THE CREDIT SUISSE GUIDE

TO MANAGING YOUR PERSONAL WEALTH

INTERNATIONAL EDITION

GET WEALTH WHEN YOU HAVE IT NOT;
GUARD WHAT YOU HAVE GOT;
INCREASE WHAT YOU HAVE GUARDED; AND
BESTOW UPON WORTHY PERSONS
WHAT YOU HAVE INCREASED.

Panchatantra (c.5th century)

THE CREDIT SUISSE GUIDE

TO MANAGING YOUR PERSONAL WEALTH

INTERNATIONAL EDITION

Publisher: *David Tait*
Associate Publisher: *Algernon Sanders*
Managing Editor: *Larissa Downes*
Production Manager: *Calum Gordon*
Editor: *Sophie Pyott*
Sub-Editors: *Linda Walsh, Susan Whitehead*
Book Design: *Calum Gordon*
Cover Design & Photo: *Hofmann & Sandmeier AG, Switzerland*
Jewellery supplied by Gay Frères SA, Switzerland
Book Layout & Production: *Koon Ming Tang*
Book Layout: *Emma Sharman, Wilson Chan*
PR & Marketing Manager: *Jacqueline Ratnavira*
Marketing Coordinators: *Jill Davies, Fraser Wilson*
Circulation & Distribution Manager: *Amanda Manning*
Printed by **Prontaprint Asia Ltd**, Hong Kong

Publisher's Note:

The opinions expressed in this publication are not necessarily those of the publisher or the institutions for which the contributing authors work. Although every care has been taken to ensure the accuracy of the information contained within the publication, the publisher, authors and their employers accept no responsibility for any inaccuracies, errors or omissions howsoever arising whether through negligence or otherwise.

Important disclaimer: *This publication is sold on the understanding that the publisher, authors and their employers are not responsible for the results of any actions, errors or omissions taken on the basis of information contained in this publication. The publisher is not engaged in the rendering of legal, accounting or other professional services. The publisher, authors and their employers expressly disclaim all and any liability to any person, whether a purchaser of this publication or not, in respect of any action or omission or the consequences of any action or omission by any such person in reliance, whether whole or partial, upon the whole or any part of the contents of this publication. If legal or other advice is required, the services of a competent professional person should be sought.*

ISBN 962-7982-07-5

Compiled and published by:

EDINBURGH FINANCIAL PUBLISHING (BOOKS) LIMITED
Rooms 1004-07, 10/F Wing On House, 71 Des Voeux Road Central, Hong Kong.
Tel: (852) 2869 8969 *Fax:* (852) 2804 6492

CONTENTS

3. INVESTING IN PROPERTY

4. ALTERNATIVE INVESTMENTS

5. MANAGING YOUR PERSONAL AFFAIRS

6. STRUCTURING YOUR WEALTH

APPENDIX

LIST OF PLATES

FOREWORD

It gives me great pleasure to write the opening words to this second, revised *Credit Suisse Guide to Managing Your Personal Wealth* and to introduce it to a wider international readership.

We all dream of being rich enough to lead a life of leisure – free to pursue our favourite hobbies and interests. Life without care!

Throughout history, people have invested a great deal of time, energy and hard work in accumulating personal wealth. Many people are willing to pay a high price to ensure that they and their families are financially secure. It is crucial, therefore, that these hard-earned savings retain their value and enjoy steady growth. Building up your own personal fortune is a challenge in itself. Making sure that it keeps its value, for use at a later date, is an equally, if not more, demanding task.

Our world is becoming ever smaller as technological advances enable us to establish communications networks that span the globe, while travelling from country to country has never been easier. 'Multinational' these days is not just another word for big business – today it is used to describe a way of life. Greater cultural diversity comes at a price, however. The increased interdependence of regional markets, offering an almost endless variety of choice, means that any decision, particularly regarding investments, needs careful consideration, and today's investor, like any multinational company, must be able to call on highly qualified advisers.

Leafing through this guide, one could be excused for thinking that personal wealth is almost more trouble than it is worth. The range of investment possibilities and the professional knowledge required is quite daunting. And this book gives but a small selection of the various aspects of wealth preservation. Each case has to be looked at individually and on a country-by-country basis. This book gives you the opportunity to learn about the many important aspects of managing and developing your personal wealth.

I would like to take this opportunity to thank the various authors for their valuable contributions to this publication. They have succeeded in presenting the complicated and broad-ranging subject of personal wealth management in a way that is straightforward and easy to understand.

Please allow me now to guide you through the fascinating world of personal wealth management. ❖

Hans Peter Sorg,
Member of Executive Board,
Credit Suisse Zurich

INTRODUCTION

In the business of managing your personal wealth, visiting a private banker may not count as one of life's pleasures, but it may be one of life's necessities. The reason for this is simple: few people, even the richest, excel at everything. A successful business executive is unlikely to be best qualified to weigh the relative merits of German equities against old master paintings, or to determine the appropriate legal structure in which his or her wealth can grow and prosper. The private banker can do just that, or at least know whom to approach for assistance. The private banker is, after all, a specialist in an area that most people know little about and would not want to spend the time learning.

The 1980s was a decade in which many people in business made their fortune. It was also a decade that left the currency of private bankers somewhat debased as the field became crowded with 'johnnies-come-lately', confident that they could compete with the services offered by those more established in the business. Private banking very nearly came to be seen as 'retail banking with a higher minimum', as one of the top-drawer practitioners recently suggested to me. As the contents of this book indicate, the private bank can offer guidance on a far broader range of issues than suggested by such a phrase. Trust and quality service are the key values of the private banker, whose role is to provide personal guidance in the matter of wealth generation and preservation.

The private bank can be the place in which to park one's excess capital and to receive investment advice. Private banking services do not, generally speaking, offer an investment opportunity that will earn supernormal returns. Typically, private banks seek to make around 10% a year for their clients over the long term. The best private banks offer a secure and very private service to help preserve the real value of your wealth, rather than a rollercoaster ride that could result in massive losses or gains.

If your investment taste is for the exotic, the articles contained in the book covering diamonds, the world of fine and decorative arts and wine might be to your liking. If past performance is any guide (and there is no guarantee that it is), diamonds might offer investment opportunities for you – Ephraim Zion points out that prices for diamonds have risen dramatically over the past 25 years. If the grape is to your taste, take John Armit's advice and consider the pros and cons of investing in fine wines. Investing in art in its many forms is covered in a series of articles beginning with Colin Sheaf's comprehensive overview of the highs and lows of the art market in recent years. Karsten Schubert offers practical advice on how to make a contemporary art purchase and, if you have ever wondered how best to safeguard your most highly prized pieces, Gordon Brennan-Jesson offers a storage solution.

A detailed article by UNESCO outlines the legal considerations that surround dealing in antiquities: provenance should be carefully investigated to guard against any challenge to title. The book caters to a range of tastes among collectors of art: from old masters to Fabergé, Judaica to musical instruments – you need not be an art aficionado to reap reward from this investment alternative.

Investors with a social conscience may like to turn to Peggy Dulany's discussion of the global social implications of investment choices, while they can rest assured that their money is reaping real benefit if the paths to charitable giving are followed according to Oliver Stanley's direction.

Property can make up a substantial part of a portfolio and the many opportunities it presents for investment in Europe and around the world are discussed by Pierre Rolin, Nigel Roberts and others. Not only can you expect a return from the purchase of commercial units in many cities in Europe, but opportunities for profit as well as pleasure may abound in the playgrounds of the world, according to Christopher Cornell.

But the serious business of being wealthy concerns the issue of how to preserve that wealth or, in the face of biological inevitability, how to ensure that your heirs receive what is due to them. The Byzantine complexity of succession issues in Europe is disentangled in articles by Coopers & Lybrand and Baker & McKenzie, offering a sober reminder of the need to keep one's affairs in good order.

The book also provides a thoroughgoing guide to money management and the complexities of the bond and stock markets, taxation issues and the use of trusts. You did not get rich reading books about how to manage your wealth but, if you do have a spare million dollars and are wondering what to do with it, you need go no further than the pages of this book. If you can't find an investment idea here, then there is no helping you. ❖

Simon Holberton,
Financial Times,
London

CHAPTER 1

PRIVATE BANKING

PRIVATE BANKING SERVICES: HOW YOU CAN BENEFIT

Max F. Günthert, Credit Suisse Private Banking, Hong Kong

Max F. Günthert studied at the University of St Gallen, Switzerland. He has been with Credit Suisse for more than 17 years. Half of that time he has spent in Asia: in Tokyo, where he was the head of Credit Suisse Trust & Banking Co. Ltd, and in Singapore and Hong Kong. Now working from Hong Kong, Mr Günthert is the regional head of Credit Suisse Private Banking East Asia, and is responsible for implementing the bank's private banking strategy in the region.

PRIVATE BANKING SERVICES: HOW YOU CAN BENEFIT

Max F. Günthert, Credit Suisse Private Banking, Hong Kong

During the 1980s and early 1990s there was an unprecedented shift towards world economic stability, accompanied by the development of local capital markets and the virtual disappearance of exchange controls.

In the past, high net worth individuals looked to private bankers simply to preserve their capital. They would often deposit their money with banks or buy foreign currency with a view to speculating on exchange rate fluctuations. They would then sit back and collect a predictable return from the bank or reap the profits generated by movements on the foreign exchange markets. However, globalisation and the volatility and complexity of financial markets have led in recent years to significant changes in asset management and the financial services industry in general. These changes have, in turn, made an increasingly wide choice of financial products and investment opportunities available to private investors.

As the market has changed, so have private investors. Their needs have become more diverse and sophisticated, and standardised products and services are no longer adequate. Investors today expect the highest standards of professional expertise and personal commitment from their private bankers; nothing less than the best in all aspects of their private banking relationship will do.

In order to ensure that they can respond promptly and effectively to the changing needs of discerning private investors, international banks such as Credit Suisse use their global networks to give private investors access to financial markets and investment opportunities outside their home country. In addition, banks have localised their portfolio management and other banking services to provide local and regional market knowledge and expertise. Private banks rely not only on their own global network but also on their ability to mobilise the resources and specialist skills of their group companies and subsidiaries. Their clients thus also gain invaluable access to global market information and to the full range of products and services important in today's financial environment.

PRIVATE BANKS

There are two types of private banking provider: the dedicated private bank, and the private banking division of a full-service bank. Dedicated private banks are devoted to the provision of private banking services only, while full-service banks offer a wide range of banking services to both corporations and individuals. In this article, we will discuss the latter type of private bank, since this better illustrates the integrated and global availability of private banking products and services.

Private banks employ teams of investment specialists equipped with the necessary knowledge and skills to help clients define their financial needs and decide how these can best be met. Of all the qualities a private bank possesses, however, a proactive attitude is perhaps the most important to its clients. Clients' needs constantly change with their personal situation. Proactive banks help clients identify their current needs and ensure that services are properly adapted and the right products delivered at the right time.

In a broad sense, the role of a private bank is to help high net worth individuals manage their wealth. A private bank assists clients in defining their financial situation and assessing their requirements, which may change over time as the 'wealth cycle' progresses. The bank will offer responses to these changing requirements through a wide range of products and services; it thereby contributes to a client's financial peace of mind.

CLIENT NEEDS

The needs of private banking clients can be separated into the tangible and the intangible. Tangible needs include:

(i) return on investments;
(ii) access to global market information;
(iii) attractive products and services;
(iv) the integrated delivery of these products and services; and
(v) a proactive, vigilant approach to service from the client's private bank.

Clients' intangible needs are somewhat more difficult to assess. In this domain, the qualities that private clients want from their private bank are:

- competence
- trust
- confidentiality
- continuity
- reliability

All of these are crucial to the development of mutual trust and understanding. Both tangible and intangible needs have to be fulfilled if clients are to gain peace of mind (see Figure 1). To some, peace of mind simply means preserving and managing their capital well enough to ensure a sense of security and control. To others, it may mean maximising returns on investments, which enables them to gain a sense of achievement and the freedom to focus on other pursuits, knowing that their wealth is being taken care of by professionals whom they trust.

FIGURE 1: PEACE OF MIND

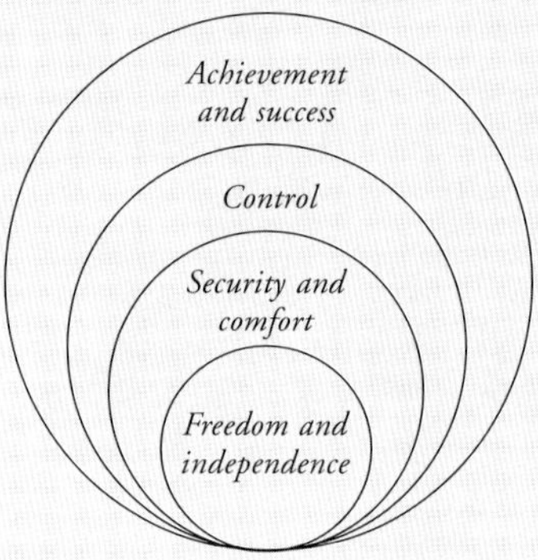

THE WEALTH CYCLE

The wealth cycle refers to changes in personal wealth over time. Typically, there are four stages in a wealth cycle (as illustrated in Figure 2). These are:

- creation
- enhancement
- preservation
- disposition

Certain events occur at different periods in different people's lives. Generally, the creation stage corresponds to the building of a career or profession, or to the setting up of one's own business. At this point an individual may begin to think about enhancing and preserving this accumulating wealth in order to raise a family or to fund a child's education. Advice on how to minimise tax obligations could also be sought

FIGURE 2: STAGES OF THE WEALTH CYCLE

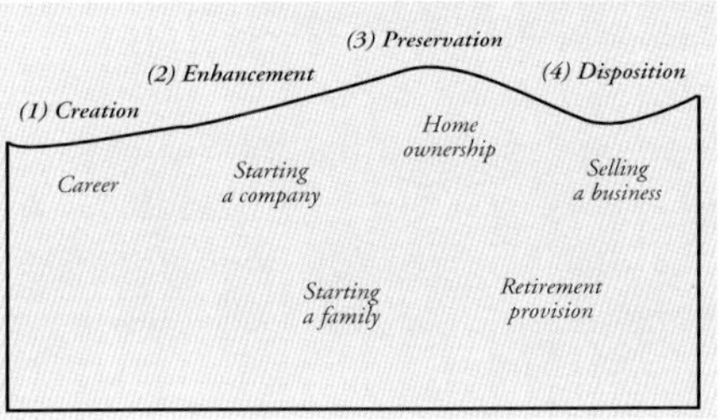

at this stage. As time passes and the children grow up, plans need to be put in place to ensure that wealth is well managed when an individual reaches retirement age. At this stage, estate and succession planning become important.

At each stage of the wealth cycle, a private bank can offer the appropriate services. A private banker will keep a close watch on how a client's wealth cycle develops, so that he or she can modify investment objectives to suit changing circumstances, and help guide the client to the best decisions.

LOCATION OF WEALTH

Wealth can be managed onshore (domestically) or offshore (internationally), or using both methods, depending on the client's situation and specific needs. Many investors will choose Switzerland as an offshore centre because of the country's reputation as a safe haven. Switzerland is well established as the world's premier private banking centre, and Swiss banks are renowned for their integrity, competence and discretion. Investment expertise, minimal taxation on foreign assets, confidentiality backed by law, and political and economic stability all add up to an ideal environment for investing personal wealth.

Location of assets may be relatively unimportant to other investors, simply because they are less concerned about sovereign risks and political and economic volatility. As a result, such investors might prefer to have their assets managed locally, allowing them to take advantage of the immediate access to information and services that locally based private bankers can provide.

INVESTMENT SERVICES

Private banks provide two principal types of investment service: advisory and discretionary. The difference between these services lies in the level of the client's involvement in investment decisions.

With an advisory service, the private bank will provide the necessary specialist advice, resources and skills, and will leave the final investment decision in the client's own hands. However, if the client would prefer the private bank to take full responsibility for an investment portfolio, then a discretionary service is required – the private bank's portfolio management service.

PORTFOLIO MANAGEMENT

Portfolio management is the core product in private banking. It provides all the benefits of active, professional, international investment management on a discretionary basis.

The investment process

Whether your primary target is the overall growth of your portfolio or a steady income, you first need to discuss your investment objectives with your private banker. Before accepting a mandate for portfolio management, your private banker will determine your investment profile. This will involve discussion of your personal needs, goals, investment objectives, risk tolerance, capital growth and income requirements, as well as legal and tax considerations. For example, it is very important for your bank to have an idea of your level of risk acceptance. While a conservative investor will accept only minimal risk, a more aggressive investor generally seeks higher returns and will accept certain fluctuations in the value of his or her assets along the way.

The investment profile is a cornerstone of portfolio management because it establishes the basis upon which the framework for 'asset allocation' in your portfolio is built. With the investment profile completed, your private bank will devise an individual investment strategy which, within the parameters of the bank's overall investment policy and strategy, is tailor-made to your needs (see Figure 3).

Asset allocation

Asset allocation – the structuring of your assets in terms of liquidity, investment instruments, currencies and countries – is crucial to the success of a portfolio. Research shows that portfolio performance is determined principally by the overall allocation of assets, rather than by stock-picking.

Investment strategy and asset allocation will be considered within the framework of your private bank's investment policy, and constantly monitored and adapted as necessary. The decisions made by your private banking specialists have an influence on the way your portfolio is established – leaving you secure in the knowledge that the management of your assets could not be in more professional hands.

Portfolio types

Generally, portfolios are divided into several categories. At Credit Suisse Hong Kong, for example, five different basic portfolio types are offered. These are:

- fixed income
- conservative
- balanced
- dynamic
- pure equities

FIGURE 3: THE INVESTMENT PROCESS AND CLIENT NEEDS

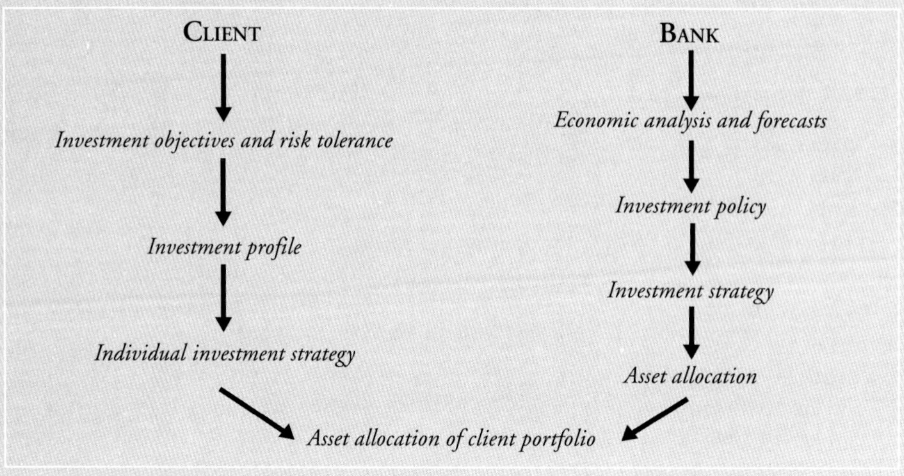

FIGURE 4: RISK/RETURN PROFILE

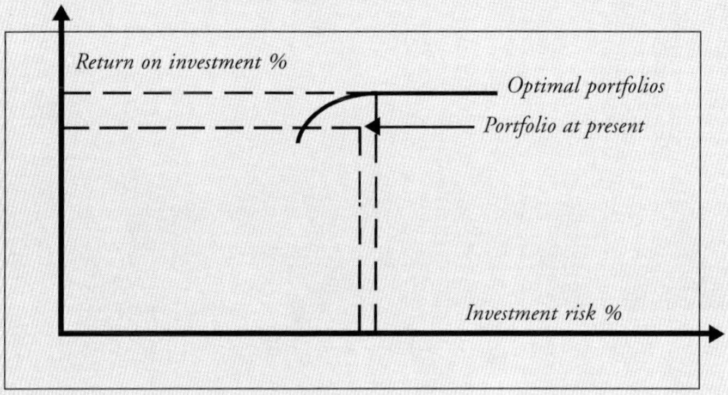

All five types can be structured to include either global securities or Asian securities only. Apart from these standard types of portfolio, tailor-made portfolios are offered to suit needs not catered for by the standard products.

Previously, most portfolios were booked and managed offshore, often in Switzerland. Today, the situation has changed significantly. Investors now have a range of choices. They are free to select where their assets are held and managed and can decide whether their portfolio management team is located onshore or offshore. Today's private banking services are very flexible and truly global, offering access to virtually any service or product you might need to fulfil your investment objectives.

Once a portfolio is established, a team of banking professionals is at your disposal and can draw on the global resources of the entire bank. This includes being able to tap into the bank's in-house research capabilities for the latest information on local and regional economies, markets, companies and securities.

A portfolio management team will be comprised of professionals (often with a multicultural background) who are specialists in their fields. They will know the intricacies of modern portfolio theory and be familiar with the more complex investment products such as derivatives. Their rich experience in both the local and international markets will enable them to recognise and seize the opportunities offered in any given market, to the benefit of their clients.

Decision making

A large private bank's investment policy is based on the systematic application of a tried and tested decision-making process. This approach leads from a general assessment of the global economic situation to the selection of specific securities by the portfolio management team. Comprehensive analysis forms the basis of the bank's global economic forecasts and determines the bank's investment policy. The portfolio management team constantly monitors the performance of your assets and continually adjusts your portfolio to reflect anticipated market changes. When selecting investments, your portfolio manager keeps in close touch with financial specialists and analysts in the securities and foreign exchange trading departments.

Traditionally, portfolio management has been based mainly on an assessment of the fundamental values of specific investments. In addition, modern portfolio theory takes into consideration the risk/return profile of a portfolio (see Figure 4). By taking such factors into account, an optimal and efficient portfolio structure can be established.

In today's complex investment environment, above-average results in portfolio management can only be achieved when a group of experts – each one of whom is a specialist committed to enhancing the performance of your portfolio – work together.

Measuring performance

Just as calculating the progress of a portfolio can be a complex matter, interpreting it can also be difficult. Portfolio managers always consider performance in relation to the risk incurred. This evaluation, made over as long a period as possible and using clearly defined benchmarks as the basis for comparison, gives you an indication of how successfully your portfolio is being managed on your behalf. Whether you decide to take a conservative or an aggressive approach to investment, it is important for you to decide on a reference currency in order to assess the performance of your portfolio.

Reporting

Your private banker will periodically discuss the development and performance of your portfolio with you. Such reporting is essential to the client-banker relationship. It will give you and your private banker an opportunity to bring your investment objectives into line with any recent changes in your personal circumstances.

It is crucial that your bank keeps you constantly and comprehensively informed on the development of your investments. Today's state-of-the-art technology should allow private banks to provide clients with regular and detailed reports on the progress of their portfolios.

OTHER PRIVATE BANKING SERVICES

In addition to investment management services, private banks also provide a broad range of other products and services. These typically include multicurrency fiduciary accounts; trading of foreign exchange and precious metals; margins and loan facilities; onshore and offshore deposit facilities and cash management services; unit trusts; and tax and estate planning through the establishment of trusts, foundations and offshore companies. All of these products and services are geared towards offering private banking clients comprehensive solutions to managing their personal wealth.

CONCLUSION

In this article, we have considered what peace of mind means to the high net worth individual. We have also examined how, by helping their clients to analyse and define their financial needs, international private bankers can tailor their products and services to meet those needs.

In view of the fast-changing economic, political and competitive environment, private banks must be flexible, proactive and innovative in their approach. They must be able to react quickly to changing circumstances in order to help investors capitalise on opportunities and redefine investment objectives to reflect new developments.

In the current dynamic financial and business environment, more people than ever before are reaching the threshold at which private banking becomes not only a viable option, but a necessity. Private bankers need to offer the highest standards in service and expertise if they are to satisfy the sophisticated and demanding requirements of today's most wealthy individuals. ❖

DETERMINING YOUR INVESTMENT OBJECTIVES

Conrad M. Eilts, Credit Suisse Private Banking, Zurich

Conrad Eilts has an MBA from the American Graduate School for International Management, Arizona and a BA in economics from the American University of Cairo. He has more than 15 years' experience in private banking, seven of which were spent at Credit Suisse London. In June 1995 he became head of overseas private banking at Credit Suisse's head office.

DETERMINING YOUR INVESTMENT OBJECTIVES

Conrad M. Eilts, Credit Suisse Private Banking, Zurich

Before making any investment decisions, you must first determine your objectives. Most people invest for the same reason: to make money. However, achieving this goal is far from straightforward, given the almost endless range of investment options and the possibility of your personal circumstances changing. Quantifying risk and return provides the framework for assessing the available alternatives. Determining your objectives will allow you to decide on a desired level of return and assess whether or not the risk necessary to achieve it is within your 'comfort zone'.

The successful management of your investment portfolio will require you to keep your objectives clearly in mind at all times and to modify them as your experience grows, the markets change, your needs evolve and, it is hoped, your wealth increases. The success or failure of a particular investment will be measured against the objectives established for it. Choosing your objectives carefully and realistically will result in achievable goals that will greatly enhance the probability of your investment being a success.

WHAT WILL THE MONEY BE USED FOR?

The ultimate use that capital will be put to is an important consideration when first establishing investment objectives. We can illustrate this point by means of a few examples.

A 45-year-old entrepreneur has spent the last 15 years building up a business and is starting to think about his eventual retirement. He might not want to continue reinvesting all his profits in the business, as he may be concerned about its cyclical nature and fearful that he might lose everything if the business folded. Over time, however, his business has prospered and grown much faster than the economy. Given this situation, the businessman should set up a retirement fund that will avoid big fluctuations in asset value. (A retirement fund that loses value at the same time as the business falters will provide him with little peace of mind.)

To take a second example: a couple in their mid-30s would like to give their two children the chance of going to university in the United States. They have inherited some capital from a relative and would like to put it towards their children's education, although they are concerned about escalating costs. To achieve this objective, the couple should measure their portfolio in US dollars, as this is the currency that they will ultimately spend. The portfolio will need to record growth at least equal to the rate of US inflation, and a portion of the money will need to be made available in 13 years' time when the eldest child is of college age.

A further example: a 40-year-old professional would like to become the senior partner of her firm; the two established senior partners are much older and could retire at any time. She wants to be in a position to buy them out when the opportunity arises, as she might not get another chance to seize control. In this situation liquidity is the individual's main concern, and this will rule out many otherwise attractive options. Another consideration is the collateral value of any investment, in case she needs to seek a loan. Some types of investment have very little, if any, collateral value.

By contrast, some people already have more funds available than they need and they should not have to dip into their invested capital at all. For this type of investor, return will be the paramount concern.

One crucial factor is the duration of an investment. The hypothetical 40-year-old professional mentioned above needs access to her funds at very short notice. Unfortunately for her, she could find herself in this situation for quite some time. Her funds will inevitably under-perform over the long term because of the short duration, or liquidity premium, she is tied to. Some people keep funds on very short deposits just in case they need the money, and are reassured by knowing that it is readily available. However, unless there is a compelling reason for maintaining immediate liquidity, it is better to take a slightly longer-term view. Since most people only need access to emergency funds once or twice during their lifetime, it is advisable to make a single break in deposits rather than have 20 years of under-performing funds. Over the short term, equity investments can be speculative and are not generally recommended unless investors are able to withstand bad timing or unexpected short-term declines.

INVESTMENT OBJECTIVES

Portfolios are structured to meet a wide range of investment objectives; a number of typical objectives are outlined below.

Maintaining income

The investor who wishes to maintain a certain income level will have a below-average acceptance of risk and will want to see virtually no fluctuation in the value of the portfolio. Earnings come exclusively from investment income, rather than from capital growth. Accordingly, investments will be primarily in the reference (local) currency with minimal foreign exchange exposure. Maintaining income might be the main objective of somebody who relies on a steady income to finance his or her daily needs.

Maintaining purchasing power

This is another conservative objective, although this type of investor will need to assume a slightly higher level of risk than the investor who simply wants to maintain income. Maintaining purchasing power requires that earnings keep pace with the rate of inflation. The 'real' interest rate is often expressed as the difference between the nominal interest rate and the rate of inflation.

A negative interest rate occurs when the rate of inflation exceeds the interest rate. During a negative phase of the inflationary cycle it is impossible to achieve the objective of maintaining purchasing power simply by earning interest. Instead, it is necessary to pursue investments with a return at least equal to the rate of inflation. Because of the higher level of risk, an investor has to accept the possibility of fluctuations in the value of the portfolio, and some foreign exchange exposure might be advisable.

If, on the other hand, the 'real' interest rate is positive – that is, it exceeds the rate of inflation – there is no need to assume any further risk, as the return will be adequate. It is, however, important to gain an overview of the trends in interest and inflation rates, which will allow the investor to choose the optimal duration with regard to the maintenance of purchasing power.

Maintaining purchasing power and achieving long-term capital growth

In order to achieve this objective, earnings will be generated through current income as well as capital and currency gains. The investor will have an average acceptance of risk and will accept some fluctuation in portfolio value. Nominal values and tangible assets will be evenly balanced and investments will generally be in the reference currency, with relatively high foreign exchange exposure possible.

Long-term capital growth

To achieve this target, earnings will be generated through capital and currency gains. The investor will have an above-average acceptance of risk and will be prepared to accept a high level of fluctuation in asset value. The portfolio will have a heavy weighting in favour of tangible assets, and high foreign exchange exposure may be appropriate.

Real capital growth

To benefit from real capital growth, the investor will have a high acceptance of risk. Real capital growth will be at a higher return over a shorter period compared with long-term capital growth, and can be attained by opportunist investment in high-risk markets, which are subject to a high degree of fluctuation.

RETURN AND RISK

In assessing the objectives set out above there can be a great temptation to focus on the return from one objective and the risk from another. For example, it would be wonderful to achieve long-term capital growth while having a below-average acceptance of risk with virtually no downward fluctuation in the capital value of the portfolio. However, the direct correlation between return and risk unfortunately does not allow such a luxury. It is always necessary to recognise that a high return will also involve a high degree of risk.

The objective of any investor is to ensure that the risk-return ratio of their investment is favourable. That is to say, a given degree of risk should have as high a return as possible or, conversely, a given degree of return should have as low a risk as possible. The quantification of risk and return is not always straightforward and there is certainly room for differences of opinion.

What rate of return is acceptable?

Everyone would like to achieve the highest possible rate of return. The problem is that the higher the return sought, the higher the risk involved. As few investors are prepared to accept a very high level of risk, it becomes necessary to think quite carefully about what rate of return is actually desired.

For example, if an investor with US dollars wants to earn interest over a three-month period, a number of options are open; the safest is a US Treasury bill, which has a minimal risk of default. A three-month term deposit with a low-rated bank would produce a higher return, but would also incur much higher risk.

Most investors would like, at the very least, to maintain the purchasing power of their funds. So earnings will primarily comprise current income derived from bonds and deposits, with the possible option of some foreign exchange exposure and earnings from dividends and capital gains. In times of high inflation this target return will not always be easy to achieve.

If the objectives are expanded to include

long-term growth, it will be necessary to accept a higher level of risk. Earnings will be generated through current income as well as capital and currency gains. Within the portfolio, bonds and deposits will be at least evenly balanced with equities, with equities perhaps even representing a greater share. A reasonably high foreign exchange exposure is also possible. Furthermore, it will be necessary to commit oneself for at least several years as the return will almost certainly vary from year to year. Three years is a reasonable time over which to measure an average rate of return.

The rate of return acceptable to a particular investor is very much a question of individual needs and expectations. The purpose of making the investment will also be a factor in determining an acceptable rate of return. For example, an elderly man who is not in a position to earn very much income might be satisfied with maintaining the purchasing power of his capital, as he cannot afford to take any chances. A doctor at the peak of her earning power is in a very different position, as the income from a successful medical practice may greatly exceed her expenses. She might be prepared to aim for real capital growth and be in a position to accept the high degree of risk involved.

Does risk really mean that you can lose?

Any investment can potentially produce a loss rather than a gain. Risk assessment is simply the identification of the factors that could lead to a loss, and the attempt by the investor to quantify each type of risk. Two different investment opportunities may each present similar expected returns, while incurring completely different types of possible risk. The better investment is the one that is considered less risky.

Standard deviation and beta analysis are two statistical measures widely used by investors to gauge statistical variability of past or expected returns. Other risks that investors seek to evaluate are market risk, business risk, interest rate risk and purchasing power risk.

TIMING

'Timing is everything' is a commonly used phrase; it may or may not be true. At its most basic level, a successful investment means buying something at a low price and selling it at a high price. If this was all that investing involved, it would simply be a matter of picking the highs and lows. Unfortunately, as discussed in connection with investment objectives, it is also necessary to have an overview of general economic trends. If the objective is to achieve long-term capital growth, isolated speculations may or may not be sufficient in the long term.

A more comprehensive approach recognises that, in the long term, markets for equities, bonds, precious metals and other investments are influenced by fundamental economic factors. It is necessary to weigh up the relative attractiveness of a wide range of possible investments at any given time, the object being to identify the one that offers the most attractive risk-reward ratio. With this approach, timing is always relevant, but it is just as important to decide what to do with the proceeds after an investment has been sold as it is to decide when to sell.

An area in which good timing is crucial is the investment of new funds. This would take place, for example, upon an investor's receipt of a large inheritance, proceeds from the sale of a business or capital from some other source. If these funds are invested in a market at its peak, whether it is in equities, bonds or precious metals, it can take some time to recover the loss let alone make any long-term gains. Care must be taken when investing for the first time in a market that is at historically high levels. For an investor who has been in the market for some time, the effects of a correction might be less painful, as the investment cost may have been somewhat lower than current values.

PAST EXPERIENCE

Everyone carries with him a certain amount of psychological baggage, accumulated as a result of his experiences. As far as investment is concerned, these experiences usually result in the individual placing a restriction on the type of investments that he or she may consider, or having a somewhat irrational predisposition to sticking with favourite proven investments.

For example, we all know someone who invested in a speculative 'can't miss' mining, oil or gas investment in a remote part of the world. It is likely that such an investment disappeared without a trace – and our friend vows that he or she will never again invest in anything but blue chips. This is not necessarily a rational conclusion but it is certainly one that our friend would be extremely reluctant to reconsider.

To take a second example: there is no doubt that precious metals represent an attractive investment opportunity and should be considered as part of a portfolio. If investors have reaped a windfall on gold or silver during one of the previous market booms, they will most likely continue to hold precious metals during an extended bear market. Although a rational assessment of the market would clearly indicate that opportunities had changed, such investors may not be inclined to remove their investment, in case the market boomed again. This is not necessarily a rational decision but it may be difficult to change tack if you have a certain degree of familiarity with a particular investment option.

In summary, it is necessary to consider every investment on its own merit. Although we must certainly learn from the the lessons of past, we should not generalise our experiences to the extent that we refuse to fully consider all future options.

DO YOU HAVE THE TIME AND ENERGY TO BE AN INVESTOR?

Ask yourself the following questions before you decide to invest your money:

- Do you enjoy the process of selecting and making investments?
- Do you follow economic trends, data and forecasts?
- Do you have the time to look after your investments properly?
- Are there occasions when you simply do not have any time for your investments, such as busy spells in your career, business travel or holidays?
- Is your temperament well-suited to investing?
- Do you make decisions by inspiration or by logic?
- Are you influenced by your emotions if events unfold much better or much worse than you expect?
- Is it hard for you to admit that you have made a mistake and to change your mind?
- Are you influenced by others, or do you prefer to make your own decisions?
- How important is it for you to be in control of the investment process?

The purpose of these questions is to highlight a few of the attributes required in a successful investor. For most individuals not in the investment industry, the biggest problems are lack of time, insufficient access to information as well as the inability to review existing investments regularly and consider new ideas. One sometimes hears the phrase 'seasoned investor', referring to someone who is experienced in both success and failure over a longer period of time. Emotional reactions become less of a factor for such an investor.

If you recognise that you need some support in the investment process, there are many ways to obtain it. Purchasing an investment fund allows you to take advantage of the expertise of a fund manager. It will still be necessary to select a fund, which is itself an investment decision, but the ongoing management will be the responsibility of the fund manager.

An alternative would be to take the advice of a private banker who can keep you posted on market developments and who will make suggestions that you can accept or reject.

One more option is to establish a discretionary managed account with your bank. This will allow you to tailor your investment decisions to your particular risk-return objectives. The bank's portfolio managers will stay on top of the markets on your behalf.

CONCLUSION

The purpose of asking these questions and providing these examples has been to assist you in determining your objectives and managing your investment portfolio. There is rarely only one correct approach. Rather, a variety of factors must be assessed to find a solution tailor-made to your individual objectives. You must be honest with yourself and communicate openly with your financial advisers in this process.

It is essential that you are comfortable with your own objectives and the risk-reward implications that are relevant to your investment portfolio. If you are comfortable with your objectives, you will achieve peace of mind and a sense of control. If the decision you make regarding risk-reward is well-informed and carefully considered, your expectations will be realistic. ❖

BANKING ON TRUSTS

Kevin Clerey, Credit Suisse Fides Trust AG, Zurich

Kevin Clerey was educated at Sedbergh in England and is a fellow of the Institute of Chartered Secretaries and Administrators. Until 1994 he was managing director of Credit Suisse Fides Trust Limited in Guernsey and he is now a member of the executive board of Credit Suisse Fides Trust AG, Zurich.

BANKING ON TRUSTS

Kevin Clerey, Credit Suisse Fides Trust AG, Zurich

Trusts have come a long way since they were first used at the end of the 11th century in feudal England. Steeped in tradition as trusts may be, their explosive growth over the last 25 years is testament to their modern-day efficiency. But why are trusts so much in the minds of today's financial planners, and for whom are they best suited? This article will seek to answer these questions and to review a solution created specifically for those private clients who choose to bank overseas.

PLANNING FOR THE FUTURE

One of the fundamental issues confronting the international private investor is that of succession planning: how to organise one's assets today so that upon one's death one's heirs can benefit in an orderly and controlled way. This issue has traditionally been addressed by writing a will, which stipulates how the assets should be allocated to the beneficiaries and names the executors who are to perform this task. However, there are many problems associated with wills:

- For the proper appointment of the executor, probate needs to be granted by the local court.
- Probate procedures vary greatly from country to country.
- Probate matters can take months and sometimes years to conclude.
- Until probate has been granted, the assets are usually frozen and cannot be passed on to the beneficiaries.
- In many countries, the wills of deceased persons are made known to the public.
- For the international investor with assets in a number of countries, it may be prudent to complete a similar number of wills.
- It is necessary to abide by local inheritance provisions and forced heirship rules in each country.
- Estate duty may well be levied by the tax authority of the country in which the assets are located.
- Not completing the appropriate will in the first place compounds some of these problems many times over!

Clearly, for the investor with an international portfolio of assets, there has to be a better way. It is for this reason that trusts feature so prominently in succession planning techniques. To many people, they are now indispensable.

TRUSTS

A trust is created when an individual (the settlor) transfers assets (the trust fund) to a person who acquires legal title to the trust fund (the trustee), the terms of the trust being defined in a document (the trust deed). The trust deed sets out the powers given to the trustees in respect of the administration of the trust fund. It also contains details of the persons for whose benefit the trust has been established (the beneficiaries). The beneficiaries in many cases will be limited to the settlor and his or her immediate family, but considerable flexibility is possible. The trust deed is drafted to accommodate the settlor's requirements both during his or her lifetime and thereafter. In many cases, and generally when trusts are used to hold bankable assets overseas, the trust deed will grant the trustees a number of discretionary powers.

For the purposes of tax planning, inheritance planning, asset protection and keeping the trust in tune with day-to-day requirements, discretionary powers are seen to be a distinct advantage to the settlor and the beneficiaries. However, by their very nature, discretionary powers give the trustee more control and greater freedom to act. The choice of trustee is therefore an issue of crucial importance.

CHOICE OF TRUSTEE

For centuries, trustees were invariably individuals and often a trusted friend or relative of the settlor. Nowadays, one has the choice of individual trustees or corporate trustees, and whilst it is not always a straightforward decision, the large corporate trustee does have some distinct advantages over the individual:

(i) it does not go absent, become unsound of mind or die;
(ii) it is likely to be financially more secure;
(iii) it will employ the skills of a number of individuals, often based in different parts of the world, with differing expertise, skills, knowledge and experience;
(iv) it will probably be located in a number of well-established trust jurisdictions;
(v) it is likely to be strictly regulated by the jurisdictions in which it operates; and
(vi) it may be better placed to deal with the unwelcome attentions of outside parties; confidentiality and the trust itself will thus be better protected.

LETTER OF WISHES

Once the trustee has been chosen, the trust deed agreed and signed, and the initial assets transferred to the trust, there remains one further matter – the letter of wishes. The trust deed names the beneficiaries, but important and detailed aspects, such as who benefits and when, can be set out in the letter of wishes written by the settlor to the trustees. The letter should provide all the extra guidance the trustees will need in the proper administration of the trust. This will become particularly important in the event of the settlor's death. A letter of wishes resembles a will in many ways, and similarly can be updated regularly, but with very little fuss or formality.

Transfer of assets

With the trust up and running, assets can continue to be transferred over by the settlor at any time. These assets come in all shapes and sizes: money, shares, bonds, property, yachts – the list is almost endless. It is also usual for payments to be made from the trust to the beneficiaries, even before the death of the settlor, and these can happen as frequently or infrequently as desired. A properly established trust can still provide a good deal of flexibility and freedom for the parties involved whilst at the same time banishing many of the worries associated with probates and wills.

DUTIES OF THE TRUSTEES

It is important to consider the duties of the trustees, for theirs is by no means a passive role:

(i) the trustees should ensure strict compliance with the terms of the trust deed at all times;

(ii) trustees must keep abreast of events. They should maintain a record of the trust's assets and monitor investment performance and security;
(iii) they are responsible for making payments to beneficiaries at the appropriate times. Trustees should therefore be in full control of the assets, enabling such payments to proceed without delay or dispute; and
(iv) trustees should be responsive to situations that arise. For example, if the settlor dies, they need to be alert and responsive to the needs of the surviving spouse and family.

TRUSTS FOR BANKING OVERSEAS

The banking trust is a solution offered by a number of financial institutions and is usually designed specifically with the private banking client in mind. A properly drawn-up trust will address the issues raised so far in this article, and offer a reasonable degree of freedom and flexibility. Desirable features to look out for in a trust:

(i) It should be discretionary in nature and be able to focus directly on the needs of the private banking client.
(ii) It should be maintained in a well-established jurisdiction and should also benefit from total tax exemption with no disclosure requirements.
(iii) It should preferably have its own underlying company integrated within the trust: the trust's bank account (or accounts) is held and operated in the name of the underlying company (see Figure 1).
(iv) It should be designed for use with one or more bank accounts around the world, the bank accounts should range from simple deposit accounts to more complex managed portfolio accounts, it should be possible to expand the structure later to meet additional non-banking needs of the client (see Figure 2). Avoid trusts which need to be dismantled to meet such needs, the personal relationship between banker and client should not be unduly disturbed, and the establishment formalities should be straightforward and the fee schedule should be simple.

FUNDING AND MAINTAINING THE TRUST

Once the banking trust is established, the client has to transfer bankable assets from his or her existing account(s).

The client authorises the payment to the trust's underlying company and, once the funds have been transferred, they immediately become part of the 'trust fund'. If the client has hitherto maintained a portfolio, there is no need for these holdings to be liquidated – the whole portfolio can normally be moved over intact.

As both the client's own account and the new trust account are usually handled by the same banking officer, the whole process is simple, fast and efficient. The client remains free to discuss the portfolio with the banking officer used in the past and also to ensure it meets the same objectives in the short and long term.

With time, more funds can be added, or they can be paid out as loans or gifts to the beneficiaries. Although the standard banking trust can exist for up to a hundred years, it is always possible to close it down sooner if necessary, although the original intention is generally for the structure to at least outlive the settlor.

FIGURE 1: A TYPICAL BANKING TRUST STRUCTURE

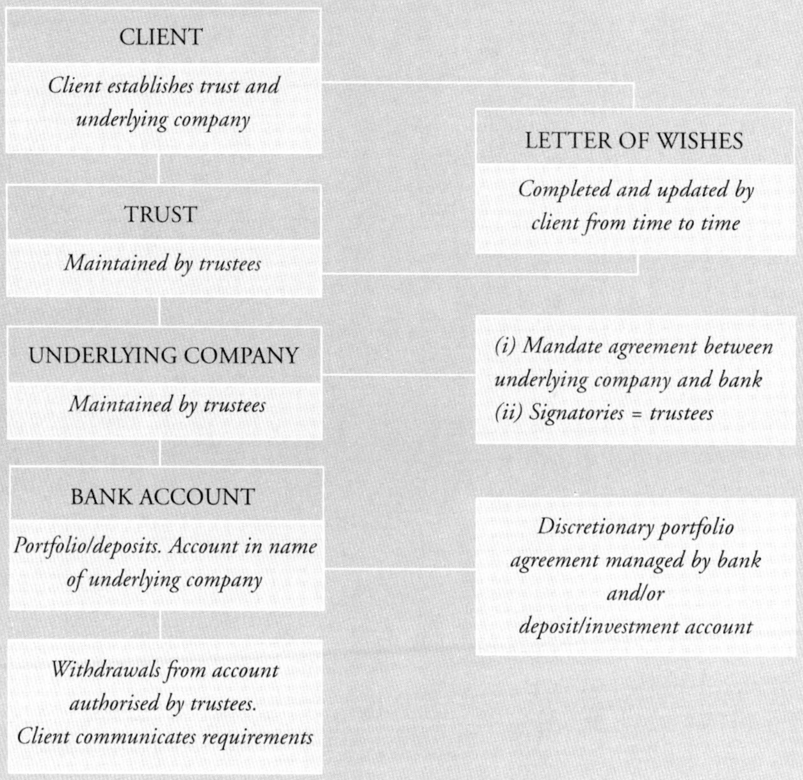

FIGURE 2: AN EXPANDABLE TRUST

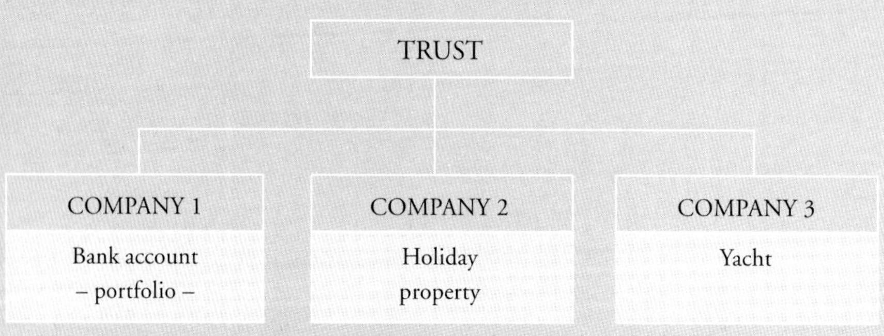

WHEN THE SETTLOR DIES

The trust, together with the assets in the trust fund, has its own legal existence and is therefore unaffected by the death of the settlor. It is at this point that some of the trust's most striking benefits come into play, neatly sidestepping the problems, delays and costs that would otherwise be faced. On the settlor's death, the trustees again consult the letter of wishes and exercise their discretion in deciding what action is to be taken. Much will depend on the contents of the letter of wishes, but a common course of action is for the trustee to contact the settlor's spouse and discuss the issues to be addressed. Normally, there is no need for any rushed decisions, particularly when one bears in mind that the assets are safely tucked away in the trust and yet remain freely available at any time.

PEACE OF MIND

The private investor banking overseas is always keen to find peace of mind: a satisfactory return on investment, security and complete faith in the discretion and competence of the institution with which he or she is dealing. Complete peace of mind will, however, prove elusive until the investor has answered the question, 'How do I ensure that my heirs benefit in a smooth, discreet and controlled way when I die ?' In many cases, a trust provides the best answer. ❖

BANKING AND INSURANCE SERVICES COMBINED

Frank G. Wettstein, CS Life, Zurich

Frank G. Wettstein is a member of the executive board of CS Life. Before joining CS Life he was in charge of Credit Suisse Monaco. He has a PhD in economics from the University of Zurich and an MBA from INSEAD.

BANKING AND INSURANCE SERVICES COMBINED

Frank G. Wettstein, CS Life, Zurich

Over the last few years, life insurance services have changed dramatically. On one hand, new products such as fund and index-linked life insurance policies have been launched. On the other – and more importantly for most customers – the advice and services provided by the insurers have been improved.

INDIVIDUAL SOLUTIONS

The buzzword here is *Allfinanz*, or 'banking and insurance under one roof'. Europe in particular is witnessing a veritable revolution in this field. The advantages of *Allfinanz* are so obvious that one cannot help but wonder why the transformation did not take place earlier. The purpose is to give customers a more comprehensive advisory service than has traditionally been available. By 'more comprehensive' we mean meeting the customers' investment and pension requirements while providing a full analysis of their income and asset situation.

Customers with a high standard of living or substantial financial commitments need a high level of insurance protection and, secondly, that they often have a high income or a large portfolio of assets. In such cases it makes sense to assess the precise scale of the risks involved, and to find out which potential losses could be sustained and which could not.

SOUND ADVICE IS ESSENTIAL

Obtaining an overview is essential. If an advisory service encompasses both banking and insurance solutions, the customer can rest assured that both aspects are being accorded equal priority. Consequently, more and more banks are incorporating insurance products into their investment counselling service as an alternative or a supplement to their traditional range of products.

Common ground

A number of life insurance products – particularly those with a savings or investment component, such as the classic combined life policies, fund or index-linked policies and annuities – have a great deal in common with conventional bank investments. The basic aim of both life insurance and financial investments is to increase or preserve customer's assets or income for themselves, for their family or for other beneficiaries. Furthermore, both bank investments and insurance products entail an element of risk. The level of risk will depend on the kind of product the customer selects and the investment strategy adopted by the insurance company.

Investing your money safely

Security oriented customers should therefore start by giving careful thought to their choice of insurance company. They will want to know about the size of the company, its position in a larger corporate group and its reputation.

At the product level, they will want to see documents and proposals that are clearly set out and easy to understand. Are the statements and figures contained in the proposal transparent, and do they make sense? Is the company promising guaranteed returns, and if so how high are they?

Another important consideration is supervision. Is the company subject to official regulation? If so, how stringent is the regulatory process? What criteria are applied to the investment of the customers' capital? Is compliance with these criteria regularly monitored?

Over the next few years, uniform standards will be applied to the insurance industry throughout the European Union. These will be concerned primarily with the solvency of insurance companies. The Swiss supervisory authority – which is one of the strictest in the world – requires additional safeguards. Each newly launched product, for instance, must undergo an authorisation procedure. Furthermore, every insurance company is subject to twice-yearly checks with regard to the structure of its investments and its ability to meet potential claims.

ADDITIONAL ADVANTAGES

Taking out life insurance is straightforward enough, but there is no guarantee that the policy will be tailored to your particular needs. Experience shows that people's investments are not just based on security risk and returns; they also take into account pension provisions, tax issues and inheritance. That is why customers often require a more detailed analysis of their financial needs, and solutions involving combinations of products. This underlines the importance of advice: only if customers receive the best possible financial advice can they be absolutely confident that the strategy to be adopted will still be the right one many years down the road.

Providing for your nearest and dearest

Life insurance offers a number of advantages over and above those offered by conventional investments. One is the freedom to nominate the beneficiary or beneficiaries (individuals or institutions) who will receive the immediate lump sum payment in the event of the policyholder's death. The choice of beneficiary, which is known only to the customer and the insurance company, can be changed quickly and easily at any time.

Depending on the legal system in the policyholder's country of residence, the funds may be protected in the event of inheritance or bankruptcy. The aim of such regulations is to safeguard the family's well-being, in other words, it is a classic method of providing for the future. Under Swiss law, for instance, creditors in bankruptcy have no claim to funds if the beneficiary is the bankrupt's wife. The situation with regard to inheritance is similar: the widow will receive the guaranteed sum even if she has renounced her inheritance.

Inheritance arrangements

The capital paid out on death is remitted directly to the nominated beneficiary. Policyholders can thus ensure that their preferred beneficiary will receive the money shortly after their death even if there is an inheritance dispute. In this way they can pay out to heirs, provide for their living costs or repay loans.

A low-cost alternative to trusts

Life insurance policies are not unlike trusts in that they are also a way of bequeathing money to individuals or institutions, either through the payment of a lump sum at maturity or death, or through the provision of a life-long income. They also offer a more advantageous cost-benefit ratio, particularly for assets up to two or three million US dollars.

A carefree retirement

At a certain time of life, planning one's retirement income becomes an important element in any long-term or comprehensive investment strategy. In this context, annuities offer an attractive alternative to conventional bank products.

Annuities free customers from the need to make investment decisions. Above all, they pay out a regular income during the lifetime of the insured person. It is very easy to take out an annuity. No medical examination is required. The financing of the policy, the starting date for payments and their nature and frequency are all largely at the discretion of the policyholder.

Providing for your old age

There is an immense variety of state and occupational pension systems around the world. In most of these systems, however, individuals still have to make some provision for themselves if they are to maintain their standard of living after they retire.

The danger of inadequate pension contributions is particularly acute in the case of expatriates, and all the more so if they have had a number of different employers and countries of residence.

Even in countries where the social security system is relatively well developed, the importance of private retirement provision is growing once again. Government budget deficits are mounting worldwide, increasing the risk of cuts in social welfare. In the medium and long term, state pension provision could be one of the areas targeted as a potential source of budget savings.

FIGURE 1: GROWTH IN STATE DEFICITS

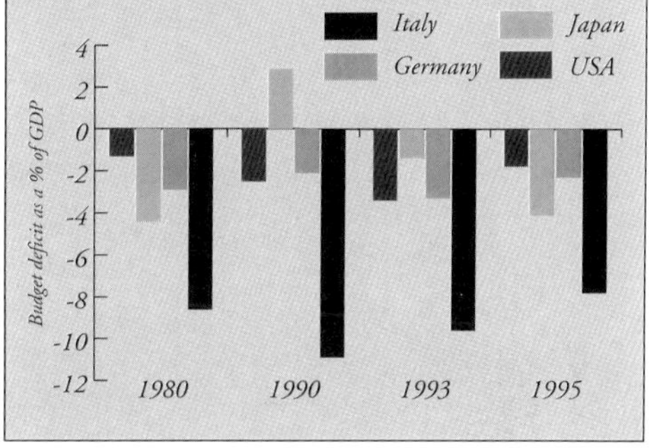

At the same time, life expectancy is increasing – particularly in those counties where health care has seen substantial improvements over the last 10 years. Consequently, state and company pensions – where they exist at all – are having to be spread over long periods. The upshot will either be higher contributions or lower pensions.

FIGURE 2: LIFE EXPECTANCY IN EUROPE

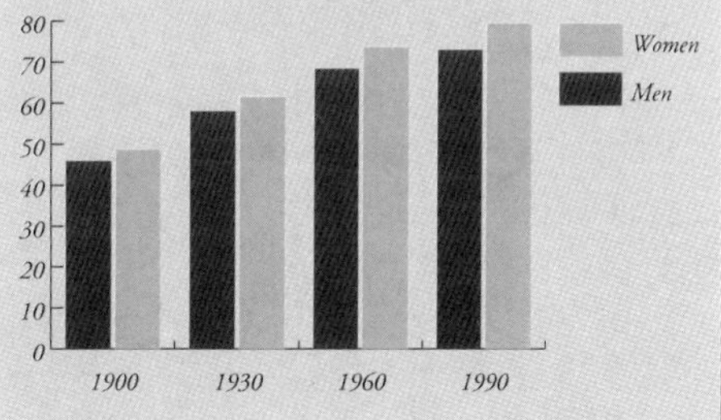

Most industrialised countries are facing the problem of an ageing population. The decline in the birth rate, and hence slower population growth, is reducing the proportion of people in employment. As a result, a diminishing workforce is having to finance a growing number of pensioners. This could have a divisive effect on society.

FIGURE 3: AGE PATTERN IN EUROPE

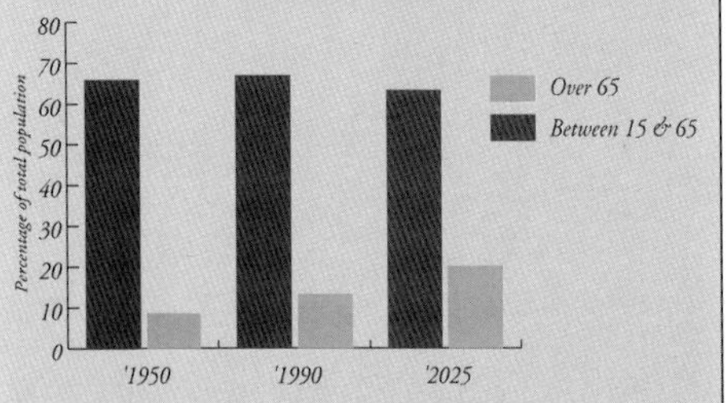

Furthermore, market deregulation is accelerating the process of industrial restructuring, which in many cases results in early retirements. More and more employees are therefore working with early retirement in mind, or are at least taking this eventuality into account. Under the circumstances, it pays to be well prepared. Only by setting aside a nest-egg or making early provision for a private pension can people ensure their future financial security.

A SWISS EXAMPLE OF ALLFINANZ

Credit Suisse was the first Swiss bank to extend its range of services to include life insurance. It did this by founding CS Life, a sister company which has developed life insurance products specifically geared to the needs of bank customers. Within the space of a few years, the success of this concept has brought many imitators onto the scene. CS Life policies are sold by Credit Suisse advisers as part of a 'one-stop' approach to financial consulting.

FIGURE 4: EARLIER RETIREMENT

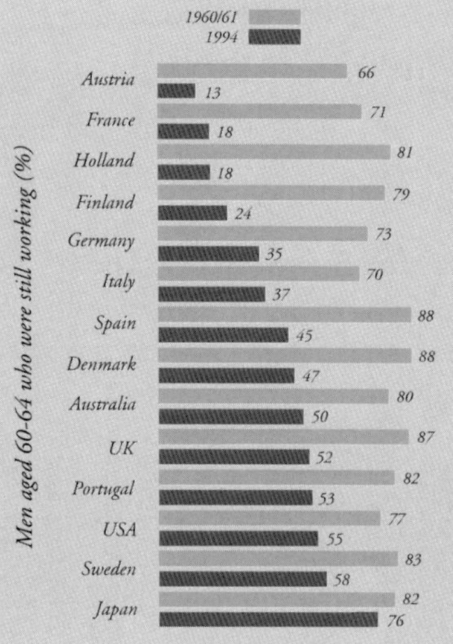

Source: ILO

Although one of the distinctive features of CS Life products is the low commission rates charged, they nevertheless meet the high quality standards demanded by banking operations. What is more, customers receive advice that gives equal weight to banking and insurance products.

CS Life products are authorised for sale in Switzerland, and the company's activities are monitored by the Swiss authorities. In all other countries, CS Life products are sold either directly or through agents in conformity with the relevant local laws.

Of particular interest to non-Swiss residents is the fact that all the life policies offered by CS Life are exempt from withholding tax. All benefits are paid out net and without deductions; nor is there any obligation to report these remittances to the public authorities. Thanks to Credit Suisse's worldwide network, moreover, investment income can be paid in the country of the investor's choice. ❖

CHAPTER 2

INVESTING IN WORLD MARKETS

INVESTING: ART OR SCIENCE?

William W. Priest Jr, BEA Associates, New York

William W. Priest Jr is CEO, executive director and portfolio manager at BEA Associates. A CPA and CFA, Mr Priest was educated at Duke University and the University of Pennsylvania's Wharton School. He co-wrote the book *The Financial Reality of Pension Funding Under ERISA*. He is on the investment council for New Jersey's state retirement funds.

Phone: (+1-212) 832 2626
Fax: (+1-212) 421 0453

INVESTING: ART OR SCIENCE?

William W. Priest Jr, BEA Associates, New York

The patron saint of the Bolsa (the stock exchange) in Barcelona is Nuestra Señora de la Esperanza – Our Lady of Hope. She is the perfect patron for all of us who desire superior returns when we invest. But competition among investors will tend to produce an efficient market. Such a market is characterised by a large number of transactions and investors, low transaction costs and transparency of the market system itself. In such a market, prices rapidly impound much if not all of any new information (e.g. interest rate changes, dividend increases, earnings surprises), and it will be very difficult consistently to earn returns superior to those of investors with similar risk tolerances. We may, indeed, hope, but all we can really expect in an efficient market is that we shall obtain a return that is just sufficient to compensate us for the time value of money and for the risks we bear.

This is where we, the money managers, come in. We seek to maximise return per unit of risk or, conversely, minimise risk per unit of return. In modern portfolio theory, this is called building an efficient portfolio, one that cannot provide more return without adding more risk.

CAPITAL MARKETS

Before we discuss how we at BEA actually build portfolios, let us look briefly at capital market history. Before performance measurement was standardised and empirical evidence collected to validate or invalidate popular investment rules, investing was all art and no science. We have learned a great deal in the past 30 years about the structure of capital markets. The concept of present value, the capital asset pricing model, the efficient market hypothesis and option pricing theory have all added to our understanding of how financial markets work. (Indeed, the authors of the capital asset pricing model recently won the Nobel Prize for Finance.) It is not our aim here to review these powerful insights in any detail, but to point out their relevance and practical effects in managing money and constructing portfolios.

A review of capital market history shows that the returns received by investors have varied according to the risks taken. At one extreme, very safe securities like US Treasury bills have provided a return over half a century of only 3.1% per year. Common stocks, the riskiest securities publicly traded, provided an average return of 11.5% for the same time period, a premium of more than 800 percentage points over the 'safe' rate of interest.

This provides us with two benchmarks for the opportunity of cost capital. If we are evaluating a safe investment, we discount the expected returns by the current risk-free rate of interest. If we are evaluating an equity investment of average risk, we discount at the expected return on the average common stock. But many, if not most, investments do not fit these simple examples and before we can deal with them we need to measure risk.

Risk is best judged in a portfolio context. Most investors do not put all their eggs in one basket: they diversify. Hence, the effective risk of any security cannot be judged by an examination of that security alone. Part of the uncertainty about the security's return is diversified away when that security is grouped with others in a portfolio.

Risk in investment means that future returns are unpredictable, and the spread of these returns is measured by a statistical term – standard deviation. The annual standard deviation of the market portfolio (S&P 500) has been about 20% over the last 70 years. Individual stocks have much higher standard deviations than this, but much of their variability represents risk unique to that stock, which can largely be diversified away. Diversification cannot eliminate market risk but it goes a long way in reducing risk related to a single security.

Intuitively, we know there is no such thing as a free lunch in life, and that is true in investing as well, with one exception: diversification pays. As we add securities to a portfolio we initially shrink dispersion (risk) far more than we lower our expected return. As long as the security added to the portfolio is sufficiently different from those already present in the portfolio, we will constantly be improving return per unit of risk within the portfolio. In other words, the portfolio becomes more and more efficient.

One more insight about risk follows. A stock's marginal contribution to portfolio risk is measured by its sensitivity to changes in the value of the portfolio. If a portfolio is efficient, as we have just described, there will be a straight-line relationship between each stock's expected return and its marginal contribution to the risk of the portfolio. The marginal contribution of a stock to the risk of the market portfolio is measured in 'beta'. If the market portfolio is efficient, there will be a straight-line relation between the expected return and beta of each stock. This is the fundamental idea behind the capital asset pricing model. It has changed the investment landscape forever because it proves that investors require extra expected return for taking on risk, and investors are concerned predominantly with the risk they cannot eliminate through diversification.

The concept of an efficient market is astonishingly simple and remarkably well supported by the facts. The suggestion that security investment is a fair game would have been regarded as bizarre 30 years ago. Today, it is not only widely accepted in business schools, it also permeates investment practice and government policy towards the security markets.

How does all this affect you? In building a portfolio, we start with the notion that the market is 'right' – that all the securities are fairly priced at this point in time. Unless we develop an insight, a theme, a valuation procedure that does not appear to be discounted in the stock market, we will own the 'market'. To provide value added above that of the market in general, we carefully 'un-diversify'. By un-diversifying, we seek to be unlike the market in a measurable way. We know we are adding risk; we also believe we are adding return.

STOCK PICKING

How do we go about the process of selecting securities given that we want to have an overall risk level about equal to that of the market? Actually the process is a straightforward one and is no different from how one runs a small business or makes capital budgeting decisions in a large corporation. Let us look at Figure 1 as it applies to equities.

We develop a model to determine the enterprise value of a company. We mark the equity of the company to market value, add all the debt with maturities beyond one year, and then subtract any excess working capital our analysis has led us to believe is present in the company. This is the market value of the company, better known as the enterprise value.

We then ask ourselves this question:

FIGURE 1: STOCK PICKING

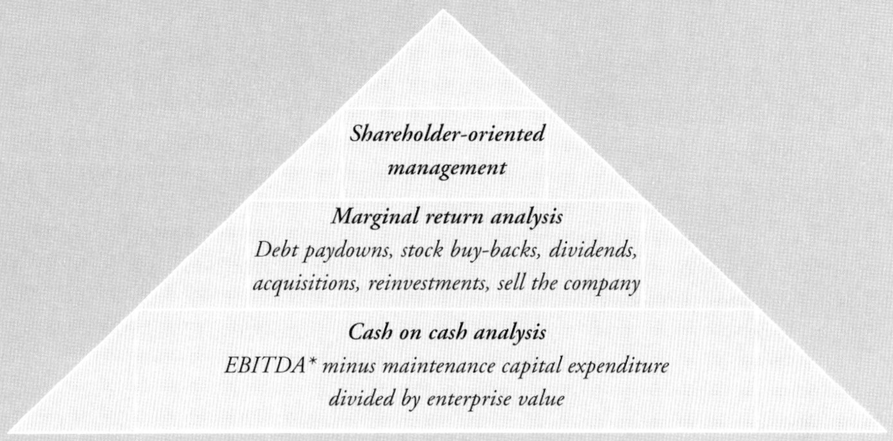

**Earnings before interest, taxes, depreciation and amortisation of intangible charges*

'Assuming we own the entire business at the price measured by the enterprise value, what are the cash flows to which we are entitled?' We calculate the earnings of the enterprise before interest, taxes, depreciation and amortisation of intangibles (EBITDA). We then deduct a sum equal to capital expenditures required to maintain the enterprise in the same state as it is today (maintenance capital expenditure). We examine the ratio of the resulting discretionary cash flow to the enterprise value. If this number is higher than the cost of capital for a security of this type, we become very interested.

Let us now look at the second level, marginal return analysis. There are only five things the company can do with its discretionary cash flow – pay dividends, pay down debt, purchase stock, make an acquisition, or reinvest in the business. There is a sixth option, and that is to sell the business to a larger company capable of exploiting economies of scale. Extensive analysis based on the public information available to us will give us an idea of what the company ought to do. In effect, putting ourselves in the shoes of the chief financial officer, we ask ourselves this question: 'What is the optimum use of the free cash flow from the point of view of the shareholder?'

Will management behave in the desired way? We move to the third box, labelled shareholder-oriented management. Does management own stock outright, purchased with its own money? We prefer management to own its company equity through outright stock purchases. This ensures that its interests are also those of the shareholders. Numerous examples, drawn from the past several years, show how well this policy works – General Dynamics, Scott Paper and Disney, to name but three.

Applying the process across all industries often yields a surprising number of stocks that are attractively priced relative to their peer groups. We seek to identify and invest in such companies in a measured and controlled way so that over time our results will exceed those of the market in general, although the risks taken are no greater than the broad market averages. We think this strategy is of lasting value and, while it is applicable to all markets, it is most relevant to the United States, where large amounts of data are available to portfolio managers and analysts.

The empirical evidence of the past 30 years – supporting much of academic theory – has become part of the investment landscape. It provides more solid foundations upon which to build portfolios than relying on Nuestra Señora. It is our task to execute a client strategy that integrates science, methodology and judgement (art) derived from experience. ❖

EUROPEAN EQUITY MARKETS

Josef Baumeler, Equity Research Europe, Credit Suisse Zurich

Josef Baumeler studied economics at the University of St Gallen (majoring in economic policy) and wrote his dissertation on 'The Economics of Water Rights in Switzerland'. He first worked as a scientific assistant in the economics department of the University of St Gallen. Since January 1992 Mr Baumeler has been head of European equity research for Credit Suisse in Zurich.

EUROPEAN EQUITY MARKETS

Josef Baumeler, Equity Research Europe, Credit Suisse Zurich

Credit Suisse's publications usually deal exclusively with the immediate future, that is the coming 12 months. The aim of this article, however, is to consider the long-term prospects for the European equity markets, and to look at the history and distinguishing characteristics of the major bourses. We hope to provide private investors unable or unwilling to keep abreast of developments in all these markets, with a general overview of events that will enable them to play a more active role in determining their investment strategies.

Let me say at the outset that the outlook for investments in European equities over the next few years is extremely promising. In fact, the climate could hardly be better: moderate growth, low inflation and thus low interest rates, and a sharp rise in corporate profits. The European markets are currently lagging behind the USA by about two years. The bullishness of the US equity market this year can also be attributed to the favourable economic climate. The inflationary fears that emerged in 1994 have proved to be unfounded: as in the 1960s, simultaneous growth and low inflation are now feasible. There has also been an unprecedented wave of restructuring and prolific growth associated with the advent of new information and communications technology. The productivity gains achieved in this economic cycle have therefore been considerably higher than in the previous two cycles. Profit margins in the USA have reached levels not seen since 1973, and the trend looks set to continue.

Europe, however, has not yet reached this stage. One of the most significant developments in recent years has been the march towards globalisation: a number of enormous, untapped markets have opened up, presenting both new competition and new opportunities. Consequently, European companies have begun to switch production to locations where labour costs are far lower than in Western Europe. The technological barriers associated with the Cold War are a thing of the past; capital can now be procured anywhere in the world. The implications of this revolution for equity investments are extremely positive: downward pressure on inflation and interest rates, appreciable productivity gains, growing profit margins, and the opportunity to tap lucrative new markets with practically unlimited growth potential.

UNITED KINGDOM

Trading in British shares began around the same time as the market for Dutch shares was founded. The two countries share a number of historical links (William of Orange, for example) and similarities (both are colonial, seafaring nations), so it is perhaps not surprising that their stock markets developed at approximately the same time and along similar lines. The first joint-stock company (where public subscription to ownership in equal shares was possible) was the Muscovy Company, founded in London in 1533. Shareholders contributed the working capital and appointed a governor and deputy governors to manage the venture for them. The initial subscription raised £6,000 in shares of £25 each. The volume of stock-trading grew rapidly in 17th-century London. This was a period of expanding international commerce, and trade stretched to the Far East and the Americas, as well as Europe. By 1698, four years after the founding of the Bank of England, business was so buoyant that dealers were forced by the other merchants to move from the Royal Exchange to Change Alley and its coffee houses. It was not until 1773, however, that the first official stock-exchange building was opened, on the corner of Threadneedle Street.

Today, the London Stock Exchange is the world's third-largest equity market, after New York and Tokyo. Operating in the European time-zone, the exchange holds a vital position in the world's international financial markets, lying as it does midway between the trading zones of Japan and the eastern seaboard of the USA. It is the largest market in the world for trading in non-domestic equities. Around 65% of all shares traded outside their home markets are traded via London: for European shares traded outside their home markets this figure rises to 90%.

FIGURE 1: LONDON TOTAL STOCK MARKET

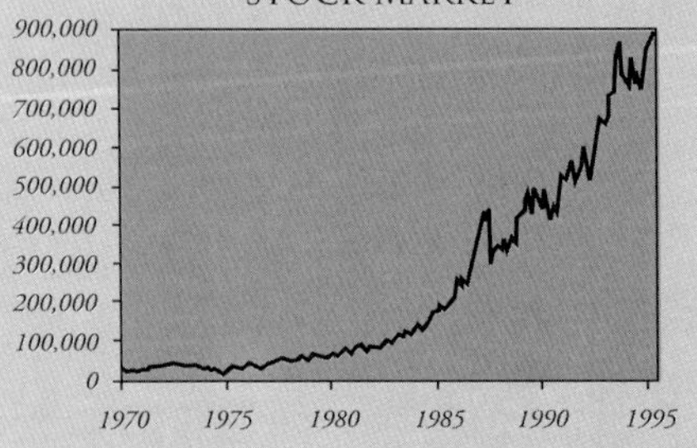

Source: Datastream

The most frequently used benchmark of stock activity in UK equities is the Financial Times-Stock Exchange 100 Share Index. The FT-SE 100 Index is a capitalisation-weighted index of the 100 most highly capitalised companies traded on the London Stock Exchange. The index was developed with a base of 1,000 as at 3 January 1984, with consumer goods weighing in as the heaviest constituent sector. Within that sector, pharmaceuticals make up the largest component, at 9.22% of the index. At the top of the list in terms of single-company weighting within the index, and part of the pharmaceutical sector, is Glaxo Wellcome at nearly 5%. Retail banking makes up around 13% of the index, and integrated oil-producers are also important at 9.3%, with BP and Shell contributing 4.6% each, giving them the second and third heaviest weightings in the FT-SE 100.

FIGURE 2: SECTOR BREAKDOWN OF FT-SE 100 INDEX

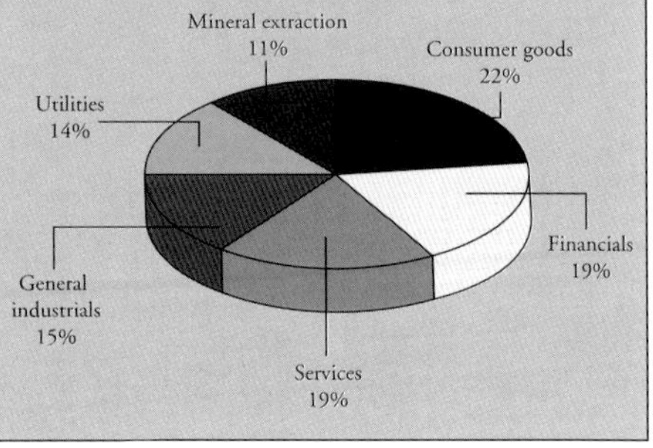

At the inception of the FT-SE 100 in 1984, total market capitalisation stood at US$225.8bn; today it stands at US$960bn. During the same period, trading volumes have grown from US$85.05bn to more than US$928bn. Those 11 years have seen many political and economic changes in the UK. Former prime minister Margaret Thatcher has come and gone, having apparently succeeded in breaking the crippling (for industry) power of the unions. Deregulation and modernisation have revolutionised the City, with the so-called 'Big Bang' in 1986. Following this came the crash of October 1987 and, most recently, Britain's entry into the ERM and subsequent ignominious withdrawal in September 1992. Throughout the ups and downs of this period, the FT-SE 100 Index nevertheless increased by

250%. Although it does not take account of differences in structure, this performance compares favourably with that of the US and Japanese indices, with the Dow Jones up 263% in the same period and the Nikkei 225 up 77%.

If the 1970s was the decade of rising inflation and industrial conflict, and the 1980s a boom period characterised by overheating in the economy, it is to be hoped that the present decade will continue as it has begun, with slow but steady growth and low inflation. British companies have implemented radical restructuring and cost-cutting measures over the last few years and now seem well-placed to keep pace with the international competition.

GERMANY

At the end of the 1980s the German stock market enjoyed an upsurge in the wake of reunification. By 1990, however, recession in the world economy had brought the good times to an abrupt halt, even though the special economic circumstances created by the absorption of the former East Germany kept the downturn at bay for 18 months or so. Apart from the various adjustment problems faced by the country as a whole – much higher interest rates, a runaway budget deficit and a swing from the traditional surplus on the current account to a shortfall – it became clear that assessments of eastern German companies

FIGURE 3: DAX PRICE INDEX

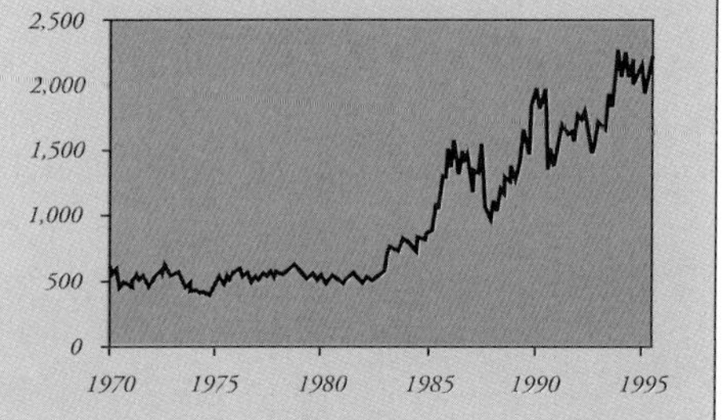

Source: Datastream

had been too optimistic. Another factor militating against the gold-digger sentiment of the late 1980s was the government's compliance with the trade unions' demand that wages in the east be quickly brought into line with those in the west.

After treading water for two years, the German equity market began to surge towards all-time highs in 1993, a year when the recession was in full swing and a number of leading German companies (e.g. Daimler Benz, Volkswagen, Thyssen, Mannesmann) posted record losses. The upswing on the bourse was fuelled by falling interest rates and, above all, by the onset of a phase of comprehensive restructuring. Inadequate cost control, massive over-capacity and a certain neglect of export markets in favour of business with the eastern *Länder*, all helped to make the process of structural adjustment in German industry even more painful than in neighbouring countries.

Germany has never been all that keen on equities as an investment medium, especially when compared with the USA or UK. There are a number of reasons for this:

- Historically and culturally, German investors tend to go for security.
- The corporate landscape is dominated by the *Mittelstand* (small and medium-sized enterprises).
- German banks tend to operate as universal banks rather than as specialist investment houses.
- The bulk of the nation's economic activity is generated by smaller businesses run for the most part as family firms. Only a small proportion of German industry takes the form of companies limited by shares (stock corporations).
- Both family firms and the majority of big businesses arrange their funding primarily through their 'house bank', with the result that industry and the banking sector are tightly interwoven.

Trading on the German stock market is heavily concentrated in a few large-scale businesses which, in many cases, are conglomerates. The broad DAX index accounts for more than 70% of total share ownership and roughly three-quarters of stock-exchange turnover. The DAX, covering 30 blue-chip companies, which the stock-exchange authority Deutsche Börse AG has updated on a minute-by-minute basis since 1988, is currently the main benchmark. Trading in options and futures on most of these stocks is handled by the Deutsche Terminbörse (DTB). Compared with the indices in use on other European bourses, the DAX is dominated by cyclical industries (chemicals, car manufacture, mechanical engineering and electronics) and the financial sector. The most heavily weighted stock on the market is Allianz, with a 12% DAX weighting. Sectors under-represented or even omitted from the DAX are food and beverages, oil, and telecommunications. The German Telekom new issue scheduled for 1996 (volume approximately DM16bn) will significantly alter the weightings.

FIGURE 4: SECTOR BREAKDOWN OF DAX INDEX

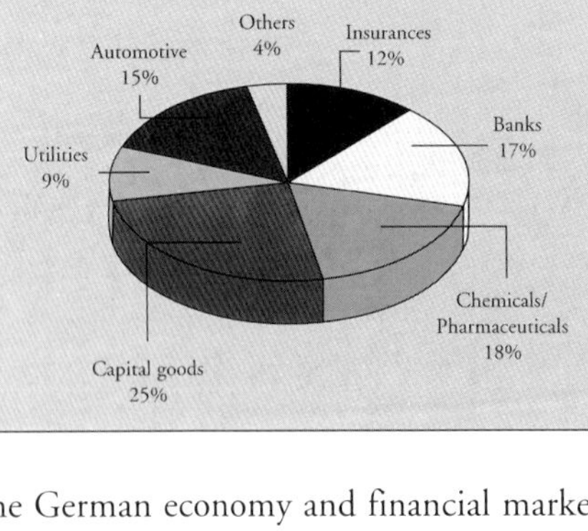

The German economy and financial markets have now largely overcome the stresses and strains caused by reunification. By 1994, corporate profits had fully recovered. However, structural adjustments continue to be made. German manufacturers, with their heavy reliance on export markets (more than 50% of sales revenue is generated abroad, mainly elsewhere in Europe), are being forced by high domestic wage costs and a susceptibility to currency movements to shift production out of the country. Against the backdrop of modest growth in the German economy, slow inflation and low interest rates, company earnings should be boosted in the years ahead by productivity increases and a more flexible attitude on the part of the trade unions.

In the wake of reunification the German stock market was revalued: we believe that the consequent over-valuation will tend to be revised downwards, with the result that the market's performance, by European standards, will be average at best, despite above-average profit increases.

FRANCE

From 1945 onwards, the French financial markets, and indeed the whole financial sector, were heavily regulated by the state, with all the principal banks in state ownership. Investor interest remained correspondingly weak. Reform did not come until the late 1970s and early 1980s, when the general liberalisation of the French economy brought with it a number of measures to modernise the financial markets. A key part of the modernisation of equity trading was the concentration of trading in Paris: hardly any major French companies are listed on the regional bourses today. Exchange efficiency was further improved by the introduction of a computerised trading system. Since then an effective, comprehensive trading system for derivatives has also been built up.

Economic reform and privatisation have done much to enhance the nation's appeal as a

FIGURE 5: PARIS TOTAL STOCK MARKET

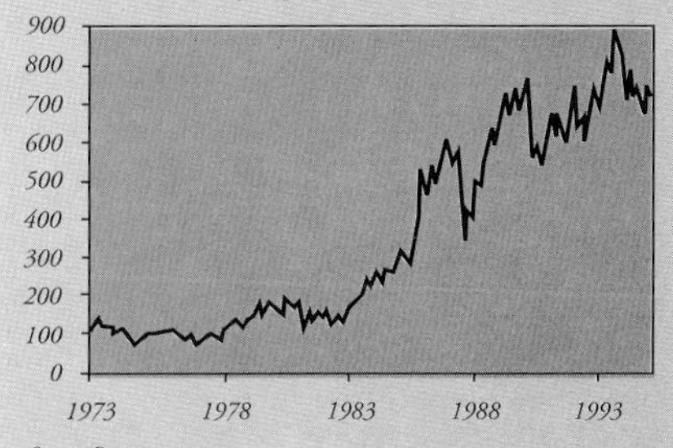

Source: Datastream

financial centre. Nowadays, more than 30% of French share capital is held by foreign investors, and another third by domestic private investors. In terms of stock-market capitalisation, Paris is now Europe's third-largest stock exchange after London and Frankfurt. The sectoral make-up of the French equity market, and in particular that of the CAC 40 index of larger companies, gives a disproportionate weighting to banks and finance companies relative to their contribution to GDP. More privatised companies (e.g. the steel-maker Usinor Sacilor), along with fast-growing enterprises from other sectors, should soon be added to the CAC, thus tightening the correlation between the share index and the economy at large.

FIGURE 6: SECTOR BREAKDOWN OF CAC 40 INDEX

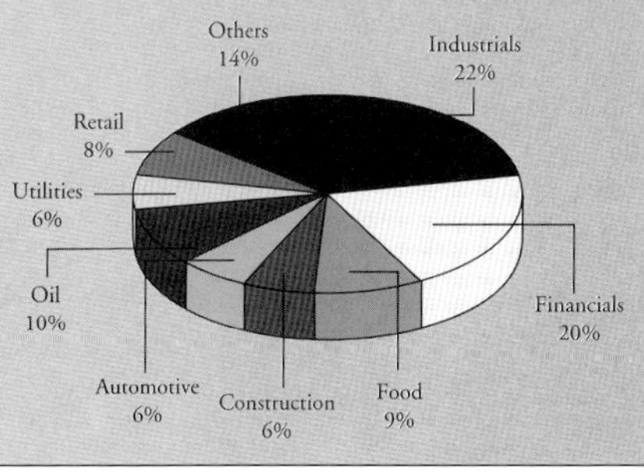

Although over the last 10 years the state has quite clearly been exercising less direct influence over French businesses, there are still strong links between the interests of the political class, the civil service and the larger companies, many of which used to be state-run. These links are to a large extent the product of personal acquaintances dating back to time spent at France's elite *grandes écoles* by future politicians, civil servants and businessmen. Furthermore, frequent job changes are generally accepted in France, not merely from one firm to another, but to and fro between industry, politics and the civil service. The links are strengthened even further by a network of mutual share-holdings amongst companies. For these reasons foreigners have always found it practically impossible to take over a French company, even after explicit legal barriers to such takeovers were dismantled. However, with the growing internationalisation of French industry and increasing European integration – and also because of a more systematic exercise of shareholders' rights, as highlighted by the last Suez AGM – it is safe to assume that, in future, not only will takeovers become more likely, but more attention will paid to the interests of minority shareholders.

The French equity market has under-performed compared with the European average in recent years. Yet the economy now stands on a very solid foundation: low inflation, a stable currency and good growth rates, reinforced by the radical restructuring of the newly privatised companies, should lift corporate profits and bolster the market's valuation to a level comparable with the other European exchanges. The fly in the ointment, and the nation's foremost social problem, is unemployment: with joblessness at such a high level, there is a danger that political considerations will delay the public spending cuts required to reduce the budget deficit and trim the national debt. This might prevent France from meeting the Maastricht convergence criteria, to the detriment of French interest rates and prices on the equity market.

THE NETHERLANDS

The Netherlands' stock exchange, the Amsterdam Effectenbeurs, was established in 1602 and therefore qualifies as the world's first 'modern' stock exchange. From the very outset, the Amsterdam bourse displayed a truly international character – the first company to sell its shares on the new exchange was the Vereenigde Oost-Indische Gesellschaft – and it remained the leading international stock exchange until the middle of the 18th century. An options market was introduced in 1978 (relatively early by European standards) and today plays a major role in continental Europe. The transition to electronic trading followed in 1988 and fixed commissions and stock-transfer tax were abolished in 1990.

FIGURE 7: AEX ALL SHARES GENERAL PRICE INDEX

Source: Datastream

Like their Swiss counterparts, large Dutch companies have a strong international orientation. This explains why Dutch shares are regarded as highly dollar-sensitive. The sector structure of the AEX Index (comprising the shares of the 25 largest companies) reveals the dominance of the financial-services sector. The food industry also plays an important role, thanks primarily to Unilever, as does the oil and chemical sector, represented by Royal Dutch. The geographical position and trading history of the Netherlands have contributed significantly to the development of the country's renowned transport sector, producing such companies as Nedlloyd, Pakhoed and KLM. The Dutch media industry is also worthy of mention, with such publishing houses as Elsevier, Wolters Kluwer and VNU, and the music and film company Polygram (a subsidiary of Philips).

FIGURE 8: SECTOR BREAKDOWN OF AEX INDEX

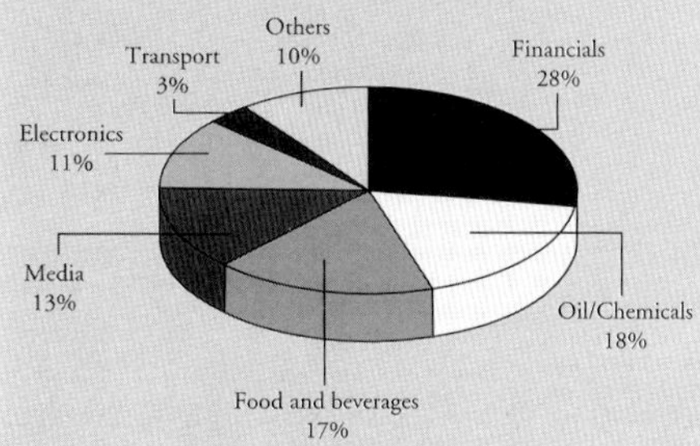

Although Dutch companies are extremely international in their approach, the law continues to allow them to erect virtually insurmountable barriers to takeovers by non-residents. This, together with the relatively high proportion of cyclicals and financials, is one of the reasons that the performance of Dutch equities has traditionally been unimpressive by European standards.

Since the early 1990s, however, the Dutch equity market has seen a significant improvement in its fortunes and is now one of the star performers in Europe. This success can be attributed to the robustness of the economy (the Netherlands managed to avoid recession); low interest rates; the stability of the guilder; sound economic, financial and monetary policy; and the international competitiveness of Dutch companies. The outlook for the second half of the decade is very similar:

moderate, virtually inflation-free growth should continue for the next few years, with recession no more than a remote possibility. The radical restructuring measures implemented in recent years will begin to take effect, resulting in higher margins and an appreciable rise in corporate profits. The relaxation of restrictions on company takeovers should boost Dutch share prices over the next few years. The proposed changes to the rules governing investments by the Dutch civil service pension scheme ABP should also boost the equity market: from 1996, ABP will have greater scope to invest its 185 billion guilders of total assets (as at end 1994) in shares.

SWITZERLAND

During the 1980s, the Swiss stock exchange was fairly inactive compared with other European bourses. From 1982 to 1989 it managed to outperform the Morgan Stanley Capital International (MSCI) Europe Index only once, in 1985 (after adjustment for currency factors). At the beginning of the 1990s, however, the situation changed dramatically. Since 1990 the MSCI Switzerland has outshone the MSCI Europe every single year. There are a number of reasons that Swiss shares have become (and should remain) more attractive to investors.

The last seven years have witnessed a plethora of far-reaching structural changes, which have had a positive effect on the valuation of Swiss stocks. These include the lifting of restrictions on the transferability of Swiss companies' registered shares (begun by Nestlé in November 1988); the simplification of capital structures (creation of unitary shares); the adoption of international accounting practices (IAS and ARR standards); and the introduction, due in 1996, of new exchange legislation (reporting requirements, rules on takeovers). In addition, many companies have realised that adopting an open information policy towards their shareholders brings long-term benefits (the strengthening of investor relations, publication of half-yearly and quarterly results, and breakdown by division are word of turnover and operating profit). Since all this applies mainly to well-capitalised firms, the equity market has in effect split in two over the past five years, with internationally active companies being the main winners and smaller ones losing out. So, for example, the Swiss Small Cap Index has shed 21% of its value since 1990. In stark contrast, the stocks listed on the Swiss Market Index posted a 68% advance over the same period. Of course, there are other reasons for this divergence: in particular, small firms were hard hit by Switzerland's 1991-93 recession.

FIGURE 9: CREDIT SUISSE GENERAL PRICE INDEX

FIGURE 10: SECTOR BREAKDOWN OF SPI INDEX

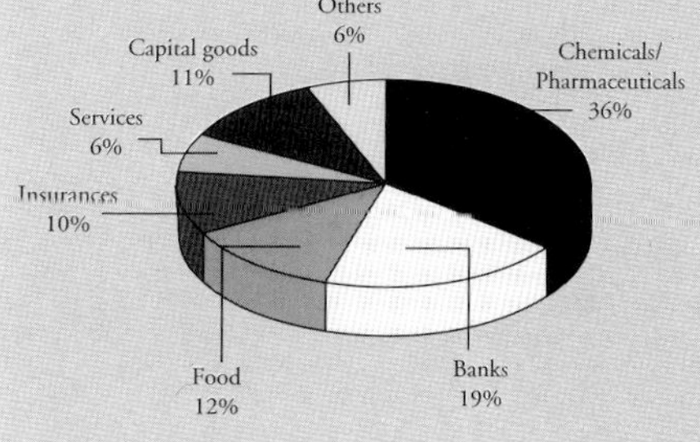

CONCLUSION

Any long-term forecast is a risky affair, since the scope for unforeseen developments is so great. However, there are a number of trends in the world economy which will be sustained or even intensified in the next few years. The main ones are: rapid economic growth in East Asia (especially China); economic reform and industrial modernisation in the countries of the former Eastern bloc, with their associated high political risks; the international boom in telecommunications technology (mobile communications, global networks); heavy investment to enhance productivity (capital goods replacing manual labour); the transfer of production facilities out of Western Europe into low-wage countries; and the expansion of international trade and transport. With their international presence and sound market positioning, the major Swiss firms are well-placed to reap the benefits of these long-term trends. This should be reflected in above-average profit growth and share price gains. ❖

UNDERSTANDING GLOBAL BOND MARKETS

Adam Beaudin and Christoph Fehr, CS Trading Strategy, Zurich

Adam Beaudin joined Credit Suisse in May 1995 as an international bond market strategist on the Zurich trading floor, after working in a similar position for Technical Data in London and New York. Before that, he was an international bond arbitrage trader at Société Générale. Mr Beaudin was educated at the University of Vermont and the Sorbonne.

Christoph Fehr has been at Credit Suisse since 1989. He now works as an analyst with the Trading Strategy Group. Before this he was employed by SBC in portfolio management, where he advised on short-term equity trading strategy. He studied at St Gallen.

UNDERSTANDING GLOBAL BOND MARKETS

Adam Beaudin and Christoph Fehr, CS Trading Strategy, Zurich

In a sense, the bond markets have been around for a very long time. Even in ancient times it was recognised that there was a price to be paid for lending money and, although the risk parameters have changed slightly, the 'time value of money' is an old and accepted concept. Although much has changed in the way the market works, the basic principles and the result of the interaction between lender and borrower are more or less the same.

Today's global bond markets handle an increasing percentage of total debt in the developed world, and their influence over both governments and commerce has expanded accordingly. Unfortunately, the growing complexity of the markets can make them difficult to understand and interpret; the ever-greater number of interest rate instruments now traded in dozens of currencies has not lessened the challenge that modern bond markets pose for investors. However, the markets are not as complicated as they may seem from the outside. And rapid advances in technology are making the international markets more accessible than ever before.

Rather than offering specific investment advice, this article will focus on general concepts and investment themes. The final section will describe the cyclical nature of bond prices and ways to exploit investment opportunities.

We will not cover derivative instruments in any detail as they are often only a way of achieving a more leveraged exposure to a market or of managing risk. Also, the implementation of any derivative strategy is very dependent on the specific needs of an investor at a particular time. In any case, a whole book would be needed to describe all the different kinds of swaps, caps, collars, floors, options, futures, swaptions and so on, in enough detail to give the reader a fair understanding of the risk/return profiles of these instruments.

MECHANICS OF THE BOND MARKET

We will assume that most bonds have a maturity of between one and 30 years. (Interest-bearing investments of less than one year are part of the money market rather than the bond market.) A bond represents an agreement between two parties and is essentially a loan made for a specified period, compensated by an agreed rate of interest. The buyer of, or investor in, a bond usually receives a fixed interest payment, known as the 'coupon', either once or twice a year.[1] At maturity the investor receives payment of the original face value from the issuer. 'Callable' bonds give issuers the right to redeem or 'call' their bonds before the final maturity. A particular characteristic of bonds is that their prices move in inverse proportion to their yields, i.e. as interest rates fall, bond prices rise and vice versa.

The bond market tries to satisfy the needs of two different groups with opposing interests. One group has a cash shortage and needs to borrow, while the other group is flush with cash and needs to invest. The first group, often

[1] *Exceptions are FRNs (floating rate notes) and zero-coupon bonds. FRNs pay a variable interest rate. Zero-coupon bonds are bought at a large discount and redeemed at maturity at face value. Annual or semi-annual coupon payments are market-dependent.*

defined by the market as 'borrowers' or 'issuers', is at present dominated by federal and state governments, supranational agencies and large companies, which have obvious and sizeable capital needs for funding or investment. The second group is made up of 'investors' or 'lenders' who have excess capital to lend – at a price. Members of this group range from private individuals to the largest corporations, banks, and even governments, that need to put their excess money to work.

The most important bond market in the world is the US Treasury market. Not only is it the largest, most liquid and most sophisticated market, but it also serves as a benchmark for global interest rates because of the dominant position of the US economy and the US dollar in world trade. In the last decade, however, the bond markets of many other industrialised countries have become much larger, more liquid and more sophisticated, largely because of the recent explosion in government debt. Lesser developed and emerging countries issue bonds as well. They tend to offer much higher interest rates because of their much higher risk: including currency, credit, liquidity and political risk.

Unfortunately, bonds have a somewhat ambiguous nature. Depending on whether they are issued by a company or a government, they can be looked at as either debt or wealth. First, the 'wealth' argument. Debt creation in the form of bond issuance has been a driving force behind virtually every modern capitalist system. The fact that private companies and individuals who go into debt are forced to produce goods and services to settle their contracts clearly helps to stimulate economic growth, thereby aiding the production of wealth.

The same cannot always be said of government debt. If a country runs up a huge budget deficit by spending more than it produces, the result is high inflation. The market will, therefore, tend to demand a higher interest rate for a country's bonds if there is a high budget deficit, and a lower one if there is a budget surplus. Furthermore, states are not often forced to produce the wealth to settle their debts. Very often, government debt is paid by issuing new debt; the burden of payment is simply shifted to future generations of tax-payers.

Another important difference between a private issuer and a state is that the holder of a bond can sue a private issuer for non-payment, while there is no easy legal means to force a state to pay its debt. Sometimes the value of a bond can be affected if a government decides to change the rules of the game: by cutting or reducing interest payments, extending bond maturities or prohibiting the export of capital, for example. This is why bond markets devote so much attention to government funding as well as to any possible regulatory changes. An investment in the bonds of the G7 developed economies risks much less in the way of regulatory surprises than is the case with emerging markets. Nevertheless, even in the developed economies, withholding tax has on occasion been imposed on domestic or foreign investors or both, with significant ramifications.

Understanding the relationship between debt and wealth creation can help investors understand why the bond markets are perhaps the most important of the main financial markets, rivalled only by the foreign exchange market. Most importantly, bond markets determine the 'price of money', i.e. a nation's interest rates. These in turn greatly influence the strain a capitalist system has to work under.[2] Sharp swings in interest rates can thus make or break any investment strategy, and often it is rising or falling interest rates that are responsible for stock market moves. The sheer size of the major international bond markets means that no government or company can afford to ignore them.[3]

[2] *Obviously somebody who has to earn a 12% interest rate to pay off a loan has to work much harder than someone who has to earn only a 5% interest rate for the same level of debt.*

[3] *This once caused a high-ranking official in the Clinton administration to say that if he could choose he would like to be reborn not as a king or a president, but as a bond market. Change in size is not neutral, as Marx once observed. At a certain level, a change in quantity can trigger a change in the characteristics of the subject.*

THE MAJOR PLAYERS AND INVESTMENT THEMES

As we have said, bond markets determine the price of money and thus, to a great extent, the overall development of an economy. However, this is not a one-way process: bond markets themselves are influenced by a combination of economic, political and technical trends. Thus, in analysing bond markets, one must get to grips with a complex system of interrelationships between numerous variables, all of which can drive prices up or down. A useful analogy might be a game of chess; the pieces have their different roles and capabilities, any move could change the landscape considerably and even pawns sometimes have the power to decide the game.

The players are:

- governments
- central banks
- banking systems
- foreign exchange markets
- commodity markets

Governments

Governments are the most important players: not because of their economic wisdom, which is open to question, or because they are in full control of the economic process, which they are not, but because they can change the rules of the game by legislation, they can influence the distribution of wealth and even confiscate assets, and they are the largest supplier of new bond issues. (Increasing levels of government debt call for higher levels of bond issuance, which tends to push market rates up.)

The goals of a government are often defined by a political process influenced by all kinds of motivations. However, it is clear that re-election is one of the primary goals of most politicians. This is achieved in most cases by promising resources to groups that might support them in an election. For a long time, the money needed to fulfil political promises has been raised by issuing debt. The burden of repayment is shifted to future generations. Only in recent years has an awareness grown that the exploding cost of servicing government debt is already hurting current voters by increasing the cost of long-term money. Governments can also have a strong influence on the next most important players, the central banks.

The central banks

Central banks shape a country's monetary policy and determine the level of official interest rates. Throughout the 1980s and into the 1990s there has been a clear move towards greater independence for central banks. The central bank now considered the most independent of government influence is the German Bundesbank, which is Europe's most important financial institution. (The US Federal Reserve Bank is the world's most influential central bank.) However, one should bear in mind that governments have the power to change laws and constitutions and alter the central bank's role. Nevertheless, the balance of power has recently shifted in favour of the central banks. This is partly because deeply indebted governments need financial systems to work smoothly in order to refinance their debt and to raise new money efficiently. Furthermore, the huge deficits piled up in recent years have virtually paralysed attempts to stimulate the economy by means of fiscal policy (tax cuts). In this environment, one can see the central bank as an intermediary between the financial markets and the government. A central bank has the following tasks:

(i) to guarantee the stability of the currency, i.e. price stability;
(ii) to support sustainable economic growth;
(iii) to act as a lender of last resort in the event of systemic crises; and
(iv) to support and supervise a settlement system.

The support of sustainable growth has become less important as the state's ability to control economic developments has been discredited. Thus the worthy goals of full employment and economic growth have often become secondary to the target of price stability. This brings with it a conflict of interests. Using high interest rates to keep inflationary pressure down makes it more difficult to stimulate growth and prevent systemic banking problems. Systemic troubles usually emerge when an important player on the financial markets fails to meet its commitments. In theory, if a major link in the chain were to fail, it could bring the entire system down with it. Thus, if there seems to be any danger of this happening, the central bank has to 'reflate' the system by providing liquidity to any needy institution. These potentially conflicting goals make central banking a tightrope walk, sometimes closer to an art than a science.

The banking system

As discussed, the bond markets determine the price of money, provide a relatively safe home for financial resources and absorb most of the distribution risk entailed in the issuance of bonds. Together with the banking system they form the core of any capitalist system and by their sheer size have an increasing influence on the economy and political situation. Despite rapid changes in the financial system, banks retain a crucial role. Only they are able to fuel speculative manias by virtue of the credit/deposit multiplier; hence the attempts of central banks to convince bankers to continue lending even if the system is in distress.

The foreign exchange markets

The foreign exchange markets determine the value of one currency in terms of another. It can be said that a strong currency helps to keep a country's bond market buoyant. On the other hand, the sudden weakening of a currency can trigger large losses in a bond market if there is an exodus of capital abroad. A weak currency also raises inflationary fears (because of the danger of imported inflation), which always unsettle the bond market. A sharp currency depreciation can also create a vicious spiral: a drop in the currency leads to losses in the bond market; if a lot of the bond selling is being done by foreigners, the large amount of money moving out of the country exerts an even stronger downward pressure on the currency. Because of this, the ratio of foreign to domestic participation in a bond market is of crucial importance and will help determine the degree to which the currency can affect market rates.

The commodity markets

It has long been a custom in the bond markets to watch the prices of major commodities very closely. Traditionally, the most important markets to follow are gold and oil. Sharp price moves these markets have often served as a leading indicator of inflation and have thus coincided with some trend reversals in interest rates. Significant volatility in the oil price can have a marked effect on the growth/inflation equation in an industrial economy. Rising oil prices are inflationary or bad for bond prices, and vice versa.

A steep rise in gold prices, besides being an inflation indicator, can also be a sign of distrust in the stability of the overall financial system. However, gold has lost some of its traditional safe-haven status to other commodities, and gold prices are also influenced by supply. Despite the fact that gold is no longer the anchor of the capitalist financial system, it continues to have a somewhat ambiguous character – somewhere between a commodity and a kind of money. It is interesting that central banks continue to hold considerable amounts on non-interest-bearing gold and have not yet replaced it to any great extent with currency reserves in the form of government bonds like US Treasury bonds or German *Bunds*. It is clear that the influence of commodities on bond markets has faded somewhat in recent years as inflation has become less of a problem than in the past.

It is useful for a bond investor to follow trends in all major markets as these can help paint a clearer picture of economic growth and inflationary pressure – the main influences on bond market performance. The dominant trends in bonds, stocks, currencies, commodities and even property prices are all important. Conflicting trends (e.g. rising commodities and rising bonds, falling bonds and rising stocks, and so on) could give early warning that something is going wrong and that one of the markets may have to give in.

How do these players and markets interact and how do they influence the level of interest rates, and hence that of bond prices?

We shall begin by analysing a 'classic' economic cycle lasting for between four and six years.[4] In a cyclical economic downturn, the central bank will provide liquidity by lowering benchmark interest rates in order to stimulate the financial system. When the economy is in recession bond markets benefit because bond prices rise as interest rates fall. As rate cuts drive market rates lower, they reinvigorate the economy by encouraging more investment, production and consumption. Once growth does start to respond to the stimulus, the central bank will try to maintain it at a sustainable rate without triggering inflationary pressure. The successful attainment of the 'holy grail' of stability is called a 'soft landing', but these are very rare.

A renewal of inflationary pressure in mid-cycle is more common than a soft landing. This will cause the central bank to drain liquidity from the system by raising interest rates to calm speculative activity. The first sign of a turn in the interest rate trend often coincides with a major downturn in bond prices. How far bond prices fall depends on the extent to which inflationary pressure rises or is expected to rise.[5] Late in the cycle, bond prices recover as growth slows down again in response to higher rates. At this time, the central bank may start easing rates again to prevent a major economic slowdown. This change in trend towards lower interest rates is usually the start of a bull market for bonds as investors bet on the risk of a recession developing.

CONCLUSION

Forecasting interest rate trends is a difficult business. As we have seen, there are numerous influences on bond price volatility. It is fair to

[4] *We do stress that this is a simplified view. Apart from cyclical factors, there are many other influences on bond market behaviour, such as financial and political crises. It is also open to debate whether there are powerful long-term cycles (e.g. 50- to 54-year Kondratieff cycles) at work that override the shorter-term cycles.*

[5] *Remember that markets are often driven as much by expectations as by reality. The market has a tendency to expect both the best and the worst until the actual outcome is known. This erratic behaviour often accounts for large swings in bond prices despite a seemingly stable economic environment; these also reflect the market's preoccupation with economic data and the fact that a single statistic can trigger a trend change.*

say that buying bonds when inflation is stable or falling is a good idea and that one should sell bonds when inflation is rising. In a rising-rate environment, the short-term money markets are usually the best refuge for capital. It is often necessary to pay as much attention to political trends as to economic ones; as the recent past has illustrated, politics can influence currency and bond markets significantly. In times of instability, safe havens like the Swiss and German bond markets, which traditionally benefit from strong domestic currencies and a stable political environment, are highly recommended. ❖

THE GLOBAL FOREIGN EXCHANGE MARKET

Mark Rees-Kohler, Credit Suisse Private Banking, Zurich

Mark Rees-Kohler is responsible for business development with Credit Suisse's private banking clients in Europe. Before this he was head of global treasury marketing. Mr Rees-Kohler was formerly general manager of a leading American bank in Switzerland. His background includes financial marketing in Europe, Asia and North America. He studied at Oxford and the London School of Economics.

THE GLOBAL FOREIGN EXCHANGE MARKET

Mark Rees-Kohler, Credit Suisse Private Banking, Zurich

Every day one trillion US dollars change hands in the most liquid and transparent market in the world – the global foreign exchange market. The market is active 24 hours a day, five days a week. It can be accessed by telephone from anywhere in the world, there are no restrictions on membership and no commissions to be paid. To some, this degree of freedom represents a threat; to many others an opportunity. In this article we will take a look at the threats and opportunities that the global foreign exchange market presents to the sophisticated investor. We will distinguish between hedging and speculation, explain the meaning of currency overlay, and look at some of the investment instruments that can be employed when treating foreign currency as an asset class.

THE MARKET

Unlike the securities market, which tends to have a central physical focus in the form of a stock exchange that has a governing body, a set of rules for trading, and defined penalties for not observing those rules, the global foreign exchange market has no single focal point and is largely self-regulating. Hence, when an investor considers entering into a foreign exchange transaction, the selection of a creditworthy and reputable counterparty is extremely important.

Some exchanges, such as the Chicago Mercantile Exchange (CME), the Philadelphia Board of Trade (PBOT) and the Singapore Monetary Exchange (SIMEX), trade currency futures contracts and options on contracts. The largest foreign exchange flows, however, are not exchange-traded: they come from the many thousands of individual transactions carried out every day between market-makers (primarily banks) and market-takers, including private individuals, pension funds, smaller banks, insurance companies and corporations.

Although the global foreign exchange market has no official focal point – trading is conducted over the telephone, via electronic dealing systems and through brokers – market-makers do tend to be concentrated in certain locations. London, which handles US$300bn of business every day, leads the world in currency trading volume, followed by New York, Tokyo, Zurich and Hong Kong.

In theory it is possible to trade any currency against any other. However, the primary quotation for all currencies is against the US dollar. A non-dollar quotation is known as a cross-currency quotation. The trading of European currencies against the Deutschmark using the so-called EMS (European Monetary System) crosses is of increasing importance.

Currencies fall into three broad categories:

- majors, such as the US dollar, Deutschmark, yen and Swiss franc;
- minors, such as the Canadian dollar, the Dutch guilder and the New Zealand dollar; and
- exotics, including all emerging-market currencies.

These categories are not strictly defined and the currencies qualifying for a particular designation may change.

The primary characteristics of a major

currency are its use as a reserve currency by central banks and, above all, its tradeability, i.e. there must be a high degree of liquidity for spot and forward trades, and minimal exchange rate controls. The major currencies are quoted by all leading market-making banks, the minors are quoted by the larger market-making banks and regional specialist banks and the exotic currencies, outside their home markets, are quoted by only a very select group of truly global market-makers.

HEDGING AND SPECULATION

As soon as we begin to invest in assets denominated in a currency other than our base currency we incur a foreign exchange exposure: the return on our investment may be positively or negatively affected by a change in the rate of exchange between the investment currency and our base currency.

In the case of two currencies that are formally pegged, such as the Hong Kong dollar and the US dollar, the risk of loss through this foreign exchange exposure is minimal in normal circumstances. However, we need only recall the dramatic depreciation of the British pound or the Italian lira when they left the EMS in 1993 to realise that it is difficult for any rigid currency alignment to survive once it ceases to reflect the underlying economic fundamentals of the participating countries. Consequently, a cross-border investment strategy must be regarded as exceedingly speculative if it does not take into account currency risks and the possible effect of exchange rate shifts – no matter how unlikely they may seem – on the overall return.

The decision whether or not to hedge a foreign exchange exposure depends not only on the estimated risk of loss from unfavourable currency movements, but also on the cost of hedging. No matter what instrument we use for hedging there is a cost attached, either a direct one (discounts or premiums on the spot rate or purchase premiums for options) or an indirect one in the form of an opportunity cost. The simplest hedge, but possibly the least efficient (except in tax terms), is to borrow the foreign currency needed to buy the asset. The foreign currency exposure is thus limited to the return on the asset at the time of repatriation of funds, minus the cost of the loan.

Variations on this theme, whereby foreign currency assets and liabilities are matched off against each other, provide more sophisticated opportunities for hedging. In particular, there has been considerable research on the correlations between the behaviour of different currencies and the effectiveness of using an asset or liability in one currency to 'proxy hedge' a position in another currency.

The historical volatility of a cross-currency relationship is used as one measure of a currency's suitability for use as a proxy hedge. Before the widening of the EMS bands in 1993, for instance, the EMS currencies could reasonably be used as proxy hedges for each other: the degree of divergence between these currencies was very small. Following the formal realignment of the EMS, however, the divergence was allowed, theoretically, to widen to as much as 30% before the central banks were obliged to intervene.

Forward contracts

The most common instrument for hedging a foreign exchange exposure is the forward foreign exchange contract. The forward exchange rate is calculated for the required maturity from the spot rate and the interest differential between the two currencies. It is simple to use and can give total protection from currency swings for any given amount and period. The disadvantage of the forward contract is that it is relatively inflexible for, once a rate has been locked in, the only alternative is to reverse the original contract. Variations on this model tend to increase flexibility at the expense of liquidity and transparency.

Futures contracts

An alternative to a forward contract is the foreign exchange futures contract, which is a standardised contract traded on an organised exchange. The contract's amount and term cannot be tailor-made to meet the client's needs and the client must pay both a margin at the outset and brokerage commission on each transaction. However, the market is highly transparent and positions are marked to market daily, with subsequent settlement of profits and losses via a margin account.

OTC options

Although forward foreign exchange contracts still account for more than 80% of all foreign exchange hedging tools, the fastest-growing product segment is in over-the-counter (OTC) options. Like forward foreign exchange contracts, OTC options contracts are made with a bank and can be tailored to the amount and term that a client wishes to hedge. Unlike a forward foreign exchange contract, a call or put option leaves open the possibility that the client may benefit from a more extreme rate move in his or her favour while offering the security of a fully hedged position.

The hedge and the added profit opportunity have their price in the form of a purchase premium which the client pays for the option. Should the client be willing to forego a part of the profit opportunity, it is possible to reduce, or even eliminate, the cost of the hedge by implementing an options strategy commonly known as a 'zero-cost option'.

For example, an investor with a long dollar position may buy a dollar put option as a hedge and, at the same time, sell a dollar call option at a strike price that is calculated to offset the amount of premium paid for the put. If the spot rate at maturity falls within the two strike prices, neither option will be exercised and the value of the underlying asset will correspond to the actual spot rate. Should the spot rate at maturity drop below the strike price on the put option, the investor will exercise the option, thereby benefiting from the hedge. If the dollar strengthens significantly, the investor will only benefit to the extent of the strike price on the call option. At levels above this strike price the investor will be relieved of the long dollar position by the purchaser of the option.

CURRENCY OVERLAY

Currency overlay is the term used to describe the proactive management of currency exposure as an asset class in an investment portfolio. In place of the simple decision of whether or not to hedge a foreign currency asset against the base currency, currency overlay is used by leading asset managers to optimise the overall return on a portfolio. It permits the asset manager to separate the basic investment decisions about what type of asset to acquire, maturity, credit quality etc. from decisions about the currency in which it is to be denominated. As a result, the asset manager can, for example, take a bullish position on the interest rate outlook for a particular security, while taking a bearish position on the currency; this allows the asset manager to hedge and unhedge the currency position based on expectations of currency rate movements.

In contrast to the asset manager who applies traditional long-investment strategies, the currency overlay asset manager has the advantage of establishing long or short positions, and can therefore profit from a market move in either direction.

The way in which the asset manager comes to decisions about currency movements is the key to the success of a currency overlay programme. This is an area increasingly dominated by mathematicians, physicists and computer experts. The basic forecasting tool is often a computer model which analyses currency trends over a long period and extrapolates them into the future. The asset manager is then provided with price targets as well as support and resistance levels; this is known as 'trend-following'.

A number of sophisticated financial institutions are now working with neural networks, a form of artificial intelligence, which can be trained to recognise trading patterns and forecast price movements. There are also a number of other forecasting methods available. These are based on fundamental economic analysis, or on the many forms of technical analysis, otherwise known as 'charting'.

Most asset managers will use a combination of forecasting techniques, with some degree of discretion left to the trader who executes the deal in the market. It is, nevertheless, clear that currency overlay entails an investment in technical specialists and computer systems that is beyond the reach of many of the smaller asset management companies.

INVESTMENT INSTRUMENTS

There are many different ways in which an investor can participate in the foreign exchange market. One of the easiest methods is to purchase units in an investment fund (also known as a mutual fund or unit trust) managed by a reputable financial institution that specialises in foreign exchange. Such funds are not to be confused with the so-called hedge funds, which may invest in currency as an asset class but often include a spread of other financial assets. A foreign exchange fund is designed to provide large and small investors with the opportunity to entrust their capital to a professional trading institution with an established track record. (One such fund is the Credis Forex Fund Global.) However, in accordance with modern portfolio theory for more-traditional asset classes, a diversification across managers, markets and trading styles will increase return and reduce volatility.

The risks and rewards of an investment in a currency fund should clearly be distinguished from those of investing in stocks or bonds. For example, the Ferrell FX Index, which monitors the performance of more than 40 funds managed in the US, the UK, France, Ireland and Switzerland, demonstrates a volatility with a standard deviation of 19.1%, compared with a 15.0% volatility for the Standard & Poor's (S&P) 500, a key stock index.

Bond indices show a still lower level of volatility with 5.0% for the US domestic bond market and 12.1% for the international bond market. The annual returns in the Ferrell FX Index over the last eight years range from a high of 56.4% in 1990 to a low of -6.3% in 1994. In comparison, the high for the S&P 500 was 31.6% in 1989 against a low of -3.1% in 1990.

One feature of currency funds that is often confused with their volatility is leverage. Whether a fund is using the futures market, the interbank forward market or currency options, only a small part of the nominal contract value is required in the form of margin or a 'security deposit'. The fund can enter into contracts with a nominal value several times in excess of its capital.

A fund will commonly leverage its capital between three and seven times. The degree of leverage should be determined by price volatility: when the market is more volatile, the fund should reduce its leverage, and vice versa.

An important indicator of a fund's success, and conservatism, in this regard is the maximum peak-to-trough loss. The aggregate of funds covered by the Ferrell FX Index shows that in the last eight years the maximum peak-to-trough loss was 14.8%, which compares very favourably with the 29.6% of the S&P 500 Index. The Lehman Corporate Bond Index showed the lowest peak-to-trough loss at 6.3%.

MARGIN TRADING

For the investor who wishes to retain control of a currency portfolio, a suitable instrument is the margin, or collateralised, trading account. In return for a pledged deposit, commonly in cash or Treasury Bills, the investor receives a trading limit from a bank or broker that can be utilised to buy and sell foreign currencies on the spot, forward or futures markets.

In the case of futures, an exchange-traded instrument, there is daily settlement of profit and loss based on a mark-to-market of the outstanding positions. In the case of spot and forward trading, the bank will generally set a level of margin utilisation, for example 40%, at which point it requires the investor to replenish the margin account to 100%.

The primary attraction of a margin trading account, apart from the direct involvement in trading, is that investors are permitted to leverage their capital 10, 20 or even 40 times, depending on the credit policy of the institution involved. Some institutions also permit different levels of leverage during local trading hours from those available overnight.

To put this in context, if the Deutschmark is trading at 1.5000 to the US dollar, a 1% move represents 1.5 pfennigs (DM0.0150). At a leverage of 20 times capital, an investor has deposited with the financial institution collateral equivalent to 5% of his potential open positions. Assuming he is required to replenish the margin account when 40% of the collateral has been lost, a negative three-pfennig (2%) move on the position can be tolerated before he is obliged to top up the account.

By dealing with a global financial institution the investor has access to the foreign exchange market 24 hours a day, either through a local night-desk or via toll-free telephone lines to a network of linked dealing rooms in the important financial centres.

CONCLUSION

Trading foreign exchange as an asset class is not just a matter of backing a hunch: successful practitioners commonly rely on sophisticated forecasting models to help them make decisions. However, still more important than forecasting is a strict appreciation of the importance of risk management. Wide swings in profitability are a sign of uncontrolled trading and will always lead to a net loss.

On a controlled basis, strategic currency investing can provide important diversification benefits to an investor; currency movements do not necessarily correlate with traditional equity and fixed income investments. At first sight, currency volatility may seem threatening to investors seeking to diversify their portfolios across borders. However, it is reassuring that instruments are available to manage currency volatility, and even to convert the threat into additional profit opportunities. ❖

PRECIOUS METALS

Norbert Schroff, Precious Metals, Credit Suisse Zurich

Norbert Schroff started work in 1960 as an apprentice at Metallgesellschaft AG before moving on to work as a commodity trader in 1962. In 1973 he became manager of a trading subsidiary and gained extensive managerial experience in Malaysia and Frankfurt. He joined the Credit Suisse bullion department in 1980 and is now in charge of Credit Suisse's worldwide precious metals trading activities.

PRECIOUS METALS

Norbert Schroff, Precious Metals, Credit Suisse Zurich

GOLD – A FADED LEGEND?

In the last 10 years gold has lost much of its appeal for investors in North America and Europe. At the end of the 1970s virtually every investment portfolio still boasted a substantial gold weighting. Then, suddenly, gold all but disappeared from the vast majority of portfolios. This loss of prestige was primarily a result of the disappointing gold price performance in the latter half of the 1980s: between 1987 and 1992 gold shed some 30% of its value. Indeed, in Swiss franc or yen terms, the slump lasted much longer (1982-95). Over the same period, the yellow metal surrendered a number of its traditional functions. Nowadays it is rarely used to hedge against the loss of purchasing power, for example.

Over the last decade, the central banks of the Western industrial nations have won impressive victories in the fight against inflation, bringing price increases down to a slower, steadier rate since the second oil shock in 1979. Other risks, such as falling financial market prices, can now be hedged systematically using new types of financial instruments; options and futures, in particular, have gained widespread popularity since the mid-1980s.

Recently, however, Western investors have taken a revived interest in gold, showing that it continues to play its traditional role as a store of value in times of political and economic instability. It should not, therefore,

FIGURE 1: GOLD PRICE (MONTHLY) IN US DOLLARS AND SWISS FRANCS

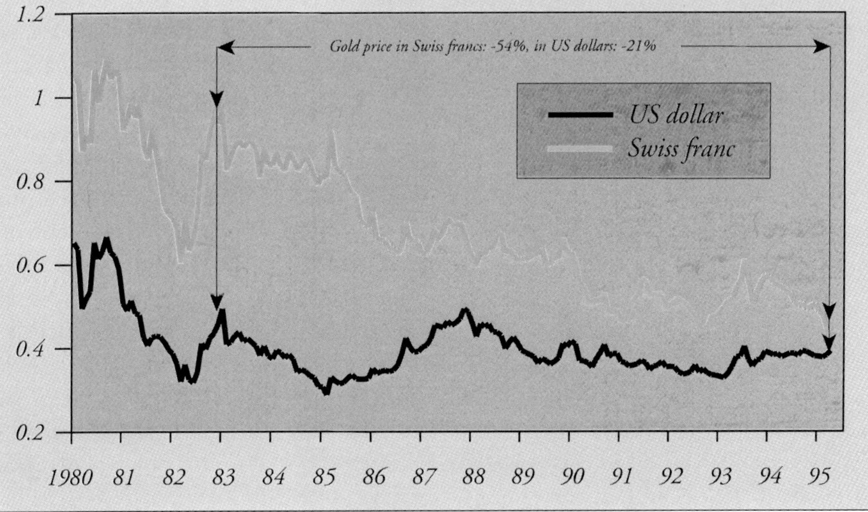

be overlooked as a potential investment medium. And, like other markets, the gold market has been transformed by modern financial instruments such as options and gold warrants, so that investors in gold, too, have a range of new investment opportunities at their disposal.

Outside Europe and North America, gold is often seen as an important investment medium, either for reasons of tradition or because the local banking system is underdeveloped. In many places, gold has retained its classic role as the ultimate hedging instrument, protecting savings against a loss of value resulting from inflation, local currency devaluation or war, for instance. In rural areas, people still use it as a strategic reserve to tide them through unexpectedly bad harvests. In all these capacities, gold serves as a 'reserve of last resort'.

The widespread habit of hoarding US-dollar notes is proof that people around the world still feel they need to keep their savings in some kind of hard currency. Just under half of all US dollars in circulation are held outside the United States: an amount equivalent to approximately 10,000 tonnes of gold. Throughout the 1980s Asia was a steady buyer of gold bars, especially when the gold price was falling. And not only bars: in Asia, gold jewellery is often purchased for investment purposes.

Any analysis of the gold market nowadays must be made from a global perspective; the emerging nations are set to become more and more significant as both buyers and producers of gold.

Gold's significance in the nineties

Autumn 1992 saw gold finally emerge from the long slump in which it had languished for most of the 1980s. Within a few months, its price climbed from slightly less than US$330/oz to nearly US$400/oz, inspiring hopes that, with the nineties, a new age was dawning for the metal. Whether this is true or not, there now seems little risk of a long-term fall in the gold price to much less than US$380/oz, given the persistent gap between the supply of mined gold and demand.

In the early 1990s the gold price sank to US$330/oz, its lowest in terms of purchasing power since the mid-1970s. At this point, gold was clearly undervalued. This under valuation led to mounting cost pressures on gold producers: in 1992, for example, roughly 20% of mining companies were no longer able to cover their overall costs (including write-downs, debt repayments and debt-service charges) by selling at the market price. By 1993, once the average gold price had recovered and cost-cutting measures had begun to bite, this proportion had fallen to around 5%. People then realised that, in real terms, gold was cheap, and this prompted an inexorable rise in demand, notably from Asia, at a time when primary supply was expanding only slowly.

Had it not been for unusually heavy selling by the central banks in 1992, when they off-loaded 626 tonnes of gold, the brisk demand could scarcely have been satisfied at such low prices. From 1988 onwards there has been a steadily widening gap between primary supply (Western mine output plus sales by the former Eastern bloc) and manufacturing demand. The largest imbalance was in 1992, when manufacturing demand outstripped primary supply by 951 tonnes, from a position of near parity in 1987. The upward adjustment in the gold price in 1993 brought a slight improvement, but the imbalance is still substantial (744 tonnes in 1994).

The growing investor interest in gold is certainly a product to some extent of oft-cited bull factors, such as the prospect of higher US inflation, America's low real-term money market rates, the downward trend on Europe's money markets and increased uncertainty regarding the outlook for bonds and equities. It is also true that the gold market often moves sharply in response to interventions by private investors, as demonstrated by its reaction to recent buying by large US investment funds. Ultimately, however, such investors have been motivated by one fundamental: the deteriorating supply and demand situation. Demand has been rising far too fast, especially in relation to mining output, for the gold price to stay where it was. Even the unusually heavy sales of central bank gold in 1992-93, when more than 1,000 tonnes came on to the market, have been unable to relieve the long-term upside pressure.

The outlook for the next few years is on the whole positive. Gold prices should remain stable, or rise gently, until the end of the decade. Although primary supply will continue to grow, such growth will be much slower than it was in the 1980s. The depth to which the gold price had sunk by the end of that decade dissuaded many mining companies from investing in exploration.

While Asian and Latin American countries are set to boost their gold production, output will stagnate in the traditional gold-mining nations. This is especially true of South Africa, the world's highest-cost producer, where production fell sharply in 1994, with every likelihood of a continuing downward slide in 1995. Supply will thus expand only gradually, while demand continues to soar.

With living standards improving in Asia, we could well see an average annual growth in demand of 2% between now and the year 2000. Furthermore, buyers in Asia and Latin America often see gold as a form of investment, partly out of tradition (as in India) and partly because of the absence of other investment options. A good example of this is provided by the former Soviet Union, where the demand for gold as a hedge against inflation was so great in 1994 that it was changing hands on the black market at roughly twice the going price on world markets.

However, although conditions are now very favourable, there is unlikely to be a repetition of the gold price boom of the late 1970s. Central banks and international organisations have amassed a huge stockpile of roughly 35,000 tonnes of gold, and investors are sitting on another 20,000 tonnes. In the future, therefore, strong advances in the gold price will be kept in check, as they have been in the past, by bouts of selling by investors as well as by forward sales of output by mines. The reserve stockpile is unlikely to be completely sold off, however, especially as gold's role in any new currency regime is yet to be clarified and the metal still serves as the reserve of last resort. In these circumstances, the likely scenario for the next 10 years is that the gold price will move in the range of US$400/oz to US$600/oz.

Investing in gold

Although the current outlook is considerably better than in the second half of the 1980s, gold still ranks as a comparatively risky investment. The gold price is always liable to temporary setbacks triggered by surprise sales of old reserves or a spate of forward selling by gold producers. Good timing thus requires an enormous amount of market information. But because gold normally shows no correlation to the equities market, its main use is as a means of diversifying portfolios, although the amount invested in gold should never be more than a relatively small percentage of overall assets.

Prospective investors in gold can choose from the following vehicles:

- metal accounts;
- gold bars and wafers from 1/4 oz to 100 oz or from 1g to 12.5kg;

- gold coins;
- call options or warrants;
- forward purchases; and
- gold-mine stocks and gold-linked investment funds.

Gold bars and coins are only recommended for collectors or investors with a longer-term outlook. In Switzerland, purchases of physical gold are exempt from VAT. Since, in the physical gold trade, only bars and coins that are free of all scratches and other damage are accepted as 'good delivery', steps must be taken to ensure proper storage. Private banks generally offer three possibilities for safe storage of gold:

- a private safe-deposit box;
- an individual safe-custody account; or
- a collective safe-custody account.

Buying call options or gold warrants is an attractive alternative to physical gold in that these instruments allow substantial investments in gold with relatively little liquidity. When the gold price is rising, much greater potential profit is offered in relation to the funds committed than in spot trading. At the same time, the potential loss is limited to the premium paid. Forward transactions in gold are not recommended for private investors: when prices move in the 'wrong' direction, an additional amount is payable and the potential loss becomes too great.

Rising gold prices also tend to bolster the shares of gold-mining companies. But because many will have already hedged their future output on the forward markets, a rise in the price of gold does not necessarily lead to a corresponding share-price gain across the board. This means that the investor needs accurate information on each company. A professionally managed gold-based investment fund such as Credit Suisse's Gold Fund may offer a more secure investment opportunity.

SILVER AND PLATINUM

The main price trends for platinum and silver run essentially parallel to those of gold, except at a much higher level of volatility. Also, because silver and platinum are mainly used in industry, they react much more strongly than gold to fluctuations in the general economy. With industrial demand on the rise, the outlook for silver and platinum over the next two to three years is even rosier than that for gold. Since physical purchases of silver and platinum are subject to VAT, investors should consider buying via a metal account or purchasing call options and warrants. However, both metals represent speculative investments entailing correspondingly high levels of risk.

In the case of platinum, the supply side will undergo a considerable squeeze, especially over the next few years. The key industrial consumers are the jewellery sector, which accounts for slightly less than 40% of total demand, and the car industry, which accounts for 35%. After the recent world recession, car manufacturers are buoyant again. With world car production expected to increase by more than 5% in 1994-95, and more and more governments adopting stricter environmental regulations, higher sales of platinum for use in catalytic converters are all but guaranteed.

The bulk of platinum jewellery is bought by the Japanese. The yen's strength means that platinum is now available at bargain prices; this will ensure that demand for platinum also picks up in the jewellery sector. Meanwhile, growth on the supply side in the next few years will probably be slow at best. This is because the 1990-93 price slump prompted mines to shelve their expansion plans, and because the profit margins of the two biggest producers in South Africa have narrowed considerably.

Silver, unlike platinum, has traditionally been seen as an investment vehicle. This means that large amounts of silver are held by private investors, much of this stockpile having been accumulated in the first half of the 1980s when the price level was between US$6/oz and US$7/oz. Although the underlying fundamentals of the silver market are excellent – the demand overhang has been growing since 1990 – the existence of this stockpile means that silver will remain rather prone to temporary price setbacks.

CONCLUSION

Although gold's importance as an investment medium has faded somewhat over the last 10 years, not least because of the plethora of new instruments launched on the international financial markets, the yellow metal has retained its traditional role as a store of value. Now and in the future, gold will remain just one of many possible investment vehicles. In the industrial nations of the OECD, it will serve as a means of diversifying portfolios or, at times, as a focus for high-risk speculation. The advent of many new financial instruments in the last 15 years has radically changed the nature of the gold market, as it has changed other markets. Today, for instance, investors in gold can buy options and gold warrants as an attractive alternative to the costly storage of physical gold.

After a long decline in the 1980s, the outlook for the gold price is bright once again. Strong demand from the jewellery industry means that modest price gains are more than possible. Rising living standards in the emerging economies of Asia will keep demand for gold at high levels. By contrast, the scope for boosting output is limited, owing to recent low levels of investment in exploration.

Considering that the downside risk in gold is limited by fundamentals and can be hedged by using modern instruments, maintaining a small percentage of gold holdings may well be considered a good way to diversify a portfolio. ❖

EMERGING MARKETS IN PERSPECTIVE

Dr Marc Faber, Marc Faber Ltd, Hong Kong
This article first appeared in *The Gloom, Boom & Doom Report*

After obtaining a PhD in economics from the University of Zurich, Marc Faber went to work for White Weld & Co. Ltd. Dr. Faber is the investment manager of Iconoclastic International Fund, a global fund specialising in unusual investment opportunities. He publishes *The Gloom, Boom & Doom Report* and is the author of *The Great Money Illusion.* He is well known for his contrarian investment approach.

Phone: (+852) 2801 5410

EMERGING MARKETS IN PERSPECTIVE

Dr Marc Faber, Marc Faber Ltd, Hong Kong

This article first appeared in *The Gloom, Boom & Doom Report*

Reading the international financial press, one might think that emerging market investing was a new phenomenon invented recently by the likes of Templeton, Morgan Stanley, Genesis or Barings. One also gets the impression that the road to emerging economies is paved with gold, based on the recent performance of these funds. Emerging market portfolio managers are as popular today as hedge-fund managers or takeover artists were in the mid-1980s.

A few years ago, I began to collect first editions of books about economics. Since books are very bulky, I soon ran out of space and so decided to limit my purchases to books relating to business-cycle theory – a subject of particular interest. The advantage of buying old books is that one gains an insight into how events were perceived at the time and not how we look at them today.

Economics is a young science. Before the 18th century practically no economic literature had been published. But following the issue of Adam Smith's *On the Wealth of Nations* in 1776, a large number of books on economics were published in the United Kingdom and France that dealt principally with issues such as the national debt, taxation, free trade, population growth, land rents, wages, etc. The first major work on business cycles, *Des Crises Commerciales et de leur Retour Périodique en France, en Angleterre et aux Etats-Unis*, was written by Clement Juglar and published in 1860 (and expanded and republished with a series of articles in a second edition in 1889). Around the same time, the German economist Max Wirth published *Geschichte der Handelskrisen* and Samuel Benner produced *Benner's Prophesies,* the first book on commodity and stock market cycles. The reason books on business cycles began to appear towards the end of the 19th century was that, although the European and American economies had grown very rapidly following the Industrial Revolution, periods of strong growth were always followed by severe commercial crises. This observation led early 20th-century economists such as Wesley Mitchell, A.C. Pigou, Nikolai Kondratieff and Joseph Schumpeter to focus on business cycles and their causes as well as their periodicity.

According to Mitchell's *Business Cycles,* published in 1920, American business had passed through a 'crisis' on 15 occasions in the preceding 110 years. The list of crisis years (1812, 1818, 1825, 1837, 1847, 1857, 1873, 1884, 1890, 1893, 1903, 1907, 1910, 1913 and 1920) reveals that, 'These cycles differ widely in duration, in intensity, in the relative prominence of their various phenomena, and the sequence of their phases.'

But what do American business cycles in the 19th century have to do with today's emerging economies? In my opinion, they provide an excellent historical precedent for today's emerging markets, simply because America was then the world's largest emerging economy. A study of past American economic cycles and the periodically recurring crises may thus provide some interesting insights into the current emerging market investment euphoria and lay to rest some of the prevailing myths about the ease of striking it rich in emerging economies.

THE EMERGING AMERICAN ECONOMY

Some readers may question the validity of comparing the economic development of 19th-century America to that of today's emerging regions. However, I have researched the subject thoroughly, and the more I read the more similarities come to light. In particular, America experienced rapid population growth and underwent a major transformation from a predominantly agrarian producer at the beginning of the 1800s to the world's largest producer of manufactured goods by 1900.

In 1790, the United States was a nation of fewer than four million people dispersed over a huge land area. (For the sake of comparison, Europe's population in 1800 was about 180 million, India's was 190 million, and China's 320 million.) Only seven cities had a population of more than 5,000, and 12 a population of more than 2,500. The remaining population of 3.7 million was rural. During the first half of the century the population grew by an average of about 3.5% per annum (largely because of immigration), but slowed to around 2% in the 1890s (see Figure 1).

By 1885, the United States (which had hardly any industry at the beginning of the century) led the world in manufacturing, producing 28.9% of the world's manufactured goods. Britain was second, with 26.6% and Germany third with 13.9%. America's rise to become the world's dominant economic power around 1885 is unique from a historical point of view. And today it is the dream of every emerging market investor to find another 19th-century America.

However, and this is the point I would like to emphasise in this article, America's road from extremely humble beginnings to an unprecedented economic dominance and prosperity was not smooth or paved with gold (certainly not for foreign investors). Rather, it was strewn with nasty and frequently, even for the skilled driver, unavoidable potholes. Furthermore, the American route to prosperity was often controlled by all manner of bandits and extremely corrupt government officials who were very clever at robbing the naive and credulous foreign or local traveller.

Economic growth in America in the last century was closely linked to the cotton industry (especially in the South), to canal and railroad construction throughout the continent

FIGURE 1: US POPULATION GROWTH 1790-1900

Census Year	Total Population	% Increase over Preceding Census
1790	*3,929,625*	-
1800	*5,308,483*	*35.1*
1810	*7,239,881*	*36.4*
1820	*9,638,453*	*33.1*
1830	*12,866,020*	*33.5*
1840	*17,069,453*	*32.7*
1850	*23,191,876*	*35.9*
1860	*31,443,321*	*35.6*
1870	*38,558,371*	*22.6*
1880	*50,155,783*	*30.1*
1890	*62,947,714*	*25.5*
1900	*75,994,575*	*20.7*

Source: *A Century of Population in the United States 1790-1900*, Department of Commerce and Labour, Washington, 1909

(which opened up the West), to the exploitation of America's vast natural resources (for example, the gold discoveries in California), to the development of new tracts of land, and to the rise of the steel industry.

The cotton industry

In the first half of the 19th century, the American cotton industry grew very rapidly and the prosperity of the South depended largely on good crops and high prices. Since the invention of the cotton gin by Ely Whitney in 1793, the South had become a highly specialised cotton-producing region and the rapidly growing income the South received from its cotton exports was used to purchase foodstuffs from the West as well as manufactured goods and services from the Northeast. Although America produced hardly any cotton before 1800, its plantations supplied five-sixths of world production by 1860.

For the South at least, the cotton industry was then about as important as the oil industry is in the Middle East today, and certainly far more economically significant than the agricultural sector is in countries such as New Zealand or Australia. Just as the South's fortunes were very much tied to the price of cotton, so too were American manufacturing output, principally in the Northeast, and imports. In the 1830s a cotton boom benefited the entire economy. Between 1831 and 1836, income from cotton trade rose from US$25m to US$71m. But when cotton prices began to fall in 1837, the South's ability to consume declined along with its income, to the detriment of industrial production in the Northeast and of imports of luxury goods from Europe. The severe depression of 1837 to 1842 was thus due, at least to some extent, to a 70% decline in the price of cotton (see Figure 2).

We introduce discussion of the American

FIGURE 2: COTTON PRICES IN NEW YORK FOR MIDDLING UPLAND COTTON 1826-1882

Year	High	Low
1826	14	9
1827	12	8
1828	13	9
1829	11	8
1830	13	8
1831	11	7
1832	12	7
1833	17	9
1834	16	10
1835	20	15
1836	20	12
1837	17	7
1838	12	9
1839	16	11
1840	10	8
1841	11	9
1842	9	7
1843	8	5
1844	9	5
1845	9	4
1846	9	6
1847	12	7
1848	8	5
1849	11	6
1850	14	11
1851	14	8
1852	10	8
1853	11	10
1854	10	8
1855	11	7
1856	12	9
1857	15	13
1858	13	9
1859	12	11
1860	11	10
1861	28	11
1862	68	20
1863	88	54
1864	190	72
1865	122	33
1866	52	32
1867	36	15 1/2
1868	33	16
1869	35	25
1870	25 3/4	15
1871	21 1/4	14 3/4
1872	27 3/8	18 5/8
1873	21 3/8	13 5/8
1874	18 7/8	14 3/4
1875	17 1/8	13 1/16
1876	13 3/8	10 7/8
1877	13 15/16	10 13/16
1878	12 3/16	8 13/16
1879	13 3/4	9 1/4
1880	13 1/4	10 15/16
1881	13	10 7/16
1882	13 1/16	10 1/4

Source: Samuel Benner, *Benner's Prophecies*, Washington, 1884

cotton industry because, although it expanded very rapidly in the first three-quarters of the last century (producing 2,300 million pounds of cotton by 1860, up from 1.5 million in 1790), from time to time it bankrupted a large number of cotton growers who had paid over the odds for land when prices were high. The South's reliance on cotton and on cheap slave labour to produce it also contributed to the outbreak of the Civil War. Furthermore, cotton fabrics were then as popular among consumers as VCRs, colour televisions and soft drinks are today. Emerging market investors should not overlook this point.

That an industry is growing rapidly in the long term does not change the fact that, when the market for its products deteriorates because of temporary oversupply or a deficiency in demand, prices fall, leading to some very painful consequences, not only for its owners, but also for its suppliers – and especially for its creditors! Today, every emerging economy wants to grow through the export of manufactured consumer goods. But that does not mean that every industry will be successful year after year. Consumer electronics, NIKE shoes, soft drinks and garments today are no different from cotton in the last century and will thus from time to time be vulnerable to periods of overproduction and falling prices.

Infrastructure growth

In view of the gargantuan appetite of investors for participation in infrastructure projects in emerging economies, a brief look at the construction of the canals and railways in America in the 19th century is in order. It may also dampen some of the prevailing enthusiasm. The impetus for the canal boom, which lasted from about 1820 to 1836, was the completion of the 587-kilometre Erie Canal in 1824. The canal proved enormously successful because it enabled grain to be transported from the Great Lakes to New York. It also linked New York to a hinterland of great economic potential and made it the financial and commercial metropolis of the Union. The Erie Canal cut to one-tenth both travel time and shipping expense for journeys from Detroit, Cleveland and Buffalo to New York.

In 1833 the Welland Canal, which bypassed Niagara Falls, was completed, connecting the Great Lakes to the St Lawrence river. The completion of these two canals was a great boon to New York, as trade routes and population growth shifted to that city. It also caused a wild canal mania as every city or state that had two or more bodies of water planned or built canals to link them, in the hope that they could replicate the success of the Erie Canal. The 1820-1836 American canal mania did not fail to arouse the interest of European emerging market investors, and the demand for American canal bonds and shares was so great that there were not enough issues to satisfy it. When the Morris Canal and Banking Company raised US$1m, most of which was offered from England, it could have been US$20m.

Property and banking

Hand in hand with the cotton and canal boom went a speculative orgy in property and banking shares. Between 1830 and 1837, 347 new banks were licensed, frequently led by men of very questionable background and no banking experience. These banks enabled speculators to purchase land in public sales at inflated prices on credit. Public land sales rose from 4.7 million acres (1.9 million hectares) in 1834 to 12.6 million in 1835 and 20 million acres in 1836 as prices for land in most parts of the East and South more than doubled.

The Chicago land boom of the early 1830s has been well documented. Until then, Chicago was a relatively unknown town and land there was much cheaper than on the East Coast. However, the planned Illinois-Michigan Canal, which was to connect Chicago's harbour and Lake Michigan with the Illinois river, began to attract great speculative interest in Chicago land. (Upon completion of this canal, Chicago became a major transportation hub.)

Within a few years, land prices rose a hundredfold. The construction of the Illinois-Michigan Canal finally began in 1836, at about the same time as Chicago land prices peaked. Following the 1837 crisis (see Figure 3), Chicago property prices plunged by 90% and it took 10 years to complete the canal. (Construction came to a standstill during the 1837-1842 depression as the State of Illinois, burdened with large debts and a weak economy, went into default – as did eight other states.)

FIGURE 3: PRICES OF SELECTED ISSUES, 1837 AND 1841

Issue	1837 High	25 November 1841
United States Bank	*122*	*4*
Vicksburg Bank	*89*	*5*
Kentucky Bank	*92*	*56*
North American Trust	*95*	*3*
Farmers' Trust	*113*	*30*
Illinois State Bank	*80*	*35*
Morris Canal	*75*	*0*
Patterson R R	*75*	*53*
Long Island R R	*60*	*52*

Source: Pratt, *Work of Wall Street* and Robert Sobel, *Panic on Wall Street*

MARKET CYCLES

How the American boom ended should be of some interest to emerging market investors: in 1836, British reserves fell by almost 50% as gold flowed to America to purchase American securities. The Bank of England was thus twice forced to increase the rediscount rate, which led to a monetary panic in Britain and the closure of several banks. Higher interest rates in Britain (in 1837 the call rate rose to around 15%) reduced the demand for American securities almost overnight, and this at precisely the time when demand for money in America was at its highest. The sudden disappearance of foreign capital had a devastating effect on the American economy: all American banks had to suspend specie payments, and between 1837 and 1839 more than 1,500 banks failed.

The 1837 crisis was aggravated by a sharp fall in the price of cotton (see Figure 2) which bankrupted many cotton speculators and led to a sharp fall in the demand for consumer goods from the South. By the autumn of 1837, nine-tenths of America's eastern factories had closed. The depression of 1837 to 1841, known as 'the Hard Time Depression', was extremely severe. It also had serious repercussions for Europe, especially England, because so much of the money that had been invested in American canals, banks and property had become worthless. Figure 3 gives an indication of the severity of the slump in financial assets which followed the earlier prosperity – a consequence of rising cotton prices, excessive monetary expansion and euphoric purchases of American securities by foreign investors.

And although before 1837 foreign investors seemed to have an insatiable appetite for American securities, after the crisis *The Times* of London wrote, 'The people of the United States may be fully persuaded that there is a certain class of securities to which no abundance of money, however great, can give value, and that in this class their own securities stand preeminent.'

From a historical point of view, the 1837 crisis and ensuing slump are interesting: the crisis was triggered by an event outside America (the increase in the rediscount rate in Britain) that reduced the flow of money from Europe to America, leading to the first wave of bankruptcies among American banks. Furthermore, the crisis and the subsequent depression (known in Europe as 'the Hungry Forties') were for the first time international in scope. Recovery began in the mid-1840s and accelerated thereafter mainly as a result of increasing railroad construction which, as Schumpeter wrote, 'played a leading role in the process of economic evolution which produced the situations that developed into crisis'.

Figure 4 shows miles of railroad in operation in America and annual completion. As can be seen, a mini-boom took place in 1835 but, in comparison to the canal craze, it was insignificant and had little impact on the economy. However, after 1848, railroad construction expanded rapidly and stimulated the economy, bringing about the first railroad mania in the mid-1850s and the crisis that followed in 1857. The economy also improved because of a gradual rise in cotton prices (see Figure 2) and an expansion of cotton production from 2.1 million bales in 1850 to 4.5 million bales in 1859.

One of the reasons for the pick-up in railroad construction in the 1850s and the attendant boom was the discovery of gold on Swiss settler Johann Sutter's property in California's lower Sacramento Valley. Gold production increased after 1847, which helped finance the economic expansion and fuel the land and railroad-construction boom.

The railroading of America led to a tenfold increase in the production of pig iron and a doubling of coal production between 1850 and 1856. Foreigners – who after the debacle of 1837 had left the American market vowing never to return – were once again eagerly buying American railroad securities and, for the first time in American financial history, speculative mining stocks, in order to capitalise on the expected bonanza from the gold discovery. By 1853, 26% of all American railroad bonds were held by foreign investors

FIGURE 4: MILES OF RAILROAD IN THE US

Years	Miles in Operation	Annual Increase
1831	95	72
1832	229	134
1833	380	151
1834	633	253
1835	1,098	465
1836	1,273	175
1837	1,497	224
1838	1,913	416
1839	2,302	389
1840	2,818	516
1841	3,535	717
1842	4,026	491
1843	4,185	159
1844	4,377	192
1845	4,633	256
1846	4,930	297
1847	5,598	668
1848	5,996	398
1849	7,365	1,369
1850	9,021	1,656
1851	10,982	1,961
1852	12,908	1,926
1853	15,360	2,452
1854	16,720	1,360
1855	18,374	1,654
1856	22,016	3,647
1857	24,503	2,647
1858	26,968	2,465
1859	28,789	1,821
1860	30,635	1,846
1861	31,286	651
1862	32,120	834
1863	33,170	1,050
1864	33,908	738
1865	35,085	1,177
1866	36,801	1,742
1867	39,250	2,449
1868	42,229	2,979
1869	46,844	4,615
1870	52,914	6,070
1871	60,283	7,379
1872	66,171	5,878
1873	70,278	4,107
1874	72,383	2,105
1875	74,096	1,712
1876	76,808	2,712
1877	79,089	2,281
1878	81,776	2,687
1879	86,497	4,721
1880	91,944	7,174
1881	101,733	9,789
1882	113,329	11,591

Source: Samuel Benner, *Benner's Prophecies*, Washington, 1884

who focused particularly on the more speculative issues.

One of the interesting aspects of the 1850-1857 railroad boom was that, for investors, it came to an end long before the 1857 crisis. Figure 5 shows the performance of the All-Inclusive Index of railroad stocks compared to the completion of railroad miles between 1850 and 1860. As can be seen, the railroad index peaked in December 1852, long before railroad construction in 1856. This was largely because capital transfers, which reached a high of US$56m in 1853, declined to only US$12m in 1856.

The reason for the slow-down in overseas buying was the Crimean War (1852-1856) fought between Great Britain, France and Turkey on one side and Russia on the other. The war drained European liquidity, and led to rising interest rates and reduced demand on the Continent for American railroad securities; and this at a time when the railroads needed to raise the maximum capital in order to finance their expansion. To put it simply, there was a large oversupply of securities in the American capital market, which depressed prices even though business conditions in America remained strong until 1857. (It should be remembered that the Crimean War was, on balance, very favourable for America as agricultural product prices rose and industrial production was stimulated.) However, emerging market investors should remember: stock prices can fall long before the economy turns down, or well ahead of deteriorating profits, simply because monetary conditions deteriorate.

The catalyst for the 1857 crisis was the failure of the Ohio Life & Trust Company (which specialised in land and railroad investments and commodity futures) and the sinking off Cape Hatteras of the steamer Central America, with its cargo of US$1.6m in California gold. (Gold shipments from California had always been a confidence booster and had provided some liquidity to the East Coast banks.) The panic reached a peak in October 1857 when 1,415 banks and numerous railroads failed (the Mining Exchange in New York even closed its doors) and the crisis spread to London and Paris, where there was a vigorous market in American securities.

An interesting aspect of the 1857 panic is that it was principally a financial crisis and relatively little damage was done to the economy, particularly in the South. American cotton worth US$128m had been exported in

FIGURE 5: ALL-INCLUSIVE INDEX OF RAILROAD STOCKS AND RAILROAD MILES COMPLETED, 1850-1860

Year	High	Low	Miles of Railroad Completed
1850	*95*	*79*	*1,656*
1851	*96*	*87*	*1,961*
1852	*110*	*89*	*1,926*
1853	*105*	*89*	*2,452*
1854	*98*	*74*	*1,360*
1855	*80*	*66*	*1,654*
1856	*73*	*68*	*3,647*
1857	*71*	*39*	*2,647*
1858	*61*	*49*	*2,465*
1859	*56*	*47*	*1,821*
1860	*74*	*48*	*1,846*

Source: Smith & Cole, *Fluctuations in Business*

1856, and cotton export earnings rose to US$192m in 1860. The strong economic performance of the cotton-growing areas at a time when the financial sector of the Northeast was in crisis gave the South confidence that, in a conflict, the Northern economy would collapse – a notion that eventually contributed to the outbreak of hostilities and the Civil War. This is another point for investors to remember: a severe financial crisis can occur for a number of reasons (including excessive speculation, which was not the case in the 1850s) and not necessarily because of a general and persistent downturn in the real economy.

For emerging market investors, the lead-up to the 1873 global crisis and the depression that followed is also of interest. The end of the American Civil War (1861-1865) and German unification, which was achieved following the Franco-Prussian War (1870-1871), gave immense impetus to American and German economic development, enabling these nations to catch up with Britain and France, once in the vanguard of the Industrial Revolution. This was especially true for Germany, whose economy was stimulated by the French War Indemnity.

Max Wirth describes in *Geschichte der Handelskrisen* (published in 1874) how a new-issue boom swept across Austria, Germany and Prussia from 1869 to 1874, designed to finance railroads, iron and steel works, property companies and banks. In Prussia 259 companies were established in 1871 and 504 in 1872, as against 34 in 1870 and only 225 in the first 75 years of the century. (The period from 1866 to 1873 became known as 'the golden age of company promoters'). People felt extremely confident about the future because the European and American economies had been expanding rapidly in the late 1860s. Between 1866 and 1873, for example, German per capita consumption of iron more than doubled. Bullish sentiment in Europe was further boosted by the laying of the first transatlantic cable in 1866 (by Cable & Wireless) and by the World Exhibition in Vienna, which was to open in 1873.

Hyndman, a contemporary economist, writes in *Commercial Crises of the Nineteenth Century* (published in 1892):

> What made matters even worse was the foundation of all sorts of banking institutions, which had little else than stock-jobbing in view. The real object of banks and companies was quite lost sight of, and men were swept into the whirl of speculation without having any other desire than to gamble and to make money in the lottery of the share market. Mortgage banks and building societies gave undue impetus to building speculations in the great cities, from which Berlin and Vienna still suffer. These building speculations were indeed among the most unsound and ruinous of all the business of the time.
>
> The price of land was run up to a purely fictitious level, and loans were made to cover the sites with houses to an extent which, when the crash came, rendered it impossible to recover even a fraction of the principal. The great object was to run up the houses in good, or what were likely to be good, situations, and put a rental upon them which, in nine cases out of ten, was never realised. To give an adequate account of this building mania in Berlin and Vienna would require a chapter to itself. But similar follies can be seen in London on a smaller proportional scale, and the speculative builder who, working on a small capital, must live continuously from hand to mouth, borrowing at illustrious rates to complete jerry-built structures, is well-known here at home.

In America, the most important event of the 1860s was the rapid territorial expansion towards the West Coast which, aided by the construction of the great transcontinental railways, was accomplished in an astonishingly short time. The building of the railways led to another speculative mania, once again financed largely by foreign investors.

Between 1860 and 1873, the number of railroad miles in operation more than doubled. Foreigners, who held US$51.9m in 1853, had increased their holdings to more than US$260m by 1872. But this time, the railroad mania was not confined to British or American railroad securities. Austria tripled her rail mileage in eight years and Russia built 12,000 miles in four years. In addition, Latin American borrowers, especially Argentina, raised funds in London and Paris for a number of rail projects. (By 1869 British investors held US$200m worth of South American railroad construction bonds; these investments led eventually to the Baring crisis in the 1890s.) The Suez Canal was completed in 1869, sparking renewed interest in ocean transport and great optimism with regard to international trade and economic growth.

Reading contemporary reports, one cannot fail to notice the parallels between the optimism of that time and today's upbeat mood about the prospects of many emerging economies such as China, India and Latin America. Then, as now, the sky seemed the limit, while risk considerations took a back seat. Everywhere fraud, manipulation of stocks, government corruption and all manner of illegal or shady schemes were common practice, but no one took much notice because, as McCulloch pointed out: 'In speculation, as in most other things, one individual derives confidence from another. Such a one purchases or sells, not because he has had any really accurate information as to the state of demand and supply, but because someone else has done so before him.'

The end of prosperity came in May 1873 when the Vienna Stock Exchange was struck by a devastating financial panic that spread like wildfire to London, Paris, Vienna, Berlin, and finally to New York.

Although the World Exhibition opened in Vienna on 1 May 1873, stock prices had already begun to drift lower in April and they totally collapsed on 8 and 9 May. Within one month, most bank shares had been halved and, according to a contemporary, 'The Vienna Stock Exchange was in a state of shock.' In September 1873 the crisis reached New York. According to a German correspondent, American bonds could not have been sold in Europe 'even if signed by an angel of Heaven'. The leading and most prestigious American investment bank, Jay Cooke & Co., was forced to close its doors and went out of business. The investment community was stunned and the ensuing panic brought the New York Exchange to a close for 10 days.

Unlike that of 1857, the 1873 crisis was followed by a deflationary depression that lasted six years. (The Wholesale Price Index fell from 133 in 1873 to 91 in 1878.) Most railroads under construction went bankrupt and, in the years following the crisis, more than 20,000 commercial and industrial failures were recorded in America alone. America underwent the most severe depression of the 19th century, a slump almost as devastating as the Great Depression of 1929 to 1932 was to be.

There were many reasons for the 1873 collapse: the railway boom sweeping Europe and America which immobilised capital and strained financial resources, over-expansion of the iron and steel industries, over-trading in stocks, widespread fraudulent practices by stock promoters, corrupt government officials and heavy speculation in property all fuelled the panic of 1873 and the subsequent depression, which was felt throughout the world.

Reading the works of contemporary economists, one notices that in 1869 some warning signals had become apparent. However, they were ignored in the whirlwind of speculation; business conditions continued to look promising because so many countries were in the process of industrialising. However, one factor in the collapse of the stock market bubble in Vienna, Berlin, Paris, London and New York should not be forgotten: a flood of new stock and bond issues came to the market between 1871 and 1873 and, sooner or later, the supply of this paper had to exceed the demand of investors and depress prices.

NOTHING CHANGES

The 19th century was undoubtedly a period of rapid economic growth in Europe and particularly the United States, then the largest emerging economy. However, in examining the economic literature of the period, one cannot help but be amazed at how much investors repeatedly lost and how severe the recessionary periods were. Foreign investors in particular were taken to the cleaners time and again by ruthless and frequently fraudulent stock promoters, swindlers and corrupt government officials. The most striking characteristic of foreign investors was that they were, throughout the 19th century, latecomers to an investment fad: they invariably bought American canal, railroad and other industrial stocks at or near the peak in the cycle. When prices were low and business conditions depressed, foreign investors were usually absent, nursing the fingers burnt during the previous boom.

Nothing has changed in the field of emerging market investment and, to my surprise, nobody seems to have learnt from the experiences that I have tried to document. From time to time, investors seem to become far too optimistic about the prospects for a new region or the potential of a new industry, about which they usually know very little. People who in the 19th century would have bought American canal and railroad shares (most of which failed) are now buying Chinese infrastructure funds and telephone companies in the remotest regions in the world. A few years ago, investors claimed a recession was impossible in Japan; today the same people seem to believe that Southeast Asia is immune to business downturns. Fortunately, our readers know that investing in rapidly growing emerging economies, like that of America in the last century, is much more complex than it appears to the casual observer.

A final point about investing in emerging regions: the centres of economic activity change with time. New England experienced a relative decline in manufacturing in the second half of the last century, while the Great Lakes region grew at above-average rates. China investors should bear this in mind: growth could, and I believe it will, shift from the south to Shanghai and the northeast, relegating Hong Kong to the status of a Baltimore or a New Orleans. ❖

INVESTMENT FUNDS

Roland Oettinger, Credis Fund Service Ltd, Zurich

Roland Oettinger studied behavioural sciences and information technology at Zurich University and the Swiss Federal Institute of Technology, Zurich. He joined Credit Suisse in 1986, where his responsibilities have included information technology, management development, marketing services and product management. In 1995 he took over the marketing department of Credis Fund Service Ltd, Zurich.

INVESTMENT FUNDS

Roland Oettinger, Credis Fund Service Ltd, Zurich

'Never marry for money, it is cheaper to borrow it': an old Scottish proverb reflecting the Scots' natural desire to get the most out of their money. Maybe this attitude explains why it was in Scotland that the idea of investment funds originated some 160 years ago.

Since then, investment funds have become established around the world. Originally conceived as a vehicle to enable small investors with limited funds to benefit from expert asset management, investment funds have become a flexible component of modern and effective wealth creation. This article describes the many advantages of funds over direct investments, how investment funds work and how to choose and use them.

RETURN AND RISK: AN INDIVISIBLE UNION

Return and risk are the twins encountered in every financial market. Return is the investor's pet, while risk is the black sheep of the family. (Risk represents the likelihood that money invested will be lost: nobody likes that.) The problem is that they do everything side by side. Together they scale the heights and plumb the depths, an inseparable pair: no risk, no return.

One of the basic tenets of portfolio theory is that the degree of risk taken is the deciding factor in the return on an investment. The founder of portfolio theory, Harry M. Markowitz, was awarded the Nobel Prize in 1990 for his achievement. The latest empirical research confirms his findings. Over a 10-year period, the average yearly return on equities came to 10.9%. Bonds brought less than half that at 4.6%. When we look at the risk involved expressed in terms of volatility, a clear pattern emerges: the risk for bonds was 4.1% compared with 17.9% for shares. So if you want a high return you will have to live with the likelihood of volatile assets. If you do not want to take risks, you will have to be satisfied with a lower rate of return.

The right combination is everything

The art of investment is to find the right mix. A Swiss would say that it is not unlike creating a cheese fondue. A good fondue depends on the right balance of wine and cheese; asset management on the right balance of return and risk. If extra risk is not going to bring a higher return, the result will be as unpalatable as a fondue ruined by too much wine.

For those who have never tasted Switzerland's national dish, here is a brief introduction. Fondue is basically cheese melted in white wine. You put a small cube of bread on a long fork and twirl it in the creamy mixture to coat it with cheese. The proportions of the fondue's ingredients are crucial: too much wine will make the mixture thin. If the wine has too little acid, the cheese won't bind with it. The flavour of the dish also depends on the blend of cheeses selected. Some people like them mild, others like them strong. And exotic variations of fondue are becoming more and more popular – with mushrooms, herbs, tomatoes or curry. Favourite beverages with fondue are black tea or white wine.

Just as there are many varieties of fondue to choose from, so your assets are invested in a way that reflects your individual preferences. Conservative Swiss would never, for instance, put curry in their fondue. Asset managers pay attention to investors' personal tastes (their 'investor profile') as different desires and circumstances will call for different solutions. Security and liquidity, steady income with limited fluctuations in value, long-term above-average growth and maximum after-tax returns – all of these need to be reconciled.

But there is at least one reason to suspect that investing money is more complicated than making fondue. While the Swiss recipe has hardly altered over the last few hundred years, the international financial markets are caught up in a whirlwind of change. Top-rate asset management calls for continual adaptation to changing conditions, and it must be flexible and favourably priced. The increasing complexity of the capital markets makes this an extremely demanding challenge.

DIVERSIFICATION REDUCES RISK

With the right recipe, anybody can make fondue. But for return and risk there is no simple recipe that works every time. Still, it is obvious that for any given return the best portfolio will be the one with the smallest risk. And how can we reduce risk? 'Don't put all your eggs in one basket!' The old proverb says it all: diversification. For example, a diversified share portfolio would be composed of shares from different regions, countries and industries, and from companies of different sizes. There would be shares from North America, Europe and Southeast Asia, and from financial-services companies and manufacturers of chemicals and consumer goods. To really spread the risk, the companies in the portfolio should be as independent of each other as possible.

Even the best investor will have trouble keeping track of everything that is happening on the international financial markets today. The amount of information involved is almost overwhelming. Opportunities vanish as quickly as they arise and risks change from moment to moment. One of the relentless developments of recent years is that even the instruments available to asset managers have undergone so many innovations that they can now be used only by specialists. It is becoming more and more difficult for investors to maintain an overview and react in time. In the end, they may just give up, and decide to settle for high-quality fixed-interest investments in their own currencies. At least they know what they will be getting – a transparent portfolio and steady returns.

An ideal spread takes a lot of capital

Even if financial superbeings existed who could process all the information and use all the available instruments, there would still be a problem: the capital available is hardly ever sufficient for a really effective distribution of risk. Portfolio experts have calculated that for the best possible diversification you need the following:

(i) at least US$10m;
(ii) investments in about 10 countries; and
(iii) 10 to 20 different securities in each of these countries.

Let us take as an example an investor with assets of US$2m. Depending on the investor's requirements, equities should make up 30% of the portfolio, and 30% of these should be European. If the European portion of the portfolio is to be well diversified it will contain around 13% French paper. This means that only US$23,400 is available for investing in French shares – not enough to get a good spread on the French equity market.

So we see that direct investing poses two basic problems: it is extremely difficult to maintain an overview that enables you to do the right thing at the right moment, and the capital available is rarely sufficient for the diversification you need.

FUNDS MAKE ASSET MANAGEMENT MORE EFFICIENT

Investment funds offer a solution to both these problems. They pool the money of a large number of investors, and each person who joins the fund gets a number of unit certificates reflecting the amount he or she pays in. Specialists invest these collective assets on the financial markets on behalf of the unitholders. The investors benefit from price gains and distributions. The large volume (typically between 50 million and several billion US dollars in any fund) guarantees the best possible distribution of risks in all relevant markets. When you buy a unit, you do not just invest in a single company, you invest in an entire market. This lets you participate in a portfolio with a diversification you could never achieve with a direct investment, and makes it possible to virtually eliminate the single-security or 'unsystemic' risk.

To illustrate unsystemic risk, when a company goes bankrupt, its shareholders lose their money. So, the better diversified the portfolio the smaller the single-security risk, as shares from this bankrupt company would only make up say 0.5% of a portfolio with a good spread. The value of the fund would decrease temporarily, but the unitholder would be much better off than a shareholder of the company. There is also a market risk, however, a so-called 'systemic' risk: when conditions are difficult on the international financial markets as a whole, the fund will also be affected.

FIGURE 1: DEVELOPMENT OF RISK

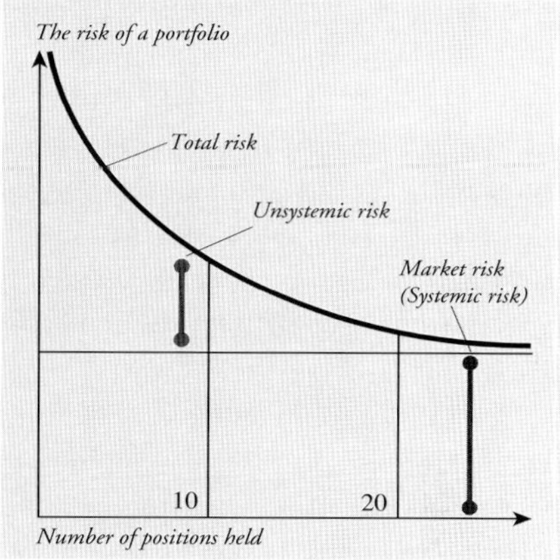

Funds have a wide range of other advantages:

(i) Fund units improve and simplify the overview of the portfolio. A single fund position covers an entire segment (sector, country, currency or region). Detailed sales prospectuses and annual and semi-annual reports provide transparency. There are also reports and comparisons from independent institutions such as Lipper, Micropal or Fund Research and from the media.

(ii) Because of their volume, investment funds enjoy better terms and conditions for securities trading than individual investors.

(iii) Unit certificates are easy to trade and can be liquidated at any time. For most funds the custodian bank is obliged to take back unit certificates. The others are traded on the exchanges. This means that small amounts can be traded at any time without incurring any price disadvantage.

(iv) The investor is relieved of the burden of difficult decisions and time-consuming administration.

USING FUNDS

Investment funds used to be used only for rounding off a portfolio; for example, apart from a lot of direct investments, you bought some units in a Japanese fund, so that you could be involved in another economy and currency. Nowadays funds are seen as flexible components for creating tailor-made solutions to an individual's needs. Instead of buying and selling separate securities, today's asset managers think more in terms of funds, so they can react more efficiently to developments on the financial markets. Funds typically serve as elements in a diversified portfolio, but they may also be used in savings or payment schemes. Fund products can also be combined with life insurance policies, an option that is becoming increasingly popular.

The wide range of funds

The range of investment fund products available is as broad as it is diverse. The biggest fund groups are equity funds, money market funds, bond funds, portfolio funds and property funds. There are also several speciality funds, aimed at a very narrow audience.

- Equity funds enable investors to tap into the economic growth of selected regions, sectors, and companies. They guarantee a broad diversification of risk, but also offer unitholders the chance to invest in markets that they could not cover effectively through direct investments. Particularly in the case of emerging markets – by the very nature of these countries – direct investments are fraught with problems and risks. Locally-managed funds offer investors an attractive opening. Equity funds are best suited to investors with an investment horizon of more than five years.
- Bond funds offer good long-term security and at the same time yield a steady return. The rapid rate of change on the world's capital markets and the wide range of bonds available have made it increasingly difficult to select the ideal mix of issuers, currencies and maturities. With funds, these tasks are performed by experts. Bond funds offer their unitholders greater security and attractive yields, although the potential for growth is not as great as with equity funds.
- Money market funds are a convenient instrument for investors who may need ready cash at short notice. The funds' assets are invested in fixed-term deposits, money market paper and other short-term securities. For private individuals a short-term exposure on the international markets is seldom to be recommended. Money market funds give you much greater flexibility than direct investments. There are no fixed terms to maturity for the unitholders, who can cash in their units whenever they choose. Money market investments offer interest rate levels in line with the market, although the growth potential is limited. These funds are best for short-term investors looking to park their money for a short period, especially to weather stock-market storms.
- Portfolio funds invest their assets on the world's equity, bond and money markets. Unitholders can buy an internationally diversified portfolio for the price of one unit, say US$1,000. They can also choose an appropriate risk strategy by selecting the equity weighting.

Property funds invest their assets in residential and commercial properties. Property is widely regarded as a secure form of investment which yields steady returns and protects the value of your assets. By purchasing units in a property fund, investors can acquire a stake in property at home and abroad with little capital outlay.

Speciality funds invest in futures and options. Speciality funds are best suited to investors who are looking for a dynamic supplement to their portfolio and are willing to accept the inherent risks.

THE RIGHT CHOICE

The wide variety of funds available means that investors can be offered tailor-made solutions. But any tailor knows that he must take the customer's complete and exact measurements if he is to achieve a proper and comfortable fit. When choosing funds, investors should start by asking themselves the following questions:

- What is my investment horizon?
- Do I want to invest for the short, medium or long term?
- What is my personal risk profile?
- What risks am I prepared to take for the returns I would expect?
- How much do I want to invest?
- Do I want to make successive payments, or invest it all at once?

Other important considerations are the investor's personal situation and intentions, as well as his present assets and the way in which they are structured.

Computer-aided fund selection

The number of alternatives available to investors is forever increasing. As financial analysis models become more and more complex, and the amount of data to be processed continues to grow, it is inevitable that traditional asset management methods will be stretched to their limits, especially when this information has to be matched with the wishes of the investors. Computer-aided fund selection systems, such as Credis Fondex, are professional tools which enable investment advisers or their clients to draw up a portfolio proposal within a matter of seconds based on the current market circumstances and up-to-date fund information. A system like Fondex also takes the investor's needs into account (see Example 1).

GOOD FUNDS AND BAD FUNDS

There is a lot of competition in the world of investment funds. They are constantly being ranked and rated by independent institutes and the financial press. They win awards. They are the subject of speculation about which will have the highest returns in the months to come. A hundred-yard dash is a much simpler affair: the winner is the one that gets to the finish line first. Funds have it harder. Basically, in order to see how well the fund is being managed it is important to make sure that only funds of the same category (just Japanese equity funds, say, or just German bond funds) and with comparable investment principles are compared. Otherwise it is like comparing apples with pears.

The most common criterion for judging funds is performance. Performance is a measure used for the return of a fund and expresses the development of its unit price. But already here problems arise. What is included in the 'performance' data? What costs are charged direct to the fund's assets, reducing its performance? Is the distribution of the fund considered in its performance? Where these things are not clear, comparisons should be made only with caution.

By now you know that not only returns but also risks play a decisive role in evaluating an

EXAMPLE 1: PROPOSAL FOR A CONSERVATIVE INVESTOR WITH DM500,000

Step 1: Define investor profile
Reference currency: Deutschmark
Investment horizon: > 3 years
Risk profile: conservative, bond-oriented

Step 2: Credis Fondex defines portfolio and currency structure

Liquidity:	*10%*	*DM:*	*77%*
Bonds:	*63%*	*Foreign currencies:*	*23%*
Equities and real estate:	*27%*		

Step 3: Credis Fondex selects funds

Number	Instrument	Currency code	Current yield in %*	Price in DM	Price as % of portfolio
1	*Savings Account*	*DEM*		*12,286*	*2.5*
21	*Credis Money Market Fund*	*DEM*	*4.0*	*39,186*	*7.9*
204	*Credis Bond Fund (B)*	*DEM*	*5.6*	*314,087*	*62.8*
170	*Credis Equity Fund Germany (B)*	*DEM*	*1.7*	*44,355*	*8.9*
37	*Credis Equity Fund UK (B)*	*GBP*	*2.7*	*11,195*	*2.2*
38	*Credis Equity Fund France (B)*	*FRF*	*2.5*	*10,621*	*2.1*
28	*Credis Equity Fund Netherlands*	*NLG*	*1.7*	*12,000*	*2.4*
54	*Credis Equity Fund North America*	*USD*	*2.6*	*23,169*	*4.6*
8	*CS Tiger*	*CHF*	*1.8*	*13,154*	*2.6*
183	*CS Euroreal*	*DEM*	*4.8*	*19,947*	*4.0*
TOTAL			*4.5*	*500,000*	*100.0*

EXAMPLE 2: PROPOSAL FOR A DYNAMIC INVESTOR WITH US$500,000

Step 1: Define investor profile
Reference currency: US dollar
Investment horizon: > 3 years
Risk profile: dynamic, equity-oriented

Step 2: Credis Fondex defines portfolio and currency structure

Liquidity:	*9%*	*USD:*	*55%*
Bonds:	*15%*	*Foreign currencies:*	*45%*
Equities and real estate:	*76%*		

Step 3: Credis Fondex selects funds

Number	Instrument	Currency code	Current yield in %*	Price in DM	Price as % of portfolio
1	*Savings account*	*USD*		*21,484*	*4.3*
12	*Credis Money Market Fund*	*USD*	*5.0*	*23,121*	*4.6*
574	*Bond Valor US-Dollar*	*USD*	*6.3*	*75,361*	*15.1*
97	*'Actions Suisses'*	*CHF*	*1.9*	*29,582*	*5.9*
326	*Credis Equity Fund Germany (B)*	*DEM*	*1.7*	*59,822*	*5.9*
127	*Credis Equity Fund France (B)*	*FRF*	*2.5*	*24,965*	*5.0*
141	*Credis Equity Fund UK (B)*	*GBP*	*2.7*	*30,006*	*6.0*
30	*Credis Equity Fund Small Cap USA*	*USD*	*0.0*	*38,794*	*7.7*
265	*Credis Equity Fund North America*	*USD*	*2.6*	*79,966*	*16.0*
410	*Credis Equity Fund Japan Megatrend*	*CHF*	*2.0*	*79,806*	*16.0*
38	*Credis Equity Fund Emerging Markets*	*USD*	*0.0*	*37,093*	*7.4*
TOTAL			*2.5*	*500,000*	*100.0*

* *The figures given for equity funds are cash yields, and those given for money market and bond funds are net portfolio yields. All figures are based on data from 2 November 1995 and are not indicative of future returns.*

investment. The comparisons mean more when the relation of return to risk is also considered. This is called the Sharpe ratio. It measures how much return the investor gets in proportion to the risk he carries. Although newspapers and magazines do not usually include the risk in their considerations, it is one of the criteria normally used in granting awards.

Choosing suitable comparison periods

Lists of funds based on performance can look very different when different periods are involved. For a long-term assessment (more than five years), especially in the case of equity funds, it is important that the management should have proven itself in various stock-market phases. Shorter-term perspectives (three years, one year or less) will not tell you much about the quality of the fund's management, but they are still worthwhile. They can allow you to observe developments that point to promising markets.

Funds are increasingly being compared with benchmarks, usually share or bond indexes. For example, equity funds that invest in particular countries are compared with the corresponding stock exchange index, such as the Hang Seng for Hong Kong, DAX for Germany, or the S&P 500 for the US. This can sometimes be misleading. Funds are often measured against a benchmark that does not match the fund's orientation. Portfolio funds, in particular, with their mixed constitution, have no corresponding index. Short-term comparisons with a benchmark over a year or less can also distort the picture because funds always hold some of their assets in cash. That is why investment funds lag behind the index when share prices go up. It must be said, however, that a fund manager who has managed to beat the index for several years at a stretch has done a very good job indeed.

DO NOT BASE PREDICTIONS PURELY ON PAST EVENTS

Regardless of statistical and scientific considerations, there is one thing that we must never forget in comparing performance: performance measures how well the fund did in the past. We would be naive to take it as an indicator of future trends. It is similar to football. A team wins the championship; it shows what it can do, and will be among the favourites next season. But there are many games to be played in the next season. Even the strongest team can go through a weak phase and lose a match. Who knows if they will be on top again next year?

The only really reliable comparison of a fund would be one with direct investments, considering the risk and all additional costs. What would the return have been? What would have been the costs for the selection, administration, purchase and sale of the securities? What would have been the cost of advice, plus the acquisition and processing of information? What would have been the risks in direct investments? What does the spread of risks cost? How quickly can the money be available? Quantifying all this is no easy matter. The fact is that in the case of direct investments everything costs extra, while these costs are included in the management of the fund. Investment funds clearly seem to be the most efficient solution.

Fund experts: Credis Investment Funds

In early 1994 the Credit Suisse Group established a specialised company for their investment fund business: Credis Investment Funds. The fund assets administered by its 350 employees currently come to around SFr50bn. More than 130 funds are available.

FIGURE 2: CREDIS INVESTMENT FUNDS
TOTAL NET ASSETS 1986-95

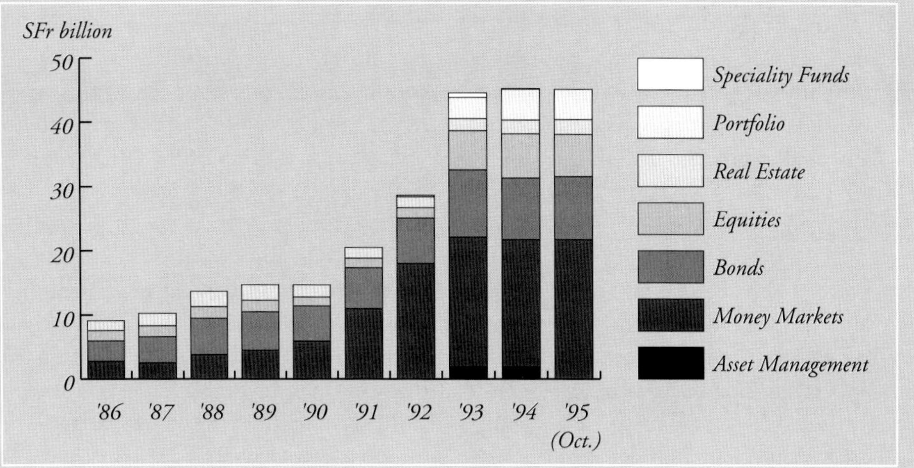

Credis Investment Funds relies on the flexibility and superior quality of its products to continue to be a successful player in the expanding investment fund business. The motto 'Think globally, act locally' holds for the fund business as well. For Credis Investment Funds this means that its dynamic international sales organisation has regional foundations, getting close to the customer through its own local sales offices, as well as through banks, insurance companies and independent financial advisers. Credis Investment Funds has subsidiaries in Luxembourg, Germany, the UK, Italy, France, Japan, the Czech Republic and Poland. A selection of the core range of funds is offered in each country according to local criteria. This is augmented by funds specifically designed for the local market. This enables them to satisfy customer requirements as fully as possible, and to take into consideration, for example, regional investment attitudes and tax regulations.

The funds' assets have to be managed by specialists with a thorough knowledge of the individual markets. Credis has selected the best of them. These are often asset management units of the Credit Suisse Group. Clear goals and performance-related reward are the foundations of this successful collaboration. ❖

LLOYD'S OF LONDON

Robert Taylor, Senior Partner, Robert Taylor Associates, London

Robert Taylor is the senior partner of Robert Taylor Associates, a firm of chartered accountants in central London, which he founded in 1987. The firm has grown considerably since then but continues to pride itself on the highly personalised service that it offers to private and corporate clients alike.

Phone: (+44-171) 491 4621
Fax: (+44-171) 495 1159

LLOYD'S OF LONDON

Robert Taylor, Senior Partner, Robert Taylor Associates, London

Of all the world's financial institutions, there can be few as well-known or as distinguished, despite its recent well-publicised troubles, as Lloyd's of London. Its internal workings and origins are unique amongst financial institutions. From its beginnings in a London coffee-house in the late 17th century, Lloyd's has developed into one of the world's leading centres for many forms of insurance, and it is the only insurer that, until recently, relied upon its members' willingness to pledge the whole of their wealth to the concept of unlimited liability in exchange for a direct share in its business.

The Corporation of Lloyd's is responsible for the centralised administration of the market, providing clerical support and the actual building where the market is located. It must be understood, however, that the Corporation does not actually write policies itself, although it does administer the Lloyd's Central Fund which would compensate a policyholder in the event of any default on the part of the underwriter.

WHAT IS A LLOYD'S NAME?

At Lloyd's it is the syndicates that actually underwrite the risk. These are groups of individuals, usually referred to as 'names', who have joined the market and who will almost always be members of several syndicates simultaneously, in order to spread their risk. Each syndicate employs a managing agent who is responsible for the day-to-day running of the business and for appointing an insurance professional, known as an underwriter, who negotiates the terms of each policy with an approved broker acting on behalf of the eventual policy holder.

For an individual who wishes to become a name, the first step is to appoint a member's agent. These are professionals with a thorough knowledge of the market, and they will recommend syndicates to the name as well as negotiating with the managing agents for the name to join a specific syndicate.

It is important to remember that a syndicate is not actually a partnership but simply a collection of individuals: each name is solely responsible for any liabilities under a particular policy and, if a name should default, the shortfall will be made up from the Lloyd's Central Fund rather than from the other members of the syndicate.

Names can reduce their exposure by taking out stop-loss policies that will cover losses after an excess has been reached. The premium payable for these policies will, of course, depend on the excess and the maximum amount payable over and above the excess. It is also possible to take out estate-protection plans to cover losses arising prior to the death of a name, which would otherwise be deducted from the name's estate.

The syndicates themselves can be divided into four main categories: motor, aviation, marine and non-marine (or general). These classifications are self-explanatory; the non-marine syndicates specialise in general insurance not covered by the other areas. In most cases a member's agent would wish to spread a name's membership across all four types of syndicates in order to avoid too much exposure to any one type of risk.

THE BENEFITS FOR THE PRIVATE INVESTOR

Why would any individual wish to become a member of Lloyd's? To answer this we need to understand how a syndicate operates. As explained above, the underwriter, appointed by the managing agent, will negotiate the terms of a policy. Once these are agreed, the premium is paid into the syndicate's premium trust funds (held in either sterling, US dollars or Canadian dollars) under a premium trust deed approved by the Committee of Lloyd's and by the Department of Trade and Industry. These funds are used to meet any claims made under the policies.

All syndicates work on a calendar-year accounting period, and most policies offer cover for a 12-month period. It follows that, if a policy was written in July 1995, for example, although the premium would be received in the accounting period ended 31 December 1995, the syndicate would still be exposed to a potential claim for an event occurring at any time up until the following July (1996). To allow for this, the 1995 account will remain open until 31 December 1997. Although this will usually permit the settlement of most outstanding claims, there may be some that remain unresolved. In these cases, it is normally possible to carry forward a reasonable estimate of the remaining exposure into the following year of account, by means of a reinsurance premium which closes the account. In rare cases this is not possible and a syndicate's account has to remain open until the outstanding matters are resolved. In the vast majority of cases, however, the accounts will be drawn up to show either a net profit or a loss. Included in these accounts will be the investment income and capital gains that the syndicate has accumulated from the asset in the premium trust funds. Any resulting net profit is distributed to the names. If, on the other hand, the syndicate should show a loss, it is at this point that the names must make up the shortfall through a cash call.

So, apart from the initial joining fee and some small fees for professional services, it has not actually cost the name anything to become a member. All of the capital invested in income-producing assets will remain invested; these assets could even be said to have been invested twice, which for a private investor is one of the greatest attractions of becoming a name.

HOW TO BECOME A NAME

To become a Lloyd's name the applicant must show that he or she has the requisite net financial worth. In order to do this he or she must complete a statement of means giving details of assets acceptable to Lloyd's up to a specified limit. The limit is determined by the Council and at present is £250,000 (US$370,000). Acceptable assets include quoted stocks and shares, gilts, UK unit trusts, local-authority bonds, national-savings certificates and, of course, cash. It is also possible to satisfy the means test with a bank guarantee or letter of credit. There are restrictions on what proportion of the total amount can be made up by certain investments and some valuable assets, such as jewellery, an applicant's main residence, an interest in a partnership or shares in an unquoted company, are not accepted by Lloyd's for the purposes of the means test.

Once an applicant has completed a satisfactory statement of means, the member's agent will guide him or her through the remaining procedures required by the Corporation and the Committee of Lloyd's.

Among these procedures is the establishment of a deposit by which the applicant assigns certain of his investments to the Corporation; these assets continue to belong to the beneficial owner, who is entitled to all income and capital gains that they earn.

The tax treatment of Lloyd's has always been the subject of great debate, and it has also been the target of criticism from some quarters in the UK. This is because losses are treated in the same way as those incurred in an active trade, in that they can be set off against other income in a variety of ways, which usually results in a tax refund. Another feature of trading at Lloyd's is the Special Reserve Fund. This system allows a name to obtain tax relief on moneys paid into his fund in years of profit, which can be used to fund any losses that may arise in later years. There are many other businesses whose trading environments are unpredictable enough to warrant such treatment, but very few actually receive it.

The tax advantages available to Lloyd's names, however, depend upon a number of factors specific to the individual. Before making an application for membership it is extremely important for the prospective member to obtain professional advice on the overall tax implications, with particular reference to the tax laws of the country in which he or she is resident.

THE PITFALLS OF MEMBERSHIP

Membership of Lloyd's can present a unique investment opportunity, but the unlimited nature of an individual's exposure must be constantly borne in mind. Although the statement of means specifies a minimum net wealth for a particular premium limit, the name's exposure is not restricted to this amount. If losses occur, the name is bound by the terms of the pledge to make available his or her total global assets to meet agreed claims. In years gone by this may have been treated as a formality; nobody expected trading losses to be so great as to involve the actual bankruptcy of a name. This is no longer the case.

There is no space here for a detailed examination of the reasons so many names have suffered financial hardship in recent years. The number and severity of these cases has forced Lloyd's to set up a hardship fund to provide some relief for those worst affected. Horror stories abound, with names apparently having to sell virtually everything they own to pay increasingly voracious cash calls from some beleaguered syndicates. The publicity has been widespread and highly critical, with some suicides even being blamed on the pressure inflicted by the seemingly never-ending losses. Suffice to say that accusations of incompetence and outright corruption have destroyed forever the reputation of a gentlemanly conduct of trade. Things got so bad at one point that many people thought the market was close to collapse.

The crisis could scarcely have come at a worse time: the Corporation had only recently fought off moves to impose independent regulators like those that have been imposed on almost every other financial services institution in the United Kingdom. However, it is now generally accepted that the measures taken by the Corporation to regulate the market and to widen its whole capital base have, at the very least, provided stability, and should form the foundations for a sound future. For instance, the introduction of limited-liability corporate capital from 1 January 1994 yielded £1.6bn (US$2.4bn) from the first 25 corporate entities to be admitted. Recent proposals to ring-fence earlier losses and indemnify new investors against judgements in legal proceedings already under way have also prompted the feeling amongst investors that now might be the time to think about joining the market.

Insurance is known to be a cyclical business, but even a brief study of the past will show that the last few years have been probably the most difficult and turbulent in the history of Lloyd's of London. Nevertheless, there appear to be grounds for optimism about both regulation and trading.

Whilst great care should be taken to ensure that membership is suitable for the individual concerned, most of the old advantages of membership still apply and, with the new potential for individuals to protect themselves from unlimited liability by the use of corporate vehicles, it would be a pity if this investment opportunity were overlooked simply because of the sins of the past. ❖

EUROPEAN ECONOMIC AND MONETARY UNION: HOW HARD WILL THE EURO BE?

Rainer Marian amd Jeanne Barras,
Credit Suisse Economic Research, Zurich

Rainer Marian trained as an economist and began his career in 1987 as a foreign exchange trader for the Hessische Landesbank in Frankfurt. In 1994 he joined Credit Suisse's economic research department where he is responsible for currency forecasts. He is now working on a study of financial markets at CS First Boston in New York.

Jeanne Barras Zwahlen joined Credit Suisse in 1989 as senior economist in charge of European integration. She previously worked for 10 years as a diplomat, particularly for regional development banks and at the Swiss delegation to the OECD in Paris.

EUROPEAN ECONOMIC AND MONETARY UNION: HOW HARD WILL THE EURO BE?

Rainer Marian and Jeanne Barras, Credit Suisse Economic Research, Zurich

The exact date of European economic and monetary union (EMU) has not yet been fixed, nor is it clear which countries will participate. Nonetheless, investors should already be thinking about how they are going to respond to this radical change in the European economic landscape. In the run-up to monetary union, we can expect to see European investors making substantial shifts in their portfolios. Given its traditional role as a safe-haven currency, the Swiss franc could well benefit from these movements.

STILL SOME WAY TO GO

Politicians and officials in Brussels are still hard at work trying to hammer out the details of the third stage of economic and monetary union. In June 1995 the European Commission published its green paper with the timetable for the implementation of this key element of the 1992 Maastricht treaty. This timetable (see Figure 1) is by no means binding on the member states, but it will soon become apparent what is actually feasible and what is essentially political wishful thinking. It is extremely unlikely, for example, that monetary union will be complete before the 1 January 1999 deadline. Even when EMU is in place, the individual national currencies will not disappear, but are more likely to be tied to each other through fixed exchange rates. The final phase, the creation of a single currency, will take another two to three years. However, if one ignores the fact that national currencies will continue to have different names, a single currency will essentially have been created through the prior fixing of the exchange rates. By this time most of the economic consequences of monetary union will already be apparent. This article looks at these consequences and at how the financial markets will behave as EMU approaches.

FIGURE 1: POSSIBLE TIMETABLE FOR EMU

Mid-1997 to end 1997	*Setting the definitive timetable and deciding who qualifies for EMU*
1.1.98 – 31.12.98	*Founding the European central bank, deciding on monetary instruments and exchange ratios*
1.1.99	*Setting exchange rates, exchange of the ecu for the Euro*
End 2001	*Exchange of national currencies for the Euro*

INITIALLY SOFTER THAN THE DEUTSCHMARK

The convergence criteria set out in the Maastricht treaty will determine which countries take part in EMU. Although only Luxembourg meets these criteria at the moment (see Figure 2), by 1998 a good number of other countries should have qualified – probably Germany, Austria, the Netherlands, France, Finland, Ireland, Belgium, Denmark and the United Kingdom. Denmark and the UK have, however, negotiated opt-out clauses which mean that they are not obliged to participate in monetary union.

Let us imagine that the exchange rates of the individual countries are fixed on a permanent basis in 1999 or 2000. This will create a currency area with a single monetary policy and a single external value vis-à-vis other currencies. The external value of the currency will reflect the economic power of the constituent member states and the markets' confidence in the European central bank. Initially, we would expect the external value of the new Euro to be lower than that of the hardest EU currency; the new European central bank will not immediately be able to command the investor confidence that the German Bundesbank – with its 40-year record of promoting monetary stability – currently enjoys. Any sort of monetary union entails some risk and the financial markets will react to this by demanding higher risk premiums.

POTENTIAL FOR PRICE GAINS

High-yielding bonds, for example, stand to benefit from the way the market develops after economic and monetary union. Once exchange rate risk is a thing of the past, European capital-market yields will – apart from considerations of creditworthiness and liquidity – be the same across the board. Initially, this will represent the average interest rate of the former national

FIGURE 2: EUROPEAN MONETARY UNION – MAASTRICHT CRITERIA 1994 (SHADED AREAS = FULFILLED)

Country	Inflation (%)	Budget deficit (%)	Gross public debt (%)	Interest rates (%)
Belgium	*1.6*	*-4.3*	*139.1*	*8.0*
Denmark	*2.1*	*-2.1*	*77.6*	*8.7*
Germany	*1.9*	*-3.6*	*58.4*	*7.2*
France	*1.9*	*-5.1*	*53.5*	*7.9*
Greece	*9.1*	*-9.0*	*112.3*	*17.4*
United Kingdom	*3.5*	*-4.9*	*52.5*	*8.6*
Ireland	*2.7*	*-2.2*	*86.1*	*8.7*
Italy	*5.3*	*-7.4*	*116.4*	*12.3*
Luxembourg	*2.3*	*+1.4*	*7.6*	*8.0*
Netherlands	*2.0*	*-3.3*	*80.3*	*7.3*
Portugal	*4.2*	*-5.2*	*70.4*	*12.1*
Spain	*4.7*	*-5.9*	*64.5*	*11.5*
Finland	*1.3*	*-4.5*	*66.3*	*9.6*
Austria	*2.3*	*-4.6*	*65.9*	*7.4*
Sweden	*3.1*	*-9.0*	*84.2*	*10.9*
EMU target	*3.1%**	*-3% of GDP*	*60% of GDP*	*10.3%***

* *Average of three lowest national inflation rates plus 1.5 percentage points*

** *Average interest rate of the three countries with lowest inflation rate plus 2 percentage points*

Source: CS Economic Research November 1995

currencies plus a risk premium. Starting from current market conditions and assuming that the 10 countries listed above will indeed participate in the economic and monetary union, we can calculate an average interest rate, weighted according to the GDP of the individual countries.

On 30 November 1995 (a randomly chosen reference date) this average rate for 10-year government paper stood at 6.7% including an appropriate risk premium (see Figure 3). Once exchange rates have been fixed, the yield on all comparable bonds will be the same, regardless of whether they are denominated in French francs, Irish punts or Dutch guilders. For a long-dated bond in a high-interest currency this would represent a price gain, and, conversely, for bonds in low-yielding currencies, a price loss. Let us look at these two extremes.

On 30 November 1995 Finnish fixed-interest government bonds due in 2005 were yielding 7.3%. If exchange rates are fixed on 1 January 1999, the price of these bonds will have risen to such an extent that the yield will have fallen to around the European average, say, 6.7%. Today's investor will, however, have secured a high yield on Finnish bonds for a further six years. By contrast, the coupon on comparable German bonds was only 6.2% on 30 November. These bonds will decline in value since the quality of the investment currency will deteriorate with economic and monetary union, at least in the initial phase. However, once it becomes clear that the European central bank is successfully pursuing a policy of stability, a drop in interest rates to the level of the hardest currency cannot be ruled out.

FIGURE 3: YIELDS ON 10-YEAR GOVERNMENT BONDS ON 30 NOVEMBER 1995

Netherlands	*6.2%*
Austria	*6.7%*
Germany	*6.2%*
Denmark	*7.7%*
Belgium	*6.7%*
France	*6.9%*
UK	*7.6%*
Ireland	*7.6%*
Finland	*7.3%*
Ecu	*7.1%*
Hypothetical GDP-weighted interest rate for the Euro under present market conditions	*6.7%*

EXCHANGE RATE RISK REMAINS

This scenario does not, of course, take account of the exchange rate risk which will remain until exchange rates are fixed. In practice, the possible gains for investments in high-interest currencies could be eliminated – or even overcompensated for – by a devaluation of the currency. This danger is all the greater since certain countries where unemployment is high (for example Finland, Ireland, France and Belgium) could well try to create a competitive advantage for themselves by devaluing before monetary union is put in place.

However, there is not a tremendous amount of time left for countries to devalue their currencies significantly. Under the Maastricht treaty, to be eligible to join EMU, countries must have kept their currency within the normal ERM bands for the two years prior to union and not devalued their currency against the bilateral ecu central rate during that time. This means that any tampering with exchange rates will have to be done before the end of 1996. Since 1992 Finland and Ireland have already seen their currencies fall by around 25% and 15% respectively in relation to the Deutschmark. Given the present interest differential vis-à-vis Germany, the Finnish mark would have to devalue by a further 12% before investors with 10-year Finnish paper would be worse off in terms of yield compared with German bonds. With only a year left until economic and monetary union, such a devaluation seems unlikely, provided Finland can meet the convergence criteria.

AND THE ECU?

The same sorts of considerations hold true for ecu-denominated paper. This is currently yielding less than investments in Finnish marks and Irish punts. However, because the ecu represents a basket of currencies it does, in theory, carry a smaller exchange rate risk. As the ecu will be exchanged for the Euro on a one-to-one basis, arbitrage will be almost risk-free (see Figure 4). Since the ecu includes Europe's soft currencies (whereas under Maastricht only the most stable countries would be eligible to join EMU), investors can be practically certain that the exchange will give them a harder currency than the ecu and therefore one with lower interest rates attached.

FIGURE 4: THE EUROPEAN CURRENCY UNIT. . .

. . .TODAY

Composition of the ecu basket of currencies on 30 November 1995 (%)

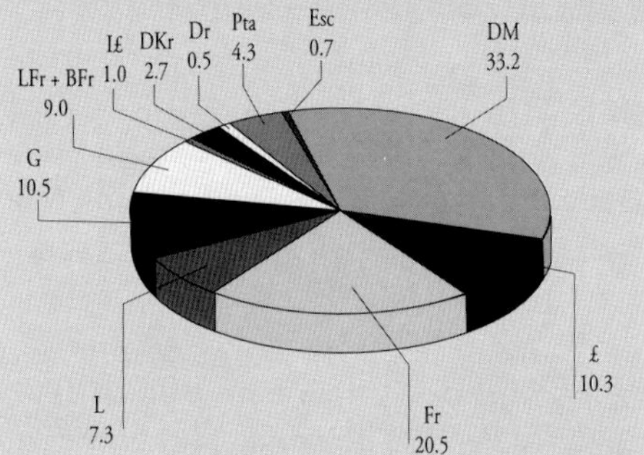

. . .TOMORROW?

Possible participants in currency union (GDP-weighted, %)

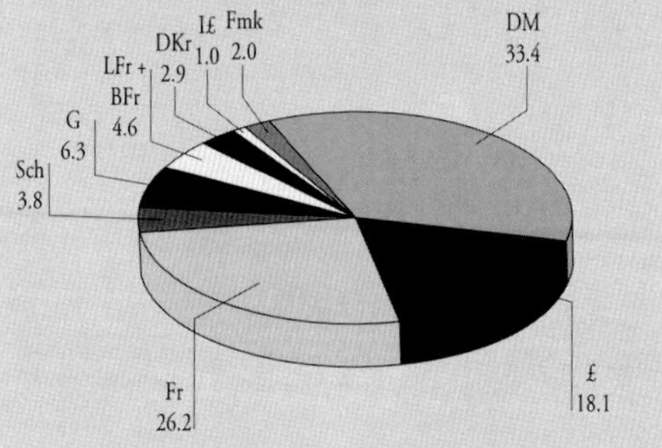

IS CAPITAL FLIGHT INEVITABLE?

If the whole situation starts to look too complicated or too uncertain, investors always have the option of shifting their money into countries which are not part of EMU. For investors in Germany, Austria and the Benelux countries, Switzerland is the obvious candidate, since the Swiss franc has tended to shadow the Deutschmark in the past. In times of political and economic uncertainty the Swiss unit has proven to be a safe-haven currency par excellence. Given that Switzerland's size bears no relation to the likely influx of capital, in 1995 we have already seen the Swiss franc appreciate considerably in the run-up to economic and monetary union. This is proving to be a tough test for Switzerland's economy.

A further factor which could encourage investors to shift funds to countries outside EMU would be the desire to diversify their risks. Most modern portfolios are managed according to the principle of risk diversification. Because there will be one currency and effectively a single exchange rate, risk diversification will hardly be possible within EMU itself. As a result, the USA and Japan (plus non-EMU European countries with solid economic fundamentals – the UK, for example, if it chooses to opt out of EMU) could well benefit.

DOUBTS REMAIN

Although the Maastricht treaty is binding on the signatory nations, there still remains a possibility that EMU will not come into being. In this case, of course, the whole scenario would look very different. Even if economic and monetary union is established, it may not be possible to force national finance ministers to exercise spending restraint. However, the Bundesbank is likely to make it a condition of its support of EMU that sanctions be tightened up in this respect. If the Bundesbank does not back EMU, it is difficult to see how the government will be able to gain the support of the majority of the German people.

Another critical aspect is the continued lack of flexibility of the European labour market. Governments will no longer be able to use the exchange rate as a way of restoring the competitiveness of their economies. This will in future have to be done through pay and mobility of labour. And even though the whole question of the viability of economic and monetary union is inextricably tied to competitiveness, it is possible that individual member states will not cooperate in dismantling the legal, linguistic and cultural barriers which are standing in the way of progress here.

Investors cannot wait until all these issues have been fully resolved. By the time the exchange rates are actually fixed, the financial markets will already have discounted many of the likely developments. Which means that it will then be too late to react to economic and monetary union. ❖

SWISS BANKS AND THE WTO: THE WALLS COME TUMBLING DOWN

Fritz Stahel, Credit Suisse Economic Research, Zurich

Following bank training, Fritz Stahel studied economics and then went on to work at the Institute for Agricultural Policy in St Gallen. He joined Credit Suisse's economic research department in 1982 and now heads the team that deals with European integration and structural issues pertaining to the financial sector.

SWISS BANKS AND THE WTO: THE WALLS COME TUMBLING DOWN

Fritz Stahel, Credit Suisse Economic Research, Zurich

On 1 July 1995 Switzerland joined the new World Trade Organisation (WTO). The inclusion of services (under GATS) will open up a whole new range of opportunities for Swiss banks and their customers. This will not happen overnight, but in stages, as more and more barriers to free trade are dismantled. The United States' refusal to extend the most-favoured-nation principle to services means, however, that the WTO agreement remains provisional as far as the financial sector is concerned.

Markets, like parachutes, work best when open. The financial sector, which has been undergoing a process of deregulation since the early eighties, is a case in point. Until now, this process has taken place on a number of different levels. On the one hand, individual states have taken their own steps – witness the lifting of the ban on interstate banking in the United States, or Japan's abandonment of the separation of commercial and investment banks, for instance. On the other, moves have been made by whole groups of countries – the creation of the EU single financial area, for example, or the OECD's guidelines on the liberalisation of capital movements.

The new WTO brings all these different initiatives under one roof. This has given fresh impetus to the globalisation of the financial services industry and, as a result, individual markets are now growing even closer together. Thanks to rapid technological progress, banks can operate all over the world and right round the clock. Foreign exchange and precious metals markets already operate on this continuous, global basis, and securities trading is quickly following suit. The financial sector is well on its way to becoming a global village.

STEP-BY-STEP LIBERALISATION

With the services sector becoming an ever more important part of the world economy, it was only natural that the Uruguay Round of GATT should seek to include services in the international process of deregulation. GATS, the General Agreement on Trade in Services, thus forms one of the three pillars of the new world trade system (see Figure 1).

Banks, securities houses and fund managers all come under the provisions of GATS and so stand to benefit from the four principles that underlie the whole WTO philosophy:

- Non-discrimination: foreign providers of financial services have to be given free access to the domestic market. Current barriers to such access include limitations on the number of licences granted, state monopolies, ceilings on the size of the stake that foreigners can take in domestic companies and restrictions on the business they can conduct.
- National treatment: the same rules have to apply to foreign institutions as to domestic ones. Limitations on currency and fund transfer operations or on the use of foreign staff, funding restrictions and discriminatory tax regimes would thus have to be abolished.
- Most-favoured-nation principle: if two WTO states agree to preferential trade

conditions, these conditions have to be extended to all other signatories of the WTO treaty, thus rendering all liberalisation universal. This principle was the keystone of GATT negotiations and now plays an equally crucial role within the WTO.

- Transparency: governments must be open about the rules that apply to their markets. Criteria for the allocation of licences, for example, should be clear.

Dismantling all the barriers that exist today will take time. Liberalisation is a slow process and it will be completed in stages rather than at one fell swoop. WTO 'country lists' provide details of the markets that member countries have made freely accessible to foreigners. These lists reveal that only a few countries have improved access to their markets since the WTO came into being. The next round of multilateral negotiations will, however, take place in five years' time at the latest. And at least the WTO will ensure that existing concessions are not withdrawn. This point is of particular significance for financial institutions: countries should not now be able to suddenly reduce the number of licences for foreign banks or to reimpose restrictions on foreigners' access to stock exchanges.

A FLY IN THE OINTMENT

Despite the WTO's aspirations to universal applicability, countries can still take advantage of opt-out clauses. Restrictions – at least temporary ones – can be applied if a country is experiencing balance of payments problems, for instance. And access to markets can be refused if a company's home country is not thought to guarantee sufficiently rigorous monitoring. In addition, free trade areas such as the EU/EEA or NAFTA can take liberalisation a stage further without having to extend rights to all other WTO members. As before, countries are given a certain amount of leeway to ensure that taxation is fair. Individual exceptions to the most-favoured-nation principle are also allowed, although only for limited periods and subject to multilateral supervision.

When the Uruguay Round was concluded, however, the United States regarded the proposals made by Japan and many emerging markets as inadequate. As a result of US concerns, when the most-favoured-nation clause for financial services came into force on 1 January 1995, it did so only in provisional form. Subsequent negotiations, which lasted until the middle of 1995, saw attempts being

FIGURE 1: THE THREE PILLARS OF THE NEW WORLD TRADE ORDER (WTO)

Agreement on goods and tariffs (GATT)	Agreement on services (GATS)	Agreement on intellectual property (TRIPS)
Lower tariffs *Textile trade* *Agriculture* *Foreign investment* *Arbitration*	*Telecommunications* *Banking* *Insurance* *Tourism* *Arbitration*	*Patent protection* *Copyrights* *Industrial design* *Brand names* *Arbitration*

Most favoured nation	*Non-discrimination*	*National treatment*	*Transparency*

made to squeeze more concessions from the 'free-loaders' and, although progress was achieved in some areas, the Americans insisted on bilateral solutions based on the principle of reciprocity. Nevertheless, under the leadership of the EU, a solution was found which should be ratified by mid-1996 and provisionally be in force until the end of 1997. At that point negotiations – in which, hopefully, the United States will take part – will begin on a definitive agreement.

ADVANTAGES FOR BANKS . . .

As Figure 2 illustrates, Swiss banks have very strong links with other countries, making the WTO an extremely important organisation for the Swiss financial sector. Swiss banks are in a position to benefit directly and immediately from the new world trade order because it presents them with an ideal opportunity to strengthen their foreign operations in a highly targeted way. These operations tend to concentrate on wholesale banking: business with major customers such as multinational companies, pension funds, government agencies, international organisations and wealthy private clients. This last group will derive particular benefit from the fact that the enormous amount of private banking expertise offered by Swiss banks is now available not only in Switzerland but in an increasing number of international locations – in Credit Suisse's case through its Private Banking International strategy.

In the industrialised world, where Swiss banks are already well established, the focus will be on tapping into new areas of business and achieving equal status with domestic banks. It is, however, clear that membership of the WTO is no substitute for participation in the EU/EEA internal market. The European single market goes much further than the WTO. It offers, for example, a single, Europe-wide licence to operate as a bank – but only to institutions from the member states. If banks in Switzerland want the same opportunities as their European counterparts, then Switzerland has to thrash out a bilateral agreement with the EU. An agreement covering the financial sector is not yet in sight.

Banks will be looking most of all towards the emerging markets for expansion opportunities. These countries expect to post above-average growth, which goes hand in hand with the need for extra financial resources (for businesses and for state infrastructure projects). Many emerging economies, especially in Asia, also have high savings rates, providing ample opportunity for private banking departments to spread their wings.

In the longer term, establishing new offices abroad will be only one of several interesting options for Swiss banks. WTO liberalisation will provide additional ways for banks to offer

FIGURE 2: SWISS BANKS' TIES WITH OTHER COUNTRIES

SWISS FINANCIAL SERVICES INDUSTRY

- *Approximately a quarter of total assets accounted for by foreign operations*
- *About half of all asset management carried out on behalf of foreign clients*
- *Net current account surplus of almost SFr20bn (banks and customers)*

MAJOR SWISS BANKS ABROAD

- *More than 200 branches, subsidiaries and representative offices spread over all five continents*
- *Around 25,000 employees, and rising*
- *Approximately one-third of group earnings generated outside Switzerland*

their services in other countries – computer access to foreign stock exchanges is a distinct possibility, for instance. New opportunities are also emerging in electronic banking, a business made all the more exciting by the inexorable march of multimedia technology.

GATT's evolution into the WTO has undoubtedly stimulated world trade and the need for trade finance is growing accordingly. The Swiss banks, which have built up a great deal of expertise in this field thanks to the openness of the Swiss economy, can only benefit from this trend.

. . . AND THEIR CUSTOMERS

New doors are opening not just for the banks but also for their investment customers. As a rule these customers are keen to increase the international diversification of their portfolios in the hope of improving returns whilst maintaining risk at the same level or, conversely, reducing risk without impairing returns. Barriers to foreign investment in emerging markets are still very high. The shares of only a limited number of companies are available for

FIGURE 3: EMERGING MARKETS: MAIN ECONOMIC INDICATORS

	GDP Real Growth in % p.a.			Inflation in % p.a.		
Asia	**1994**	**1995**	**1996***	**1994**	**1995**	**1996***
China	11.8	10.0	9.0	24.2	17.0	15.0
Hong Kong	5.5	5.0	5.0	8.0	9.5	8.8
Taiwan	6.9	6.6	6.3	4.1	4.5	4.2
Singapore	10.1	8.1	7.8	3.8	2.5	2.3
South Korea	8.4	9.0	8.0	5.6	5.0	6.0
Malaysia	8.7	8.5	8.0	3.7	4.2	4.5
Thailand	8.5	8.7	8.7	5.0	5.2	4.8
Indonesia	7.0	7.5	7.0	9.2	8.6	7.9
Philippines	5.1	6.0	6.0	9.1	7.5	8.5
India	6.2	5.8	5.5	8.9	9.5	9.0
Latin America						
Mexico	3.5	-5.0	2.0	6.9	49.0	25.0
Brazil	5.7	5.5	4.5	2310.9	27.0	20.0
Argentina	7.4	-0.5	1.0	4.3	2.3	3.0
Chile	4.2	7.0	6.5	11.4	8.2	8.0
Venezuela	-3.3	0.0	1.5	60.8	50.0	60.0
Eastern Europe						
Czech Republic	2.6	4.2	5.0	10.0	9.0	8.0
Poland	5.0	6.0	4.7	32.0	27.0	20.0
Hungary	3.0	1.5	2.7	19.0	29.0	22.0

* *Forecast*

FIGURE 4: THE WTO AND THE SWISS BANKS

- *Opening up of markets provides new opportunities for internationally oriented banks and for investment funds*
- *New opportunities for international portfolio diversification*
- *Success will not come overnight but through a long-term process of liberalisation*
- *One short-term advantage, however, is that there will be no going back on current commitments*
- *Increased competition for the Swiss financial centre in the longer run*
- *Largely positive indirect effect via corporate customers; selective reassessment of credit risks*

direct investment. The consequent narrowness of these markets entails a high degree of risk and price volatility.

The Mexican crisis of 1994 to 1995 proved once and for all that enthusiasm about emerging markets must be kept in proportion. Stock markets in all these up-and-coming countries were hit hard by the fallout from Mexico, prompting the question of whether it is worth investing in emerging economies at all. The answer is a resounding *yes* – after the initial overreaction, it became clear that investors were in a position to discriminate between the different countries in this bracket. The economic dynamism of Asia's emerging nations is being sustained by strong domestic demand for consumer and capital goods and by continued rapid expansion in China. Although the region is due to grow a little more slowly in 1996, it will remain a major contributor to world economic growth as a whole (see Figure 4). Following the collapse triggered by events in Mexico, high growth rates can also be expected from Latin America in 1996. The former communist states of central Europe will also see accelerating growth.

Given these prospects for future expansion, investments in emerging markets must still be seen as worthwhile, provided investors are prepared to take a relatively long-term view. The liberalisation of capital flows foreseen by the WTO should gradually be implemented. Investment funds provide a good alternative to direct investment wherever securities markets are still not broad and deep enough. Funds are also useful as a means of entering unfamiliar markets in industrialised countries.

For the Swiss banks' corporate customers, the WTO will not only mean enhanced export opportunities; they should also find it easier to improve their direct presence in foreign markets. The increased volume of local currency issues, foreign stock market listings and mergers and acquisitions with local companies that this will bring will provide rich seams of new business for the banks to mine.

SELECTIVE REALIGNMENT OF CREDIT POLICIES

More indirectly, the existence of the WTO will also prompt banks to reassess credit risks. On balance, most Swiss exporters and their suppliers will find that the WTO has a positive effect on business. The challenge for banks will thus be to hunt out the customers who are quick enough and flexible enough to recognise and exploit new opportunities ahead of the pack.

On the other hand, some sectors will find that the positive effects of the WTO are counteracted by other facts. Until now, for instance, the Swiss textile and clothing industry has been protected by the quota system of the Multi-Fibre Agreement. When this is scrapped, manufacturers are bound to lose market share to low-cost producers in other industrial countries. Liberalisation of public tendering might give the Swiss construction industry additional opportunities abroad, but it will simultaneously expose it to fierce competition for business at home. The food industry will certainly feel the cold if export subsidies are abolished.

The biggest changes of all, however, will probably be in agriculture. The current realignment of Swiss agricultural policy was necessary anyway and the inclusion of farming issues in the WTO remit will serve to speed things up. With structural adjustment gathering momentum, banks will have to take a long, hard look at their agricultural credit portfolios, and should bear in mind that businesses on the periphery of the industry – purchasing cooperatives at one end and sales organisations at the other – are also facing increased competitive pressure.

NEW COMPETITION FOR SWISS FINANCIAL SERVICES

Internationally oriented Swiss banks stand to gain the most from the WTO. But at the same time the whole of the Swiss financial services industry will find that it is facing new competition from abroad. Financial centres in other countries will become more attractive as restrictions are removed. Switzerland's traditional advantages are becoming less and less distinct and centres all over the world – not only in other industrial nations but also in the emerging markets – are increasingly vying with the Swiss for certain areas of business. This is not to say, however, that Switzerland will not continue to enjoy its status as a safe haven in times of uncertainty. ❖

GLOBAL SOCIAL INVESTMENT

Peggy Dulany, President and Founder,
The Synergies Institute, New York

Peggy Dulany is president and founder of The Synergies Institute. She has travelled and worked in Africa and Latin America and lived and worked in a squatter settlement in Rio de Janeiro. For six years she headed a Boston-area public programme for high-school dropouts. Dr Dulany has been a consultant to the United Nations and the Ford Foundation on health care and family planning in Brazil, the USA and Portugal, and to the National Endowment for the Arts on non-profit management and planning. As senior vice president of the New York City Partnership, she headed its youth employment and education programmes for five years. Dr Dulany is an honours graduate of Radcliffe and holds a doctorate in education from Harvard University.

Phone: (+1-212) 517 4900
Fax: (+1-212) 517 4815

GLOBAL SOCIAL INVESTMENT

Peggy Dulany, President and Founder, The Synergies Institute, New York

It is in the best interests of investors to invest part of their capital in ensuring social stability and equitable development because, otherwise, we will find ourselves living in an environment that will not even permit us to retain our capital, much less see it grow.

So what can be done to avoid such a catastrophe? And what can the individual investor do? Look at the main elements of any society: the state, or government; the market, or the economic forces that generate wealth; and civil society, the expression of human interests organised into formal or informal groupings which help meet the needs of citizens. The nature of civil society has an important role to play in protecting our future security.

The market produces goods and services and thus creates employment and generates wealth. Salaries paid for work rendered enable families to meet their basic needs, buying goods the production of which leads, in turn, to the creation of further jobs.

The state generates laws for the protection of its citizens and regulations to restrain the market from running amok and from serving the interests only of those who are able to succeed by its rules.

Civil society has the unenviable task of helping people to meet all the needs that are left unfulfilled by the market and the state and, to a certain extent, that of holding the state and the market accountable for their behaviour. Until recently civil society has been the least emphasised and understood in this triad – although Alexis de Tocqueville believed it was the basis of the success of American society in the 19th century. Because of its rich diversity, civil society is hard to grasp as a single concept; it includes groups as disparate as religious organisations, trade unions, professional associations and community groups.

In a democratic society, the people elect the government which, in turn, enacts the laws. But it is the people's organisations that affect their opinions, that give them strength and that are capable of producing the kind of orchestrated action that can apply pressure for changes in the law and even bring about changes of government, when abuses are perceived.

Research shows that the principal factor contributing to the strength of civil society is social capital: the bonds of trust and reciprocity which grow up between people when they interact on a regular basis around some shared interest. These interests provide a way for people of otherwise very different backgrounds and views to come together in a way that builds social capital, which leads in its turn to increased productivity and more effective political institutions.

The relative size and influence of the three sectors varies from country to country. In communist societies, the state tried to fill the entire space, arguing that it was best equipped both to generate wealth and to look after the needs of the people, a theory now universally discredited. But with the death of communism, the market has moved into much of the space vacated by government, although often without taking on the roles that government ceded. Civil society also varies greatly in strength, although where government was formerly strong to overpowering, civil society tends to start out weak, leaving an even larger vacuum for the market to fill, at least in the short term.

THE NEED FOR SOCIAL INVESTMENT

Looking around the world, one can see that the countries with the highest standards of living and the lowest levels of social conflict are usually those where the three elements are most evenly balanced, such as the Scandinavian countries and Chile. Furthermore, a recent study demonstrates that 'high income inequality creates social unrest and pressures for heavy taxes on the rich and, in turn, economic uncertainty which discourages investment.' Apart from balance between the sectors, the most successful societies are those where there is not a wide gap between the incomes of the richest and the poorest citizens. The United States is an exception but, if the trend towards income inequality is not soon reversed, it will inevitably face increasing social unrest and suffer a consequent economic downturn.

Given the dominant role of the market in much of the world, it is probably unrealistic to assume that governments and civil societies alone will be able to restrain it to any great degree. But, as shown by the correlation discovered in recent research findings between economic growth and social stability on the one hand and inequality on the other, it is clearly in the interests of members of the private sector to invest greater energy and resources in ensuring human and social development.

Why social investment? Firstly, it is the *modus operandi* of the private sector. Secondly, it is an effective way of allocating resources to produce the greatest return – whether human or financial. The difference between social and financial investment lies in the nature of the return. When choosing an investment purely for financial return, the likelihood of substantial profit is the deciding factor. When investing for social return – even if the decision to invest is based on probable financial return – one takes other factors into account, such as strategies that will help balance income inequalities in the long term, thereby contributing to social stability and economic growth.

Forms of social investment

One form of social investment is to become involved in activities that increase the earning potential of those at the bottom of the social spectrum by helping them to become better nourished, housed and educated. While this is generally considered to be the role of government, some governments are so weak or inefficient that they cannot, or do not, give it a high enough priority. In that case, a significant investment of time and energy by the private sector would begin to mobilise popular support to encourage the government to be more effective – or to change. The private sector as a whole has shown itself to be an effective lobbyist on behalf of causes about which it feels strongly.

Latin America provides an example of private social investment . The Americas Society and the Latin American Business Association are documenting cases where businesses have played an effective part in improving the quality of and access to basic education, and conducting business seminars on such cases. They are motivating businesses in Latin America to become more involved both in tangible projects in their own communities and in putting pressure on their governments to allocate additional resources and provide better education for young people.

According to Roberto Mizrahi, president of the South North Development Initiative, a second and more direct form of social investment by the private sector involves allocating resources directly to economic activities that will have a high social impact. Mizrahi refers to 'new socially responsible economic organisations' (NEOs), as emerging institutional vehicles well suited to private sector participation and necessary for success. These NEOs might be community banks, local venture-capital companies, marketing facilities or trading companies. The difference between them and the traditional institutional forms from which they draw their names is that, as well as producing a financial return, they have other goals: increasing employment; providing much-needed services; broadening ownership; increasing access to information and the market for those traditionally left out; and instilling business skills which will generate further economic returns. In short, they are born with a social goal which they pursue just as seriously as the financial one.

The role of private investors in supporting such institutions can range from technical assistance with business skills, equity participation, the donation of seed capital or agreement to purchase their products. These are practicable and appropriate ways for individuals and companies to apply their skills and resources to generate both financial and social returns.

A third way for private investors to use their skills and resources to invest socially is by helping to strengthen the civil society organisations that dedicate themselves to improving the economic and social conditions of the worst-off among the population. Two exemplary methods come to mind. First, business has outstripped both government and non-profit sectors in advances in management and organisational techniques. Private-sector individuals skilled in these processes can make a big difference by investing some time with such non-profit organisations. Second, private-sector individuals can assist such groups with their finances either by contributing to their funds or by helping them to develop ways of earning their own income. This too, would be money well spent, for the work these non-profit organisations do will contribute to a more stable society and a more productive economy.

THE NEED FOR GLOBAL INVESTMENT

Looking at the ways in which various phenomena cross national boundaries and affect different societies, it becomes apparent that social investment on a global scale is very much in the interests of individuals, societies and market forces all over the world.

At the most basic level are the 'people flows' that are changing the dynamics within and between countries. If economies are crumbling at home, people will move to places where the economy is more stable.

With the movement of people comes the spread of disease. The ease and frequency with which large numbers of people move between countries greatly increases the danger of diseases becoming pandemic.

Environmental damage is also no longer a local phenomenon. If we do not stem the emission of chlorofluorocarbons (CFCs) into the atmosphere and if we continue to cut down the forests, we are guaranteeing cancer-producing sunshine everywhere – not just in the offending countries. The same is true for the effects of toxic wastes, both airborne and waterborne.

Information flows, too, affect trends and events across national boundaries. The availability of radio and television in the most remote places lets people know that theirs is not the only way of life. It influences rural-to-urban migration as well as transnational migration.

Finally, capital no longer circulates primarily within its country of origin; instead, it follows opportunity, as revealed by information flows. But, all too often, it follows opportunity in a very short-sighted way – focusing on the immediate return and paying scant attention to the long-term consequences for development. Fluctuations in the world market may have more effect on certain subregional or national economies than government tax policies and foreign aid flows combined.

Market forces seem to have complete freedom to act and react on a global basis. And they do. But their actions have an echo effect which will eventually bounce off some distant shore and come back to their place of origin in distorted form. Therefore, unless we think socially as well as economically about the investment of our money and time – and unless we think not only locally but also globally when planning these social investments – we are sowing the seeds of social breakdown.

With information flows serving as interconnections, one might speculate that it will soon no longer be possible to maintain vast income differentials, either within countries or between rich and poor societies, without damaging the health of the world economy. As long as we are looking beyond the returns for the next quarter, therefore, it is in our collective interest to pay attention to what is going on in other parts of the world.

CONCLUSION

What does this have to do with the investor, and how can the individual make a difference when the problems are so many and so vast? As a first step, one can add a global social perspective to one's investment decisions. By doing so, one will reap both financial and social returns in the long run.

Obviously investors want to maximise financial returns on most of the portfolios they control. But as a hedge – against the kind of social, political and economic instability that could ultimately undermine even the apparently safest financial investment – a portion of each portfolio should focus on investments designed to bring social as well as financial returns. ❖

CHAPTER 3

INVESTING IN PROPERTY

REAL ESTATE: GLOBAL ASSET DIVERSIFICATION AND THE PRIVATE CLIENT

Pierre N. Rolin, Credit Suisse London

❖

Pierre N. Rolin is based at Credit Suisse London and is head of CS International Real Estate Investment Advisory Services. Established in 1993, this unit has global product responsibility for real estate investment counselling throughout Credit Suisse Private Banking. Mr Rolin has 10 years of international real estate experience advising private and institutional investors on global real estate diversification. Before joining Credit Suisse, Mr Rolin worked in Frankfurt, New York and Toronto for the real estate units of major international banks and specialist real estate investment advisers.

REAL ESTATE: GLOBAL ASSET DIVERSIFICATION AND THE PRIVATE CLIENT

Pierre N. Rolin, Credit Suisse London

An individual making a well-structured and conservative direct investment in carefully selected property markets has a holding in one of the most prudent and attractive classes of asset. It offers the international private investor the triple benefits of global diversification, long-term asset growth and stability.

Most private investors seeking to buy real estate have certain objectives in common:

- A physical stake: many investors are attracted to real estate because of its concrete nature.
- Preservation of capital (this is often the main objective).
- Inflation protection: this is really an extension of the second objective, since private investors want to preserve the real purchasing power of their investments.
- Inheritance and tax planning tend to be private investors' primary motivations for including long-term high-quality real estate assets in their portfolios as part of their global asset diversification strategy.

International banks often have little experience in advising private clients, whom they consider of secondary importance. As a result, they do not take the time to understand the individual, personal needs of the international private client who is investing in real estate.

The investment banking or brokerage approach to real estate investments is product-driven and transaction-oriented and focuses on selling properties. This approach generally creates a conflict of interest by representing the seller rather than the buyer. In contrast, CS International Real Estate Investment Advisory Services has a client-driven approach, which represents the buyer and focuses on each client's particular objectives in order to create a tailor-made solution to his or her real estate investment requirements.

US AND UK INVESTMENT PROPERTY

The UK and US property markets regained their fundamentals in 1993 and continue to attract a growing share of the dedicated global equity and debt. The main reasons for this include a substantial decrease in the volume of non-performing loans, debt foreclosures and sales in the corporate real estate sector, acceleration of debt repayments and higher occupancy levels. These have brought an end to the worst crisis in the UK and US property markets in 60 years (see Figure 1).

The United States . . .

Let us take a closer look at the US market. A primary cause of the recent real estate crisis in the US was the easy availability of funds after the economic recession in 1982 and 1983. However, this was itself the result of a number of factors, including aggressive lending by banks and insurance companies, a significant increase in property investment by domestic pension funds and the deregulation of the savings and loan industry, which opened up a large pool of new funds to real estate developers. All these factors combined to generate an

FIGURE 1: UK AND US PROPERTY RETURNS

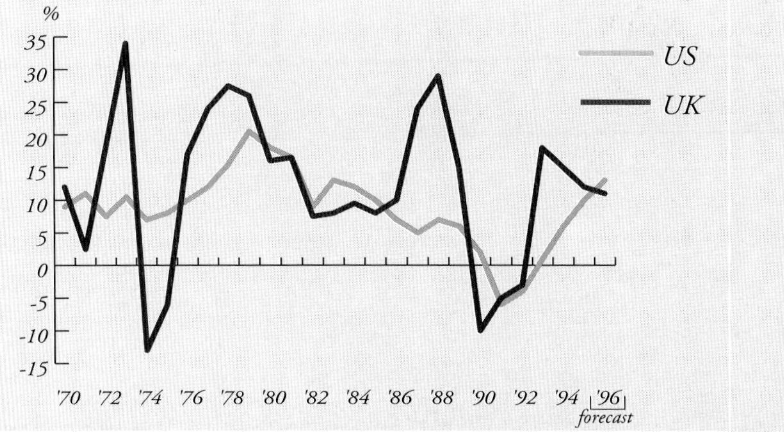

Source: *US: Evaluation Associates Inc. data 1970-1977 / NCREIF data 1987-1994 / LaSalle forecast*
UK: Jones Lang Wootton 1970-1980 / Investment Property Databank (IPD) 1980-1994 / BH&S 1995-96 forecast

enormous over-supply of capital for US real estate development and investment. During the following 10 years, tremendous over-building and a slow rental market resulted in high vacancies, declining rents and falling real estate values throughout the USA. Between 1988 and 1992, inflation-adjusted capital values declined by an average of 8% a year and by early 1992 the USA was experiencing the worst real estate crisis since the Great Depression of the 1930s.

Why should international private investors buy US real estate today? The overseas investor entering the US real estate market can obtain initial cash returns of between 8% and 10% for institutional quality, long-term leased, credit-tenanted commercial properties – the highest level in 20 years. In addition, real estate investments are available at prices below original construction cost or today's replacement cost, offering the possibility of excellent capital gains as the market improves. On an unleveraged basis, these bond-type investments currently provide private investors with income returns 200-300 basis points higher than those on 30-year US Treasury bonds, and substantially higher returns on a financed return-on-equity basis.

The fundamentals are also very positive. US commercial office vacancy rates, 18% in 1993 and 15% in 1994, were down to 14% by December 1995. In addition, with its good inflation outlook and sustainable economic recovery, the US will remain an attractive investment market and this may help to strengthen the dollar. There are few markets or asset classes in the world where all of these conditions are so favourable.

So why should German private investors, for example, diversify into US real estate at the moment? There are several reasons:

- US long-term-leased credit-tenant commercial properties give initial cash returns in the region of 9%, compared with 7% for similar properties in Germany.
- The Deutschmark's strong purchasing power.
- A favourable double-taxation agreement with the USA.
- A top US marginal tax rate of 39.6%, compared with the top German income tax bracket of 61.75%.
- The real possibility of a reduction in US income and capital gains tax rates.

In these circumstances, Credit Suisse (Deutschland) AG believes that the time is ripe for German-domiciled private investors to acquire US real estate as part of their global asset diversification strategy.

. . . and the United Kingdom

Why are international investors currently investing in selected European property markets? By way of example, let us look at London, the first European city to emerge from the recent recession. With current office vacancy rates in central London at 11.6%, compared with 14.6% in 1993, fundamental rental growth in all UK property sectors is finally happening for the first time since 1990. It is estimated that foreign investors acquired more than £5bn-worth of UK commercial property between 1991 and 1994, producing a total rate of return higher than 50%.

Speculative property developments are currently in extremely short supply for a number of reasons: the banks' pre-let funding requirements, the tighter building and environmental regulations imposed by various levels of government, and rationalisation and consolidation by white-collar employers. Consequently, Credit Suisse believes we are beginning to see a very different property cycle taking shape in the UK and US markets – one that is more orderly and stable.

The 50-year pattern of sustainable expansion driven by the white-collar employer ended in 1990-91. This new climate on the UK and US property markets is fuelled by two main trends: large institutional users are rethinking their long-term property requirements, and a general process of down-sizing is currently under way. This new reality has forced the prudent property investor – who is motivated primarily by the prospect of fundamental rental growth – to pay close attention not only to location (traditionally the primary consideration) but also to individual building specifications. As a result, the most prudent property investment is one that contains the dynamics of a committed tenant investment.

DIRECT INVESTMENT STRATEGIES

In designing a direct investment strategy one must adhere to a set of very conservative qualifying criteria, with the greatest emphasis on purchasing the highest-quality real estate in a particular market. The most important criteria for purchasing quality investment property in the UK and US markets are that the land is freehold, that the property is fully leased, that it contains institutionally acceptable tenant covenants with net lease structures, produces secure long-term rental cash flows, offers competitive investment returns and is controlled through 100% direct ownership. Two of the most important considerations in the UK and US property markets are the unique UK 'institutional lease' and the US 'triple net lease' structures, which have bond-type characteristics.

With respect to the actual lease structure, the UK institutional FRI (full repairing and insuring) lease provides private investors with a number of very attractive benefits. Lease terms of up to 25 years can be negotiated, with upward-only rent reviews every five years (although the more common lease will have a tenant break clause after 10 years). The FRI lease generally stipulates that the landlord pays nothing towards the maintenance and upkeep of the property investment: the tenant or tenants pay for the cost of maintenance, repair and renewal and for the insurance of the building.

In the US market, Credit Suisse is focusing on properties with the 'triple net lease' structure. In a triple net lease, the investor receives the rent without any deduction for operating expenses. Apart from the rent, the tenant is obliged to pay all charges with respect to the property, including property taxes, repairs, maintenance and all other expenditure incurred in the use and operation of the property.

WHAT KIND OF PROPERTY?

In both the UK and US markets there are essentially three different types of income-producing properties that Credit Suisse considers for purchase: office buildings, retail units or shopping centres, and industrial warehouses. Generally, multi-tenant properties are more prudent investments if they use the aforementioned lease structures. One of the advantages of multi-tenant properties is that the risk of vacancy is spread more evenly than with single-tenant properties.

Credit Suisse is also focusing on three avenues through which the private investor can invest directly in the UK and US property markets. First, by adopting a 'passive' investment strategy, investors can acquire existing income-producing institutional-grade investment property within a set of very conservative investment parameters. The second avenue is a 'proactive' investment strategy, whereby the investor provides a forward funding facility to a property development company in return for the right to purchase a 100% pre-let development on completion. This can give the investor a higher initial yield and generally a longer-term lease structure. Adhering to the above investment parameters, foreign investors should currently be aiming for an annualised internal rate of return of more than 12% over a seven- to 10-year period.

The third avenue Credit Suisse focuses on is its 'created value' investment strategy, an approach designed specifically for the sophisticated property investor. This involves acquiring existing properties where the present owner is in a distressed position or under pressure to sell and has not fully realised a number of features of 'hidden value' inherent in the property. With this type of investment, Credit Suisse recommends several UK and US blue-chip property developers who can provide a wealth of experience and would act as strategic partners, helping Credit Suisse clients to identify and co-invest in these added-value opportunities.

In addition, through close cooperation with sister company CS First Boston, Credit Suisse can provide its international clients with access to a fully integrated US real estate products group providing a complete range of investment banking services. This includes, for example, assistance in structuring off-market corporate sale and lease-back commercial properties leased to credit tenants on long-term leases.

Property investments meeting the above criteria generally range between US$5m and US$25m, with US$10m being the typical transaction size. This translates into a minimum equity commitment by our private clients of around US$3m before financing.

TAX PLANNING

The international investor has come to realise from experience that tax planning plays as crucial a role as the actual 'bricks and mortar' investment strategy. It is essential that private investors retain a team of specialists who can provide detailed tax planning advice in connection with the ownership vehicle, and with income tax, capital gains tax and inheritance tax issues regarding investments in UK and US property.

The UK and US are countries of Anglo-Saxon origin and these countries share not only a common language but also common law. The Anglo-Saxon trust has been employed as an effective means of estate and inheritance planning in both these countries for centuries. Together with Credit Suisse Fides Trust's office in Guernsey, Credit Suisse has spent some considerable time designing a number of ownership and tax planning structures tailored to fit the specific needs and objectives of each of its clients (see Figures 2 and 3).

UK tax attractions

Investments in UK real estate by non-UK residents have several unique tax attractions. The principal one is that rental income received on the property (after the deduction of running costs and maintenance expenses, tax depreciation and qualifying interest) will only be subject to UK income tax. The benefit there is that the rate of tax applying is only the basic rate of income tax, at present 25%. Furthermore, deductions for tax depreciation on plant and machinery within the property mean that the effective tax rate on the cash return from the property will be significantly less than 25% and can be as little as 10% in the early years of ownership.

The financing of UK property is another area that must be tailored to each specific property investment. Under the UK Finance Act in effect since 5 April 1995, any interest paid on loans taken out wholly and exclusively for the purposes of the 'business' of property investment will be deductible from rental income, not just interest payable on the loan taken for the actual acquisition as previously was the case. This means it may be possible to make such deductions in respect of borrowings in excess of the original purchase price. Since 5 January 1995, UK property investors have also been able to make use of non-UK-resident bank financing.

As a result, an individual investing through an offshore property holding structure can secure non-recourse property financing from a non-UK bank, the interest on which can be deducted against rental income. This is usually subject to the condition that the country in which the lending bank is domiciled has a double-taxation agreement with the UK to eliminate withholding tax on interest. The effect of these changes is that non-UK residents investing in UK property have much wider choice with regard to lender, sum borrowed and timing of the loan.

VAT may be charged on the purchase price of the property at a rate of 17.5%. If the purchaser registers for VAT then this VAT can be recovered; alternatively, to avoid the cashflow disadvantage of paying and then recovering VAT, it may be possible to avoid paying VAT on the purchase price by treating the purchase as a 'transfer as a going concern'.

As long as it is clear that the property is held as an investment, capital gains tax does not apply to non-UK-resident private investors who acquire property through an offshore property holding company. In addition, there is no inheritance tax exposure if the beneficial owner controls the UK property through shares held by an offshore trust of which the owner is a beneficiary. Inheritance tax does apply if the offshore trust is set up by a non-UK-domiciled individual and the trust owns only non-UK property, such as shares in the offshore property holding company.

Minimising US tax

Turning to the US market, the best way for a non-US resident to maximise the return from a US real estate investment will generally be to hold it through a US corporation incorporated

FIGURE 2: EXAMPLE FOR A £10M DIRECT INVESTMENT IN UK REAL ESTATE

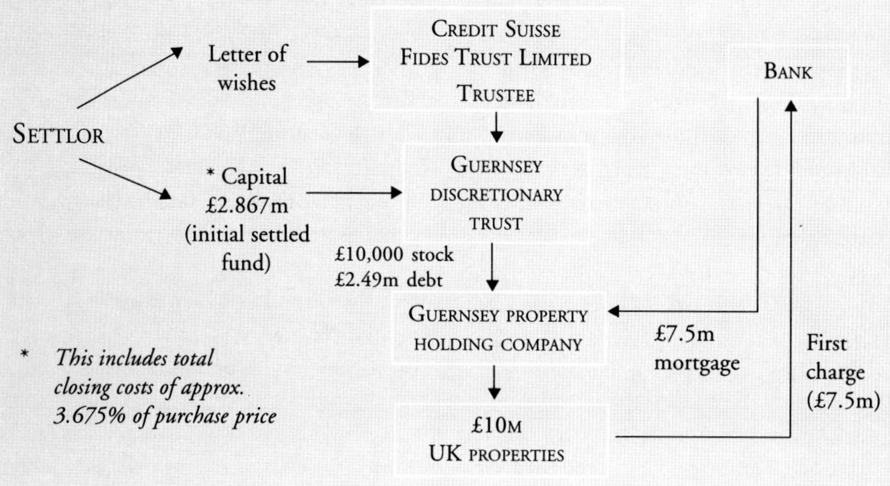

FIGURE 3: EXAMPLE FOR INVESTMENT IN US REAL ESTATE

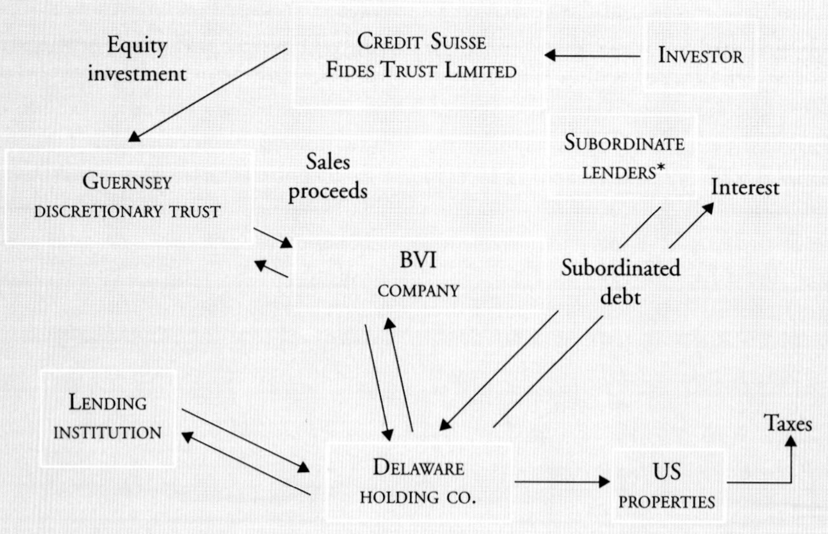

* *No US withholding tax if subordinated debt is not directly or indirectly held by shareholders of the BVI corporation or their spouses, children, grandchildren or parents. Otherwise, 30% withholding tax is levied on these interest payments.*

either in Delaware or in the state where the property is situated. The US company is, in turn, held by a non-US company typically incorporated in one of a number of offshore financial centres, such as the British Virgin Islands. In many cases, the shares in the non-US company will, in turn, be held by a Guernsey trust. The chief advantages of this structure include exemption from US estate, inheritance and gift taxes, the maximum legally permissible protection of the investor's confidentiality and the potential for maximising income and capital gains tax savings.

In addition, since the Republicans won a majority in the US Congress in January 1995 the centrepiece of their campaign has been the so-called 'Contract with America'. One of its most important items is the promise to reduce capital gains tax for both individuals and corporations. This would reduce the tax from the current top rates for individuals and corporations of 28% and 34% to 19.8% and 17.5% respectively.

There is one limited but important case in which America's flat-rate withholding tax on investment income does not apply: qualified 'portfolio' debt investments are not subject to US withholding tax. The use of portfolio loans usually allows the investor to effectively 'strip' some of the earnings from a real estate investment out of the US free of all US tax. There are, however, a number of conditions that must be met for a loan to qualify as a portfolio loan; what is needed is a well-thought-out and sophisticated structure, which must be approved by a qualified US accountant.

CONCLUSION

It is essential that any international private investor contemplating a direct investment in US or UK commercial property obtain the assistance of a full team of property specialists including real estate specialists, solicitors, accountants, tax planning and trust advisers, and property managers. A property acquisition coordination team is crucial to the acquisition of a good-quality investment, and to the minimisation of the overall potential tax exposure, thereby preserving rental income and maximising overall investment returns. In this way, the international private investor will achieve passive global diversification into a secure asset class.

By broadening the range of services and assisting private clients with their real estate investments, CS Real Estate Investment Advisory Services aims to reinforce Credit Suisse Private Banking's position as a leading provider of private banking services. ❖

EUROPE'S COMMERCIAL PROPERTY MARKETS

Nigel Roberts, Partner and Head of European Research,
Jones Lang Wootton, London

Nigel Roberts has more than 20 years' experience of property market analysis, the last 15 with Jones Lang Wootton, which today has 33 offices in Europe. As head of research he coordinates a team of 20 analysts across Europe who monitor market activity, forecast future real estate trends, and advise a wide range of investor clients on market prospects and investment strategy.

Phone: (+44-171) 413 1319
Fax: (+44-171) 493 9539

EUROPE'S COMMERCIAL PROPERTY MARKETS

Nigel Roberts, Partner and Head of European Research, Jones Lang Wootton, London

This article reviews the characteristics of Europe's main commercial property markets and their performance during the quite extraordinary market conditions experienced by real estate investors, developers and occupiers alike, during the first half of this decade. Its focus is the commercial markets of Europe: office, retail and industrial. It reviews the underlying economic conditions that created the boom-to-bust environment of the early 1990s; comparative investment performance, in terms of risk and return, between real estate and equities and gilts; relative investor returns between the main real estate sectors; and today's main markets.

THE ECONOMIC BACKDROP

Increased activity in national and local economies will generally precede events in the real estate markets by at least one year. Imrpoved economic output, for example, does not immediately boost demand for space. Occupiers are usually unable to switch premises because of existing leasehold or freehold obligations, and they will also need to satisfy themselves that any improvement in trading conditions can be sustained before taking on new space and committing themselves to increased overheads.

FIGURE 1: PROPERTY CYCLES: A CONCEPTUAL FRAMEWORK (BASED ON THE LONDON MARKET)

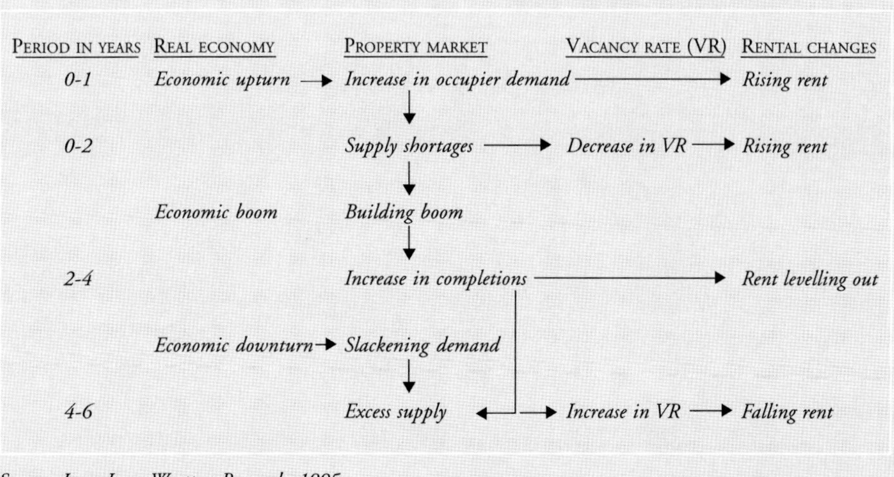

Source: Jones Lang Wootton Research, 1995

The increase in demand usually comes up to 12 months after an economic upturn and, for the next two years at least, the quantity and quality of available space are likely to be reduced. This can lead to a noticeable rise in the cost of space as vacancy rates fall. At the same time the confidence of both developers and financiers (the supply side) will improve sufficiently for new building developments to start. The two-year-plus time-scale required to secure building permits and carry out construction, however, means that new space often does not become available in time to benefit fully from the best of the demand conditions, often coming on to the market after the economy has slowed. Expected lead and lag times in a typical property market cycle are shown in Figure 1.

An understanding of the dynamics of the property market's response to the economy, helps us to interpret property market activity across Europe in recent years. Figure 2 illustrates both the rapid growth of the European economy in the late 1980s, which peaked in 1988 with overall growth of 4.2% per year, and the lead that the UK had on other major European economies (France, Germany, Italy). (The UK peaked in 1986-7 and entered recession in 1991, while post-reunification Germany was recording annual average growth rates of 5% at the time). These differing economic growth rates have since had a direct impact on European real estate markets, including the commercial property sector.

THE OFFICE MARKET

The largest commercial real estate markets in Europe are plotted in Figure 3, along with an estimation of their office stock. These markets represent an important focus for investor activity in Europe, and most are tracked quarterly by the Jones Lang Wootton European Office Index. This is a barometer of prime office performance, measuring rental and capital value changes across 14 markets (eight countries) weighted according to their office

FIGURE 2: GDP IN EUROPE (% P.A.)

EUROPEAN UNION MEMBER STATES AND THE FOUR LARGEST MEMBER STATES

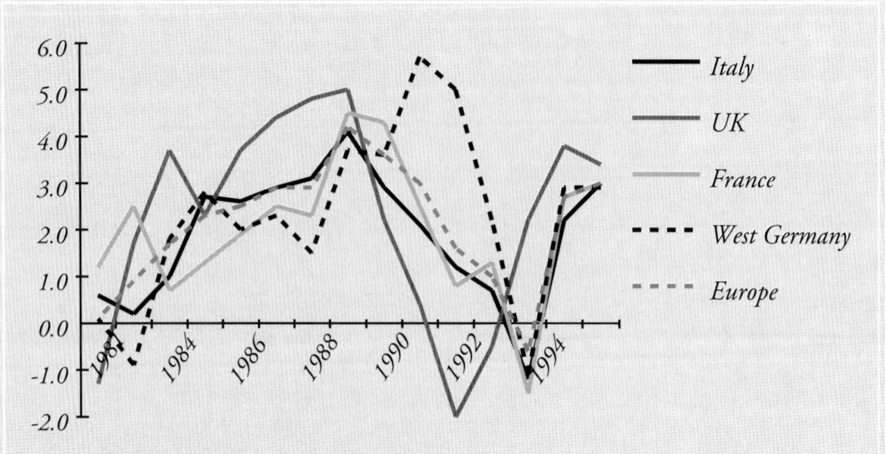

Source: OECD Economic Outlook (57), June 1995

FIGURE 3: EUROPE'S MAJOR OFFICE MARKETS[1]

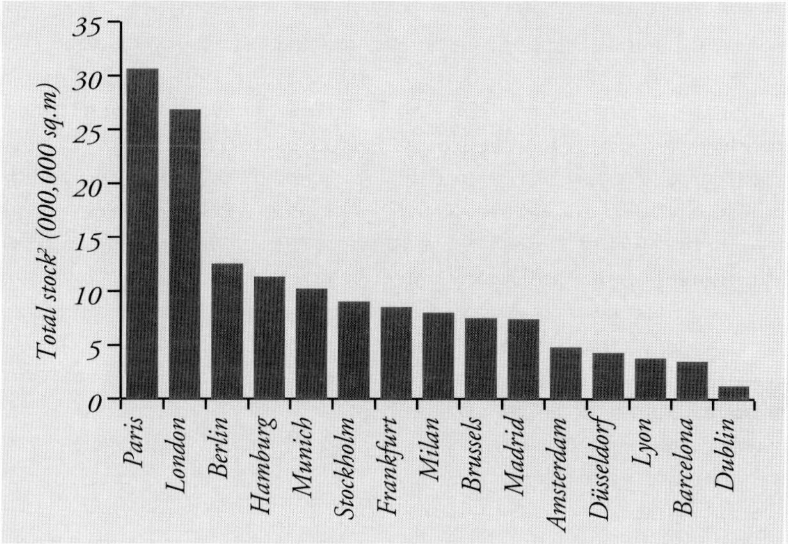

[1] The 15 cities selected are among the largest office markets (according to stock) in Europe.
[2] 1995 estimate of total office stock in each city's urban area.
Source: Jones Lang Wootton Research, November 1995

FIGURE 4: JONES LANG WOOTTON EUROPEAN OFFICE INDEX[1], SEPTEMBER 1995

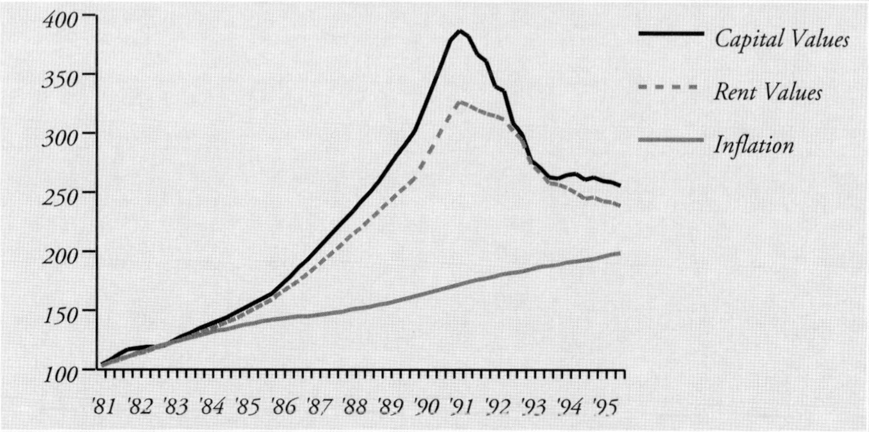

[1] The Jones Lang Wootton European Office Index tracks changes in capital and rental values in European prime office property since 1980. The indices are composed from the weighted average change in prime office capital and rental values in 14 European cities. The Inflation Index is derived from the consumer price indices (CPI) of the eight countries covered by the Index, weighted according to total GDP.
Source: Jones Lang Wootton International Research, September 1995

stock. The overall rental and capital indices peaked in the first quarter of 1991, two years after the peak in gross take-up, and three years after the peak in Europe's economic output. Since then both indices have fallen more or less continually, and are now some 30% below their peaks (see Figure 4). Over the same period vacancy rates have risen from an average of about 3% to around 10% today.

The significant overall downturn did not affect all markets at the same time, each being subject to its own particular local economic influences. Because of the timing of the macro-economic cycle, the London market was the first to peak in early 1989, benefiting from rapid economic growth in 1986 and 1987 and the deregulation of financial services affecting many of the capital's service companies. However, by the third quarter of 1993 the London prime rental market had reached bottom, prime rents having fallen from more than £70 per sq.ft in 1988 to about £30 per sq.ft in 1993. Vacancy rates in central London increased from around 2% to more than 15% over the same five-year period.

By contrast, in the early 1990s Germany's real estate markets were still feeling the benefit of the strong demand conditions that accompanied reunification, and for this reason rental growth in offices did not peak until 1991-92. Figure 5 shows the timing of rental peaks and troughs for a selection of the largest European office markets. These markets peaked over a four-year period, and we estimate that some are still in a downward phase. Others, however, have passed the low point of their rental cycles but are still experiencing relatively little upward movement.

CURRENT OFFICE MARKET CONDITIONS

European office markets have remained relatively subdued during 1995, even though the European economies have moved out of recession. Occupiers remain cautious, activity levels are low and high vacancy rates are a persistent feature of most markets. The European Rental Index fell by 1.8% during

FIGURE 5: RENTAL PEAKS AND TROUGHS IN EUROPEAN OFFICE MARKETS*

	Rental peak	Fall in rents from peak %	Rental trough	Rise in rents from trough %
London (City)	*Q1 1989*	*54*	*Q3 1993*	*15*
Paris	*Q4 1990*	*34*	—	—
Madrid	*Q4 1990*	*53*	*Q4 1994*	*0*
Milan	*Q1 1991*	*40*	—	—
Frankfurt	*Q3 1991*	*32*	—	—
Brussels	*Q2 1992*	*20*	*Q2 1994*	*3*
Hamburg	*Q2 1992*	*13*	—	—
Amsterdam	*Q2 1992*	*6*	*Q2 1995*	*0*

**Based on prime achievable office rents*
Source: Jones Lang Wootton Research, November 1995

FIGURE 6: EUROPEAN OFFICE RENTS, THIRD QUARTER 1995

	CURRENCY UNIT AND MEASUREMENT	PRIME RENT[1] PER UNIT	PRIME RENT[2] (US$/SQ.M /YEAR)
Amsterdam	*G/sq.m/year*	*425*	*267.3*
Barcelona	*Pta/sq.m/month*	*2,300*	*225.6*
Berlin	*DM/sq.m/year*	*55*	*465.0*
Dublin	*£/sq.m/year*	*172*	*275.4*
Brussels	*BFr/sq.m/year*	*8,250*	*282.7*
Düsseldorf	*DM/sq.m/month*	*40*	*338.1*
Frankfurt	*DM/sq.m/month*	*65*	*549.5*
Hamburg	*DM/sq.m/month*	*48*	*405.8*
London - West End	*£/sq.m/year*	*455*	*708.9*
London - City	*£/sq.m/year*	*405*	*629.2*
Lyon	*FFr/sq.m/year*	*900*	*183.9*
Madrid	*Pta/sq.m/month*	*2,700*	*264.9*
Milan	*Lire/sq.m/year*	*450,000*	*281.7*
Munich	*DM/sq.m/month*	*50*	*422.7*
Paris	*FFr/sq.m/year*	*2,950*	*602.8*
Stockholm	*SKr/sq.m/year*	*2,600*	*389.2*

[1]Represents the top open-market rent achievable for prime-quality office space greater than 500 sq.m.
[2]Exchange rate as at 13/11/95.
Source: Jones Lang Wootton Research, November 1995

the first nine months of the year. The index is now 27% below the peak achieved in early 1991. Investors are gradually returning to the market, but still remain hesitant, seeking well-let income-producing investments. The European Capital Index has fallen by a further 2.7%.

Figure 7 shows the rental cycles of the major European office markets at the end of 1995 and their current vacancy rates. It gives only a general representation of market conditions, and for this reason should be used with care, although it does provide a convenient tool for comparing the relative positions of the different markets in their short-term cycles.

The position of each market on the 'clock' (see Figure 7) is determined by rental movements for top-quality, centrally located offices in each market. At 12 o'clock rental values are at a peak (e.g. Warsaw) and, although strong rental growth has been recorded in recent years, prime rents now appear to have reached a plateau. As markets move past 12 o'clock, they are characterised by gradually accelerating rates of decline in rents. By three o'clock, in theory, the rate of decline should be at its maximum. Most European office markets are located in the bottom quadrants, bunched around the six-o'clock mark; they are at or around the bottom of their rental cycles with growth potential apparent in the next few years. London, Luxembourg and Dublin, followed by Scandinavia, are most advanced in their cycles.

FIGURE 7: SHORT-TERM RENTAL CYCLE, END 1995

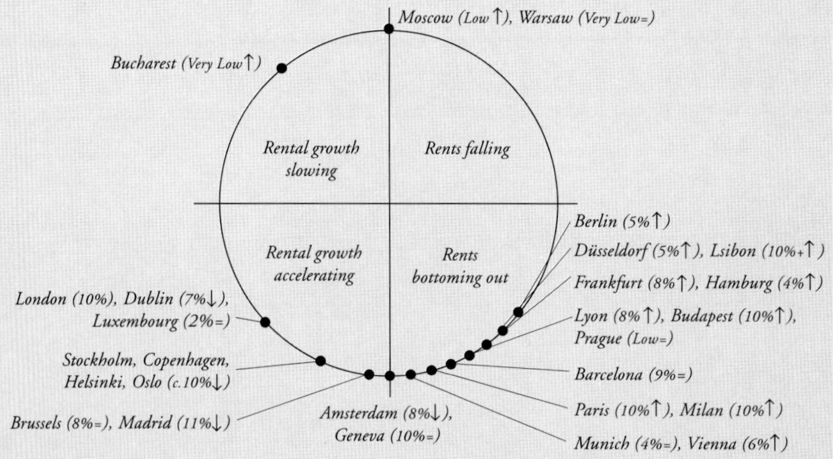

Notes: Each market's position in its cycle is based on prime rental values. Markets can move around the 'clock' at different speeds and directions. The position of a market is not necessarily representative of investment or development prospects. The figures in parenthesis refer to the current vacancy rate in each office market, and its direction over 1995.
Source: Jones Lang Wootton Research, November 1995

The UK and Ireland

The London office investment market and the leasing market have been moving upward since 1993 and 1994, respectively. However, the evidence suggests that the upturn has been less pronounced than expected. Prime rental values are under some upward pressure, reflecting falling vacancy rates (particularly for grade-A space) combined with a gradual strengthening in occupier demand. With almost no new space coming onto the market, development activity is at its lowest level since the early 1980s, and supply shortages for grade-A space are likely to become more acute during 1996-97. Nonetheless, there is still a substantial amount of second-hand space and overall vacancy rates in central London remain at around 10%. Investment activity has become more moderate since the high levels of 1993-94, reflecting the shortage of quality investments, keener yields and lower levels of activity on the part of some UK institutions.

The market in Ireland is at a similar stage, although it does appear to have sustained some of its 1994 impetus, with vacancy rates dropping from 10% in 1994 to around 7.5% in mid-1995. The strengthening occupier market has been dominated by banking, computer and financial services companies. Prime rental levels have remained unchanged, even though little new development is taking place.

Scandinavia

The markets in Copenhagen, Helsinki, Oslo and Stockholm are in a similar cyclical position to that of the UK. Following deep and prolonged recessions, 1995 has provided more evidence of an upturn in these markets.

Moderate rental growth has occurred in all four markets, but has been strongest in Stockholm. Vacancy rates are estimated to be in the region of 10% in each market. An improvement in demand and the virtual absence of new office development activity have resulted in some downward movement in vacancy levels. However, the speed with which overall vacancy rates are falling continues to be affected by weak net absorption. Only in the central areas, notably Stockholm and Helsinki, is there any evidence of more substantial falls in supply. The investment market remains dominated by domestic investors, but with emerging international interest in the larger markets.

Benelux (Belgium, Netherlands, Luxembourg)

The Brussels leasing market has been relatively active, with take-up maintained at high levels. This has been underpinned by demand from the European Union (EU), the Belgian public sector and international companies. Prime rents are under upward pressure, with a modest gain recorded during 1995. Overall vacancy rates are stable at 8% but, with limited speculative development, vacancies are of increasingly poor quality. The grade-A space vacancy rate stands at just under 3%. Investment activity has improved but remains dominated by local institutional investors seeking well-let, high-quality buildings.

Prime office rents in the main Dutch markets of Amsterdam, The Hague, Rotterdam and Utrecht have remained characteristically stable. The overall vacancy rate stands at 8% – ranging from 6% in The Hague to 10% in Rotterdam – having fallen a little from a peak of 9% in 1992-94. There is a moderate downward trend in vacancies in all four markets due to the continuing withdrawal of office space, low levels of speculative development activity and steady take-up. The 1995 investment market was less active than in 1994, with large Dutch investors adopting a more cautious approach. A number of major Dutch investors are concentrating on converting direct investments in property into other forms of investment to improve liquidity.

Luxembourg is one of the few office markets in Europe to have recorded significant rental growth during the past couple of years, with rents increasing by approximately 20%. Prime rents still remain under upward pressure. This reflects supply shortages in the prime areas, with vacancy rates as low as 2%, combined with steady demand from financial institutions.

Spain

The Madrid office market appears to have reached the bottom of its rental cycle. Prime monthly rents have stabilised at about Ptas2,700 per sq.m per month. With political uncertainty dampening confidence in the short term, there is no substantial evidence of upward pressure on rents. Fewer new completions and steady levels of take-up have translated into a fall in vacancy rates, from 15% in 1993 to 11% in 1995, with shortages of prime office space emerging in Madrid's central area.

The property cycle in Spain's second-largest office market, Barcelona, lags slightly behind Madrid's. Barcelona's monthly rents are beginning to stabilise, at about Ptas2,300 per square metre per month and vacancy rates have levelled off at around 9%. Both cities are experiencing increasing competition in investment markets. Interest from domestic and international investors is at its highest level since 1991, with the view prevailing that property is historically 'cheap' and well positioned for rental growth. Madrid and Barcelona are among the few European markets to have recorded capital-value growth during the past year.

Italy

The Milan office market has been severely affected by the property recession in recent years. However, during 1995 demand from both international and national occupiers recovered to reasonable levels; although difficult to quantify, this largely consists of companies taking advantage of low rentals to relocate. Vacancy rates are high, but concentrated in poor-quality space; good-quality modern office space is in relatively short supply. Prime rents are stable, following falls of 40% from the peak in 1990. There is a general dearth of 'institutionally acceptable' property in the market, and the investment market therefore remains quiet; any investment activity is centred on property exchanges between institutions.

Switzerland

Although the Swiss economy is recovering it has yet to have an impact on the country's two largest office markets, Zurich and Geneva, which appear to be firmly positioned at the bottom of the rental cycle. Demand levels are subdued and overall vacancy rates remain high, in excess of 10%, although lower in Zurich's central area. Levels of activity in the investment market are difficult to assess, as the majority of transactions do not take place on the open market, but are negotiated between principals.

Austria

This market has reached the bottom of its cycle, with rents stabilising after dropping by as much as 30% since 1991. Demand is improving, particularly from overseas companies, as a result of entry into the EU on 1 January 1995, but with new supply continuing to come onto the market the vacancy rate is expected to increase further from its current level of 5.5%. Within the city centre supply is more constrained and vacancy rates are expected to remain low.

France

The Paris office market has been badly affected by the recession in France and this has been compounded by the property lending crisis. Gross take-up levels have nonetheless held up comparatively well, with occupiers taking advantage of favourable leasing terms. Net absorption rates, by contrast, have been very low. Vacancy rates have risen from 3.5% in 1990 to almost 10% by 1995, although they are now expected to level off as development activity continues to decline. Prime rents remain under downward pressure, with no change expected until 1996. International investors, however, are focusing on good quality multi-let office buildings in central locations, taking the view that the market is approaching a turning point. Relative illiquidity, combined with the high transaction costs and limited availability faced by international investors entering the Paris market, have not assisted in the market's recovery.

Germany

The German real estate market is still feeling the effects of recession, with vacancy rates continuing to rise and prime rents under downward pressure. The overall vacancy rate in the five largest markets, Berlin, Frankfurt, Munich, Hamburg and Düsseldorf, stands at around 5.3%, ranging from 8.5% in Frankfurt to 4.9% in Berlin. Availability is expected to level off in most major markets, with the exception of Berlin. In Berlin, more than 2 million sq.m of new development is due on the market in the next two to three years; this will inevitably lead to an increased vacancy rate. The German investment market has been quieter during 1995 because of reduced liquidity in open ended funds and a decreased number of international investors.

Portugal

Prime office rents in Lisbon have stabilised at about 40% below the peak recorded in 1991. The leasing market is active, but this is because of lower rents encouraging occupiers to move, rather than any underlying strength of demand. Net absorption is low, with limited demand for large new office units. Vacancy rates are high and rising, indicating further downward pressure in rents. New commercial property leasing legislation was introduced mid-year, enabling both landlords and tenants to freely negotiate the terms of their leases with no restriction on rental levels. The investment market remains relatively shallow and is dominated by local players, although new leasing legislation may encourage greater international interest.

Czech Republic

During 1995, there was development of the Prague office market and there are now more predictable prices and definable sub-markets. Prime rents have been stable for a year, following falls of around 30% from the artificially high levels of 1990. A shortage of internationally acceptable office accommodation continues and vacancy rates remain low, despite many building completions over the last few years. The investment market has yet to develop, and no significant transactions have been completed on standing investment properties.

Hungary

The Budapest office market is going through a period of consolidation, in part reflecting the short-term uncertainties facing the Hungarian economy. Prime rents have stabilised and are expected to remain constant in the short to medium term. Vacancy rates for modern stock stand at around 10%, high by Central and Eastern European standards. Future supply is expected to more than match demand, and vacancy rates could rise further. There have been few true investment purchases in the Budapest market, with investment activity largely connected to development.

Poland

Warsaw's office market has been expanding rapidly, although it is not as mature as either Prague or Budapest. Occupier demand is outstripping supply, vacancy rates have not risen above 2% for five years and there is a serious shortage of high-quality office space. Despite this, rents have remained largely unchanged over the last few years, reflecting a reluctance on the part of occupiers to pay rents that are among the highest in Europe. The investment market has been slow to develop; foreign individuals and companies need a permit from the Ministry of Internal Affairs to buy property.

Russia

Now that office supply and demand are largely equal, rents in the Moscow market have stabilised. Rent prepayments, which until recently were of the order of three or more years, have now come down to between 18 and 24 months. During the next six months expanding multinational companies will have a wider choice of office accommodation as approximately 150,000 sq.m of 'Western-quality' space is expected to come onto the market. The development of the investment market is still at an early stage, although foreign nationals may now hold property without the need for a Russian partner, through wholly owned subsidiaries registered in Russia. Land is normally held on a ground lease from the regional government and freehold titles are unusual. Investment markets in general are constrained by a severe shortage of long-term capital.

Romania

The Bucharest office market is relatively undeveloped and rents have spiralled, with prime office rents increasing by about 30% in the first nine months of 1995. In view of the shortage of high-quality office accommodation, rents are expected to continue to rise in the short term. However, with proposals for 20 new office schemes, some of which are expected to come onto the market in mid-1996, rents should stabilise.

COMPARATIVE SECTOR RETURNS

The office market has had the lion's share of both local and cross-border investment in recent years. However, with the downturn in the office market, the retail and industrial sectors have been attracting more interest from investors. Figure 8 compares total returns (measured in local currencies) from prime office, unit shop and industrial investments in five of Europe's most transparent markets.

FIGURE 8: REAL ESTATE TOTAL RETURNS[1] 1981 - 1994

	OFFICES		SHOP		INDUSTRIAL	
	1981-94 % p.a.	*1989-94 % p.a.*	*1981-94 % p.a.*	*1989-94 % p.a.*	*1981-94 % p.a.*	*1989-94 % p.a.*
Europe[2]	13.5	8.3	12.8	12.1	13.0	11.9
Paris	13.0	2.9	14.9	12.1	15.0	9.4
London	9.6	-4.2	6.6	1.8	16.2[3]	17.6[3]
Frankfurt	11.1	7.6	11.8[4]	12.1[4]	10.8	10.6
Amsterdam	9.6	8.0	7.0	10.8	11.6	8.2
Brussels	13.2	5.4	12.7	17.7	12.0	9.6

[1] Total returns are measured in local currency and represent the sum of the annual income and annual capital appreciation (or depreciation) for a notional office/retail/industrial building of the highest quality in the best location in each city. These have been calculated from top prime rents, prime yields and prime capital values data and are intended to provide a benchmark for the comparison of performance between locations. They illustrate notional returns at the top end of the market and are not based on individual properties. This method of calculation will tend to overstate returns as it assumes annual reversion to prime rent every year, and makes no allowance for depreciation, or the differences in lease terms between countries. Annual average total returns (1981-94 and 1989-94) have been calculated on the basis of the geometric (compound) mean of each year's total returns.
[2] Based on the appropriate Jones Lang Wootton Index, each sector index comprises a different blend of markets.
[3] UK, based on Birmingham, London and Glasgow.
[4] Hamburg.
Source: Jones Lang Wootton Research, November 1995

FIGURE 9: COMPARATIVE RETURNS AND VOLATILITY FOR OFFICES, EQUITIES AND BONDS IN EUROPE, 1977-1994

	Returns % p.a.			Volatility[2]		
	Offices[1]	Equities[3] (national)	Bonds[4] (national)	Offices	Equities (national)	Bonds (national)
Amsterdam	*9.6*	*16.3*	*7.9*	*0.9*	*1.1*	*0.8*
Brussels	*12.5*	*17.3*	*9.1*	*0.8*	*1.2*	*0.6*
Frankfurt	*11.4*	*10.5*	*6.9*	*1.1*	*1.8*	*0.8*
London	*12.4*	*19.1*	*9.9*	*1.4*	*0.7*	*0.8*
Paris	*14.6*	*16.9*	*10.8*	*0.9*	*1.3*	*0.7*
Average[5]	*12.1*	*16.0*	*8.9*	*1.0*	*1.2*	*0.7*

[1] Total returns are measured in local currency and represent the sum of the annual income and annual capital appreciation (or depreciation) for a notional office building of the highest quality in the best location in each city. Total returns have been calculated from top prime rents, prime yields and prime capital values data and are intended to provide a benchmark for the comparison of performance between locations. They illustrate notional returns at the top end of the market and are not based on individual properties. This method of calculation will tend to overstate returns as it assumes annual reversion to prime rent every year, and makes no allowance for depreciation, or the differences in lease terms between countries.
[2] Volatility is measured using the coefficient of variation, which is the standard deviation of returns around the mean for each asset class, adjusted to reflect the mean size, and hence allow comparison between asset classes.
[3] Arithmetic mean.
[4] Total returns, based on national equities indices.
[5] Returns on 10-year government bonds.

Sources: Jones Lang Wootton Research, UBS Phillips and Drew, Morgan Stanley Capital International and Salomon Brothers, November 1995

We can see that in Europe as a whole, total returns on office investments have exceeded those in the other commercial sectors over the period. More recently (1989-94), investors have enjoyed better returns on shops and industrial property than on offices. In each sector and in most markets investment performance over the past five years has been below the overall average. Only UK industrials and shops in Hamburg, Amsterdam and Brussels have bucked this particular trend.

OTHER ASSET CLASSES

There are two principal reasons to invest in any specific market or asset class – for performance and to diversify risk. We have chosen five cities (London, Paris, Frankfurt, Brussels and Amsterdam) to assess for real estate investment on the basis of past performance. The cities represent real estate markets that have experienced significant falls in value in their latest real estate cycles (e.g. London and Paris), as well as markets that have seen less of a fall (e.g. Brussels and Amsterdam). Figure 9 shows that gross returns on office investments in these cities over the period 1977-1994 averaged 12.1%. This performance, over a period which included two peaks and two troughs in Europe's economic and property cycle – including a significant real estate downturn in the past four years – was ahead of the return on 10-year government bonds, but significantly lower than the return from equities. Figure 9 also shows that the volatility in returns on offices (1.0),

FIGURE 10: PERFORMANCE CHARACTERISTICS OF REAL ESTATE

ADVANTAGES	DISADVANTAGES
• *Total return is higher than bonds*	• *but lower than equities.*
• *Income yield is higher than equities*	• *but lower than bonds.*
• *Income growth is predictable and frequently inflation-linked*	• *but asset can depreciate.*
• *Owner can enhance value*	• *but dealing and management costs are higher.*
• *If income flow defaults, owner has the building*	• *but sales are often taxed and market can be illiquid.*
• *More predictable than equities*	• *but more risky than government bonds.*

Source: Jones Lang Wootton Research, 1995

measured using the coefficient of variation to reflect risk, was greater than for bonds (0.7) but less than for equities (1.2). These figures reflect real estate's investment characteristics, somewhere between bonds and equities (see Figure 10).

Investors should focus as much on future as on past performance, and an appropriate tool for assessing the future performance of real estate against bonds is the forward-looking gross redemption yield. This assumes that the rational investor expects a return from property determined by the following formula:

property yield + rental growth = bond yield + premium

The premium added to a bond yield compensates the investor for covenant risk, irrecoverable outgoings, depreciation and illiquidity, and a 2%-3% figure is often used as a rule of thumb. The difference between the bond yield plus premium and the initial property yield should represent a moderate investor's rental growth expectations over the medium term (say five years).

Figure 11 shows the gross redemption yields for 10-year bonds in the five markets in November 1995, compared with the likely income yield on institutional office property purchased in these markets. In all markets the implied rental growth rates are limited, reflecting the position of most European markets in their rental cycle; however, with

FIGURE 11: CURRENT YIELD PROFILE FOR BONDS AND REAL ESTATE

	OFFICES [1] %	BONDS [2] %
Amsterdam	*6.5-7.0*	*6.4*
Brussels	*7.0-7.5*	*6.8*
Frankfurt	*5.0-5.5*	*6.4*
London	*5.0-5.5*	*8.3*
Paris	*6.0-6.5*	*7.1*

[1] *Estimated representative core yields for prime offices.*
[2] *Gross redemption yields on 10-year bonds maturing in 2005 (8/11/95).*

Sources: Datastream / Jones Lang Wootton Research, November 1995

FIGURE 12: EUROPEAN REAL ESTATE LEASING PRACTICES AND ACQUISITION COSTS

	Main costs of purchase				Leasing practice			
	Restrictions on foreign ownership	Transfer tax % of purchase price [1]	VAT %	Agent's fee [2] %	Lease length (yrs)	Statutory right to renew lease	Frequency of rent review (yrs)	Indexation to cost of living
AUSTRIA	*Yes*	*3.5*	*None*	*2-3*	*5-15*	*No (post 1967 buildings)*	*Rare*	*Annual*
BELGIUM	*No*	*12.5*	*20.5 (on fees)*	*1-3*	*3/6/9*	*Yes, shops No, other*	*9 Shops 3 Other*	*Annual*
CZECH REPUBLIC	*Yes*	*5.0*	*5.0 (on purchase price)*	*3-5*	*3/5/10/15*	*No*	*3-5*	*Annual*
FRANCE	*No*	*18.2*	*18.6 (on fees)*	*1-3*	*3/6/9*	*Yes*	*None*	*Every 3 years by law*[3]
GERMANY	*No*	*2.0*	*15.0 (on fees)*	*3-5*	*5-10*	*No*	*Very Rare*	*Cost of living trigger*
HUNGARY	*Partial*	*10.0*	*25.0* [4]	*1-3*	*3-5*	*No*	*None*	*Annual*
ITALY	*No*	*10.0*	*19.0 (on fees)*	*2-4*	*6*	*Yes*	*None*	*Annual*
NETHERLANDS	*No*	*6.0*	*17.5 (on fees)*	*1.25-2*	*5-10*	*Yes*	*5 or 10*	*Annual*
SPAIN	*No*	*6.0*	*16.0 (on purchase price)*	*2-3*	*3-10*	*No*	*Rare*	*Annual*
UK	*No*	*1.0*	*17.5 (on fees)*	*1*	*25*	*Yes*	*5*	*None*

[1]Varies in some cases depending on the age of the property and the time-scale of resale. [2]In addition, legal fees will vary between 01.% and 2% of purchase price.
[3]Indexed to building costs. [4]Property often traded in single asset companies to avoid VAT and transfer tax.
Source: Jones Lang Wootton Research, November 1995.

implied annual growth rates of between 2% and 6% (on the basis of the above formula), there is a good chance of rental growth exceeding these implied growth rates. We believe that now is an appropriate time for investors to be considering European market opportunities, after a period of limited investment activity.

The practicalities

Market analysis alone will not be of very much use without the ability to secure the product in a local market and to deal effectively with local conventions and landlord obligations. It is also imperative that investors are aware of local costs likely to be associated with purchasing and leasing commercial property.

Figure 12 summarises the main real estate acquisition costs and leasing practices that may confront a European property investor in Europe; it is intended to be a guide to the operation of individual markets, rather than being applicable in all situations. The table serves to illustrate the variation in market practice that exists in Europe, and shows that movement towards a convergence in leasing practices and acquisition costs associated with European Union has so far been limited. In Eastern Europe, in particular, the situation is further complicated by constraints on foreign nationals purchasing land, different forms of ownership practice, and evolving investment markets.

To ensure that the necessary market know-how is readily to hand, it is advisable for investors intending to purchase commercial property in Europe to seek professional advice before proceeding. ❖

THE GERMAN REAL ESTATE MARKET

Wolfram Gelsner, Blumenauer International Beteiligungs GmbH, Frankfurt

Wolfram Gelsner took over as managing partner of Blumenauer International in September 1995. Mr Gelsner has been with the Blumenauer group since 1990 and was responsible for the national expansion of the company. Blumenauer International coordinates and superintends the cross-border consultancy and broker services of its international partners.

Phone: (+49-69) 9591 960
Fax: (+49-69) 9591 9620

THE GERMAN REAL ESTATE MARKET

Wolfram Gelsner, Blumenauer International Beteiligungs GmbH, Frankfurt

Germany's real estate markets felt the pinch of recession in 1994. Worst hit was the commercial real estate sector, where a glut of vacant space reflected weak demand and sustained supply-side pressure.

Demand for residential real estate was curtailed primarily by a decline in disposable incomes. This, coupled with a surge in construction activity, meant that vacant properties were attracting fewer and fewer potential tenants; this was in contrast with the housing shortages of preceding years. Landlords could no longer impose high rent increases, and the upper end of the market in particular began to see rents crumble.

On the other hand, the market for retail real estate was largely stable. The general market situation and the moderate upturn predicted for German industry over the next few years raises questions regarding the future of the real estate sector. The following report endeavours to provide the answers.

RISING UNEMPLOYMENT AND CHANGING STRUCTURES

Although the economic upturn has been stronger than anticipated, western Germany's commercial real estate market has not yet been able to reap the benefits. This is mainly because, despite the recovery, jobs were still being axed in 1994, putting paid to hopes that office payrolls (and with them demand for office space) would expand. A rise in the number of unemployed is not simply one of the phenomena attending an economic downturn: in Germany in particular it is mainly the result of large-scale economic re-engineering allied to a strategy of lean management, which aims to increase competitiveness by improving efficiency all along the value-added chain. Manufacturing companies do not have a monopoly on lean management: more and more banks, insurers and other service suppliers are seeing the potential for rationalisation by shortening decision-making procedures, flattening hierarchies and out-sourcing.

Regardless of cyclical developments, this structural shift will have repercussions on the scope and nature of employment throughout the economy. Despite the upswing anticipated for western Germany over the next few years (with projected GDP growth rates of some 2%+), any increase in employment will be slight. The number of employed is expected to pick up very slightly from 29 million in 1993 to 29.5 million in 1998. Manufacturing will see its share of the total workforce continue to decline (down from 37% to 35.5%), while the service sector will witness corresponding increases. Many service suppliers are benefiting from the general trend towards out-sourcing.

Having exhausted most of the potential for streamlining production processes, companies are coming under mounting pressure to cut back on administrative and office staff. This explains why the number of office employees is unlikely to grow much despite the bright economic climate.

Compared with the crisis-ridden markets in many other countries, the market for western German commercial real estate has come through the latest recession relatively unscathed. More and more office space is becoming available, but rising vacancy rates show how scarce demand is: 7.8% of real estate in Frankfurt is vacant, 4.7% in Munich, 4.0% in Düsseldorf, 3.6% in Hamburg and around 2.3% in Berlin. With space still at a premium, around 1.5% of properties are standing empty in Leipzig and Dresden. These figures relate to properties within city limits; the percentage of vacant properties in surrounding areas is much higher.

Rents are around 15% to 25% lower than the record sums paid at the height of the boom, when tenants were prepared to pay up to DM90 per square metre in some areas. Many landlords are now trying to attract new clients with rent-free periods and other incentives. A lot of companies are taking advantage of the current tenant's market to move from old-fashioned properties in locations with poor transport connections to well-equipped buildings in attractive sites. The relatively high turnover of office space has little to do with expansion drives and additional demand; it is more a sign that tenants are on the lookout for better alternatives.

Demand for additional office space after the expected economic upswing will be modest, because any increase in the number of people in office jobs is likely to be slow over the next few years and more firms are reducing per capita space occupancy in an effort to cut office rental costs. Future demand will focus on expensive properties in prime locations; with the general fall in rents these are now a much more attractive option although further rent cuts in this sector are unlikely. Vacancies in downmarket properties in less salubrious locations, however, will depress the market. Things are likely to improve over the medium term only if there is a drastic reduction in new developments which, in turn, depends on numerous building projects being postponed indefinitely.

MARKET FOR RETAIL REAL ESTATE STABLE

The decline in retail sales has not yet had a negative impact on western Germany's market for retail real estate. The fact that the number of prime locations for retail outlets is obviously restricted lends the market additional stability. However, the general upward trend in rents has either flattened out or halted at a high level. In cities such as Munich, Frankfurt and Hamburg top rents of around DM300 per square metre are being paid for preferred locations. Despite a predicted retail slump in 1995 (with sales down between 1% and 1.5%), rents are unlikely to come under pressure, the only possible exceptions being rents for large-scale developments or for shops in secondary locations or cities with high unemployment.

HOUSING MARKET EASING

The situation in the western German housing market has eased considerably over the past few months. With the average number of people applying for vacant apartments on the decline, landlords could suddenly no longer enforce the kind of rent increases that had been common for upmarket properties in preceding years. This trend was due primarily to a sudden expansion in new developments, and also to a decrease in migration and shrinking disposable incomes. In the period from 1988 to 1993, the number of housing starts doubled to 432,000. In 1994 some 500,000 new residences were built. However, our estimates show that housing demand is hardly going to soften over the medium term.

For the foreseeable future we are predicting an annual need for some 500,000 to 550,000 new dwellings. The demand for these residences will be particularly strong when expected wage increases begin to affect market behaviour.

Although the eastern German construction industry is buoyant, the process of providing the greater quantity and quality of housing required will be a gradual one. We still expect problems with properties designed more to lure mainly western German investors with attractive tax conditions than to entice tenants in the new *Länder*. In many cases it will be impossible to command profitable rents over the next few years given that the spending power of the local population is certain to remain weak.

THE LEGAL SITUATION

Essentially, there are no restrictions on non-Germans buying real estate in Germany.

Real estate purchase contracts can be lawfully concluded only in the presence of a notary. The latter is responsible for having the change of ownership recorded in the Land Register. Details of the real estate's owner are recorded in the Land Register kept by the local district court.

As a rule, notary and registration fees amount to around 1% to 1.5% of the purchase price.

The purchaser of a piece of real estate is required to submit to the tax authorities a land transfer tax equal to 2% of the notarised purchase price.

Local municipalities charge a yearly real estate tax to be levied separately for the land and the building(s). Tax rates are fixed by the local authorities themselves.

If real estate is sold within a blocking period of five years, any profit realised is subject to income tax. Profits are exempt from income tax after expiry of this period.

Estate agents liaise between sellers and buyers. Varying from region to region, buyers can expect to pay a commission of between 3% and 6% and sellers up to 3% of the purchase price. ❖

THE SWISS REAL ESTATE MARKET

Peter Baur, Göhner Merkur AG, Zurich

Peter Baur is a member of the executive board of the Göhner Merkur Group, where he is in charge of real estate management. In this capacity he oversees a property portfolio worth SFr14bn. Göhner Merkur offers investors comprehensive assistance and advice, not only on technical and market-related matters but also on legal issues.

Phone: (+41-1) 384 8484
Fax: (+41-1) 382 0433

THE SWISS REAL ESTATE MARKET

Peter Baur, Göhner Merkur AG, Zurich

Real estate is an inflation-proof form of capital investment with long-term profit potential. But if it is to be successful, the investor should stick to certain basic rules to ensure that the property retains its value and provides a steady return over a long period of time. This article offers general advice as well as a brief description of the main features of the Swiss property market. It also lists the key criteria for assessing a real estate investment and discusses the legal requirements governing the acquisition of Swiss real estate by foreign nationals.

Real estate plays an important part in a balanced portfolio and can offer considerable attractions. It belongs, with equities, in one of three distinct investment categories: capital investment offering a regular return and an appreciation in asset value. While investment in equities may bring high levels of return, the potential for profit is balanced out by the risk of losses, and the shareholder normally cannot influence the share price, which is subject to cyclical fluctuations and psychological factors. With real estate, on the other hand, the value of the property can actually be increased if it is well managed by professionals. Moreover, real estate returns can be higher if the investment is part of a long-term strategy.

For the sake of completeness, we should mention the other two types of investment, namely the non-interest-bearing capital investment with value appreciation, e.g. precious metals, art or undeveloped land in prime locations, and the interest-bearing capital investment with no potential appreciation in value, such as bonds or bank deposits.

CHARACTERISTICS AND DEMANDS OF REAL ESTATE INVESTMENT

Apart from the characteristics of the property itself – quality, duration and location – the most important factor in real estate investment is the investor's strategy. Real estate has no place in a short-term investment strategy, since it represents a long-term investment characterised not by spectacular short-term gains but by a regular return. Investors buying real estate in safe havens such as Switzerland have always had their patience rewarded by seeing the value of the property appreciate and their assets accumulate.

At the same time it should be remembered that a real estate portfolio entails not only regular income but also expenditure, in the form of financing costs (mortgage interest) as well as operating and maintenance costs. For newer property, these costs can amount to around 20% of the return.

Because real estate is a long-term investment, the investor needs to examine carefully all the features of the property and to draw up a list of objective and subjective criteria. Of course, there will inevitably be an element of uncertainty. For instance, one cannot predict all the socio-economic developments that will take place during the life of the property. This makes it all the more important for the investor to look to past trends for guidance and to base his or her strategy for the future on tried and tested principles.

QUALITY BEFORE QUANTITY

In real estate investment, the golden rule is often 'quality before quantity'. Since the investor is making a long-term commitment, he or she must look carefully at the overall business climate. Careful consideration needs to be given to factors such as political stability and security, the state of the economy and conditions in the job market. Population structure and demographic trends are also significant, as are a whole range of socio-economic factors. In Switzerland, the general conditions are excellent by international standards. Swiss property therefore generally represents a very good investment.

That said, it is still, of course, essential to look very carefully at each individual property and to assess its potential to provide a steady return and to maintain or increase its value. The main consideration is its location. This is not to say that only properties in prime locations are worth considering, but only by weighing up the advantages and drawbacks of the location is it possible to make an informed decision whether to invest. A comprehensive list of strengths and weaknesses will help the investor to gauge the risks and opportunities involved, and to forecast a likely rate of return. Normally, a real estate investor will accept a relatively low return if the risks associated with the location are also comparatively low. A property can be assessed on any number of criteria; the objectives and strategy of the investor must determine how relevant each factor is. But some criteria are of primary importance to any investor (see Figure 1).

The success of an investment is also determined by the property's own inherent features, such as its architecture and the quality of construction. It can be assumed as a rule that the construction standards of Swiss properties are high.

FIGURE 1: LOCATION CHECKLIST

Offices

- *Easily accessible by public and private transport: near airport, commuter and long-distance rail links, and motorway*
- *Parking*
- *Suitable setting for users (industrial, dynamic, urban, historic)*
- *Development of area (flourishing, or are other companies moving out?)*
- *General image of location, e.g. key location for industry, major international companies based in the area, etc.*
- *Synergy with other users*
- *Job market, highly skilled workforce available*
- *Wage levels, cost of living*
- *Tax levels*
- *Political stability, law and order situation*
- *Financial services industry, currency stability*
- *Health service*
- *Social and economic factors*

Flats

- *Schools and workplace within easy reach*
- *Good shopping facilities, restaurants, sports facilities, etc.*
- *Situation: town or country*
- *Good transport links, especially commuter rail links*
- *Quiet area*
- *Daylight amenity*
- *Rent levels*
- *Tax levels*
- *Type of local residents, e.g. mainly families*
- *Reputation of community, potential for development*

The high quality of Swiss property means that the average condition of buildings is good. But to avoid a possible drop in the rate of return, it is essential to draw up long-term renovation and maintenance plans at the outset with a projected investment schedule and profit and loss account. This is the only way to ensure that in the long run the property remains attractive, keeping its value in terms of price and rate of return. To run a property competently requires considerable expertise. The complexity of tenancy laws, among other things, makes property management a daunting task for the layman. Bringing in specialists to manage one's property portfolio is a sure way of protecting oneself from the pitfalls of a more amateurish approach.

A VERY SAFE INVESTMENT

For the reasons described above, Switzerland has a long tradition as an attractive location for real estate investment. Because of the country's very limited settlement area, land is a scarce, valuable and important commodity. Thus the Swiss property market has traditionally been characterised by consistently strong demand and steady appreciation in value for all types of real estate. Switzerland's local advantages and its political and economic stability make real estate a very safe investment. Other points in Switzerland's favour are the generally recognised high quality of life and the high standards of construction in its residential and commercial property.

In general, therefore, all types of Swiss real estate are suitable for investment. Consistency and particularly low risk make flats an excellent investment. The best place to buy into offices is probably Zurich and the surrounding area. Zurich is one of Europe's prime locations for services and industry, offering great advantages for property investors. Hotel and tourist facilities also offer investment opportunities,

FIGURE 2: ZURICH'S POSITION IN EUROPE

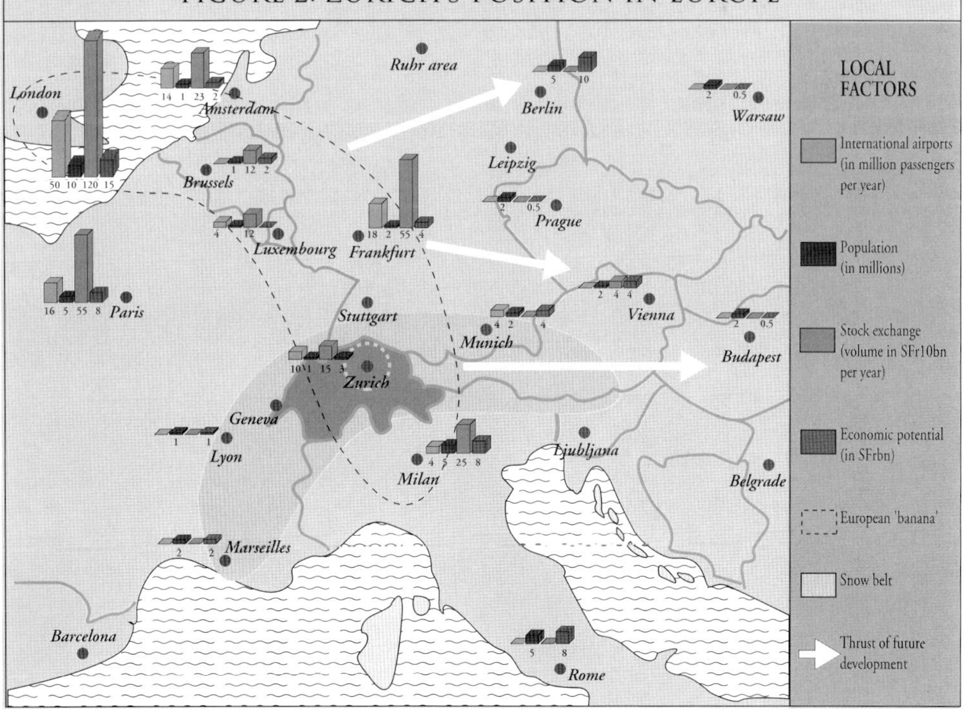

but are only suitable for investors with an excellent knowledge of this particular kind of real estate.

Solid returns

Of course, the appeal of investing in real estate essentially depends on the projected rate of return. In this respect, real estate has to compete with many other types of investment. But despite the competitive environment and the difficult climate on the property market, real estate investment has remained rock solid. This is due partly to the fact that it is a long-term, inflation-proof form of investment, but mainly to the reliable return it offers.

Although types of property and investment strategies vary widely, quite a number of general truths can be discerned. For instance, at current interest rates, investors buying into flats should ideally aim for a gross return (rental income) of 6.5%. If this looks unattainable in the medium term, such an investment should be discouraged from the outset. This means that the potential net return essentially depends on the investor's leverage. For investments financed with borrowed funds, Swiss financial institutions offer expert advice and a wide choice of financial arrangements.

LEGAL CONSIDERATIONS

One particular feature of Swiss law is the tight restriction on ownership of Swiss real estate by foreign nationals. Usually, those wishing to buy land must first obtain a residence permit, a requirement which obviously places strict limits on investment. Efforts are being made to have these laws relaxed and thus to make the market more attractive to foreign capital. But given the complexity of the legal requirements, we would strongly recommend that investors enlist the services of experienced professionals to help them make the most of existing opportunities. For instance, it is easier for investors to acquire property if they are setting up a business and thus creating jobs. ❖

EUROPEAN REAL ESTATE INVESTMENT: TAX ISSUES

Paul Kearney, Senior Tax Manager, and David Brown,
Tax Manager, Arthur Andersen, London

Paul Kearney is a senior tax manager with Arthur Andersen in London. He is an economics graduate from the University of Liverpool and a chartered accountant. He advises extensively on the UK and international taxation aspects of investing in and financing real estate, and is a frequent writer and speaker on the subject.

David Brown is a tax manager with Arthur Andersen in London. He holds an MA in history from Jesus College, Cambridge and qualified as a chartered accountant in 1993. He advises several international real estate investors on the structuring of their UK investments. He has also worked in the real estate services group of Arthur Andersen in New York.

Phone: (+44-171) 438 3989/2358
Fax: (+44-171) 438 3694

EUROPEAN REAL ESTATE INVESTMENT: TAX ISSUES

Paul Kearney, Senior Tax Manager, and David Brown, Tax Manager, Arthur Andersen, London

An important issue in any international tax planning for the acquisition of assets generally, and of real estate in particular, is the interaction between local and foreign taxes, and the income and gains arising from the investment. It is of the utmost importance to ensure that a tax deduction is available for the costs of financing the investment and, at the same time, to minimise the tax costs of the lender through withholding taxes.

This article outlines some of the main differences between the ways in which real estate investors are taxed in a number of European countries with markets for investment real estate. It provides a necessarily superficial overview of a complex subject, and potential investors should always take the advice of a qualified professional. European real estate is often regarded as an attractive investment for international investors, both because of the underlying nature of the assets available in mature investment markets, and because Europe's sophisticated and well-developed tax and legal systems allow investors to operate with a degree of certainty in the treatment of their income flows. The effective exploitation of these conditions depends upon many factors, including the domestic tax law of the investor and the terms of any double taxation treaties agreed between the state where the real estate is situated and the investor's country of residence.

In the interests of brevity, this article considers only the following jurisdictions: Belgium, France, Germany, Ireland, Italy, the Netherlands, Portugal, Spain, Sweden, Switzerland and the United Kingdom.

STRUCTURING THE INVESTMENT

The choice of structure will depend mainly on the investor's requirements for the repatriation of income and gains and the tax rates applicable to each in his or her domestic jurisdiction.

There are essentially two ways of structuring an investment:

(i) the real estate may be owned by an entity resident in the investor's country (a company incorporated specially for the purpose of acquiring foreign investment property, or an existing one); or

(ii) it may be acquired by a non-resident entity, owned by the investor either directly or through one or more intermediate companies (and countries).

A real benefit of holding the investment via a company is the ability to choose between selling the property itself, therefore realising any capital gains or losses in the company, and disposing of the shares in the company and crystallising the gain at the shareholder level. This decision will obviously depend upon the particular circumstances of the investment.

The tax guidelines set out below are based on the assumption that the real estate is treated as an investment for tax purposes, i.e. as a long-term holding for generating income rather than as an asset acquired for the purpose of making a profit upon resale. Whether a property holding constitutes an investment is determined differently across the various jurisdictions. In Germany, for example, gains are taxed when realised on a holding of less than two years (speculative gains) or if the purchaser is considered to be a real estate dealer. In the UK, in contrast, there is no safe-harbour holding period and the nature of the holding is assessed according to the purchaser's motives for acquisition.

Furthermore, if the investor takes an active role in managing the property, the country where it is located may tax the income on the basis that the investor is trading in that country; careful planning should prevent this issue arising. In the Netherlands, gains on 'business assets' are taxed, whereas gains on 'passive investments' are not, for either individuals or corporates.

For individuals investing their personal wealth, another important consideration in structuring the investment is the existence and scope of estate taxes (death duties) in the jurisdiction where the real estate is located, which may constitute a substantial cost in the case of direct investment. Such investments are typically taken outside the scope of estate taxes by investing through a limited liability company or other distinct legal entity.

The choice between direct and indirect investment or any other type of holding structure will depend upon the investor's particular circumstances. Certain generalisations regarding overseas investment can be made for a number of countries. Such generalisations are based on current practice, and up-to-date advice is recommended.

The most common vehicles for acquisitions by overseas investors in Germany and the UK are non-resident companies. Resident limited liability companies are commonly used for investments in Ireland, Sweden, Belgium and Portugal, although more complex arrangements may be required in Switzerland, Italy and Spain. Real estate investment in France is usually undertaken through a partnership.

It should be noted that investment in Portugal by non-resident individuals or companies is subject to regulations which require formal prior declarations to be made to the fiscal authorities. Legislation is, however, due to be enacted to liberalise foreign investment in Portugal for both European Union and non-EU residents.

FINANCING

Clearly, it is crucial to ensure that a tax deduction can be obtained for the cost of financing any investment in European real estate. Interest will generally be deductible against rental income if the loan advance is used directly for real estate acquisition.

The rules regarding the deductibility of interest costs are generally quite complex: it is usually necessary to show that the debt to which the interest relates comes from real estate investment and the rental income arising from that investment. Most jurisdictions will also seek to disallow deductions for interest in excess of a commercial level of financing. Italy and Spain deny a deduction when the acquisition is made directly by a non-resident investor.

When the international investor intends to pay interest to an overseas lender, more rigorous structuring may be required. Certain European countries (for example, the UK, Spain and Ireland) have special rules to avoid profit stripping by interest deductions on loans to associated entities. In Spain, deductible debt from related parties is limited to twice the amount of the equity of the local subsidiary company, and Portugal is proposing to introduce a similar limit on debt-to-equity ratios. Generally, the country where the real estate is situated will not allow a full tax deduction if the interest paid is higher than a commercial level, or if the amount of the loan is more favourable than it would be with an unconnected entity. There are, however, certain structures that may be used to avoid these restrictions, as the rules generally apply only to 'connected' companies and not to other entities.

RENTAL INCOME

Foreign investors in European real estate are generally liable to pay income tax in the country where the property is situated. Profits are typically calculated as the gross rent less expenses related to the rental of the property. Deductible non-interest expenses normally include (but are not restricted to):

(i) repairs and maintenance;
(ii) management fees;
(iii) the insurance of premises, fixtures and fittings;
(iv) certain property taxes; and
(v) rents and other land-related periodic payments, e.g. ground rents.

Capital expenses, and those not exclusively incurred for the purposes of the investment business, will usually be disallowed.

The Spanish system is an exception to this general rule. If the rental income accrues to a non-resident investor, tax is charged at 25% of gross rental income and no deduction for expenses is allowed. If the income accrues to a resident company, however, a 35% tax is charged on the net surplus arising.

Belgium and Italy also depart from the rule. For individual direct investors, the amount of tax payable in Belgium will generally be calculated on a government-determined notional rental value, subject to a 40% lump-sum deduction. Similarly, in Italy, either a notional amount will be subject to tax, or the annual rent less a fixed-percentage deduction will be charged, whichever is the higher. In addition, if the Italian property is not let, a penalty is charged by increasing the notional income by one third.

In France, if the investor is a resident of a country that does not have a double taxation treaty, he or she is taxed on notional income equal to three times the rental value of the property. This rule does not, however, apply to non-resident investors whose actual French

FIGURE 1: RATES OF TAX ON RENTAL INCOME (AS AT 15 NOVEMBER 1995)

	Higher corporate tax rate for non-resident(%)	Highest individual tax rate (%)	Highest corporate tax rate (%)
Belgium	*39.0*	*55.0*[1]	*40.17*
France	*33.33*	*56.8*	*33.33*
Germany	*42.0*	*53.0*[2]	*30/45.0*[3]
Ireland	*40.0*	*48.0*	*40.0*[4]
Italy	*36.0*	*51.0*	*37.0*
Netherlands	*40.0*	*60.0*	*40.0*
Portugal	*25.0*	*40.0*	*39.6*
Spain	*25.0*[5]	*56.0*	*35.0*
Sweden	*28.0*	*65.0*	*28.0*
Switzerland[6]	-	-	-
United Kingdom	*25.0*	*40.0*	*33.0*

[1]*Plus 3% crisis contribution.*
[2]*Plus a 7.5% surcharge.*
[3]*45% on retained income, 30% on distributed income (plus 7.5% sundry).*
[4]*Closely held companies are liable to a 20% surcharge on after-tax non-trading profits and certain service income not distributed within 18 months of the year end.*
[5]*No deduction on tax expenses.*
[6]*Tax rates in Switzerland vary greatly between cantons.*

income is higher than the notional amount. These rules also apply to an investment held through a French resident partnership, which is the most common investment vehicle. Figure 1 shows the rates of tax chargeable.

TAX DEPRECIATION

Tax depreciation can generally be deducted for expenditure on certain eligible assets, including:

(i) expenditure on certain commercial buildings;
(ii) expenditure on buildings used for industrial purposes; and
(iii) expenditure on plant and machinery fitted in a building, e.g. lifts and air conditioning.

Certain countries, such as the UK and Ireland, allow privileged tax depreciation treatment and rates to buildings in areas targeted for economic regeneration. Land and houses do not generally attract tax depreciation, except in Switzerland where land can be depreciated in certain circumstances.

Most European countries depreciate commercial and industrial buildings at rates varying between 1.5% and 8% per annum, most commonly between 2% and 5%.

To claim tax depreciation, it is not usually necessary to be a resident of the country where the property is located. Spain, however, allows tax depreciation for Spanish residents only, although overseas investors are able to claim tax depreciation if they invest through a Spanish company, in which case the company can depreciate commercial or industrial buildings at a rate of between 2% and 3% per annum.

Certain countries also allow tax depreciation on plant and machinery included within a building; these tend to be structural assets such as lifts, air conditioning and heating systems, and fire-safety apparatus. A portion of the purchase price of the building may be attributed to plant and machinery, and tax depreciation is claimed on this. It is often worth making a detailed cost segregation study to maximise the expenditure on which depreciation may be calculated. The rate of tax depreciation in these circumstances is generally higher than for buildings, as the expected useful life is shorter. In the UK, for example, the rate is 25% per annum (on a reducing-balance basis) and a rate of between 20% and 30% per annum is allowed in Sweden, depending upon the specific property.

WITHHOLDING TAX

Withholding tax may apply to either rents or interest paid to the non-resident investor or to the lender.

Certain European countries, including the UK, Ireland, Portugal and Spain, impose withholding tax at rates varying between 15% and 27% of the rent paid to a non-resident landlord. In the UK and Ireland, if the rent is payable to the non-resident investor's agent (typically the letting agent) who is resident in the country where the real estate is situated, no withholding is required to be made by the tenants as the agent is liable for the income tax. In addition, the agent will usually be responsible for handling the tax assessments relating to the rental income. Any excess tax withheld over the eventual liability will be repaid when the investor settles his or her tax affairs in the country where the real estate is situated. In the UK, a non-resident investor may obtain exemption from the withholding requirement by following a formal procedure. The other countries considered here do not at present withhold tax on rents paid to non-resident landlords.

Interest paid on borrowings to a person or bank outside the European country where the real estate is situated will invariably be subject to withholding tax unless the lender is resident in a country with an appropriate tax treaty. Most European countries have negotiated tax treaties that reduce the rate of withholding tax on interest, in some cases to nil. This is in line with EU policy on intra-EU investments.

In the absence of an appropriate tax treaty, the rates of withholding tax on rents or interest are typically nil in countries such as France, Germany, the Netherlands and Sweden, 20% in Portugal, and 27% in Ireland. In Switzerland, the rate varies from nil on inter-company and mortgage interest, to 35% on bank and bond-like debts.

In the UK, it may be possible to structure debt in such a way that the interest falls outside the scope of the normal 25% withholding tax on interest, even when paid to a resident of a non-treaty country.

The withholding of tax on either interest or rents can have a significant impact on the cash-flow profile of an investment; we would therefore reiterate that professional advice must be sought in order to keep cash-flow liabilities to a minimum. The structure of the investment is also important, as in certain European countries the interest paid can be deemed to have a source in the country where the property is situated. As such, the investment will be tax-deductible against the rental income, but will be assessable for tax on the basis that it has a source within the European jurisdiction.

Outlined in Figure 2 are the rates of withholding tax applicable to rents and interest paid to non-residents. This does not reflect the benefits available under double taxation treaties in force between the relevant jurisdictions.

When investment takes the form of a local company, it may also be necessary to consider the effect of withholding taxes on dividends.

FIGURE 2: WITHHOLDING TAXES ON INTEREST AND RENT PAID TO A NONRESIDENT[1] (AS AT 15 NOVEMBER 1995)

	Interest (%)	Rent (%)
Belgium	*13.39*	*13.39*
France	-	-
Germany	-	-
Ireland	*27*	*27*[2]
Italy	-[3]	-
Netherlands	-	-
Portugal	*20*	*15*
Spain	*25*	*25*
Sweden	-	-
Switzerland	*35*[4]	-
United Kingdom	*25*	*25*[2]

[1] Excluding relief available under tax treaties.
[2] Nil if an agent is appointed to be assessed to tax.
[3] If paid to a bank.
[4] No withholding on inter-company or mortgage interest.

VALUE ADDED TAX

Value added tax (VAT) is a transaction-based tax and is likely to substantially affect the economic analysis of any investment.

Generally, land and buildings are exempt from VAT. However, it may be possible to elect to charge VAT on certain commercial properties. The purpose of the option to tax in relation to a particular property is to allow the recovery of VAT (both on the acquisition of the property and in respect of other services), which might otherwise be denied.

If an option to tax is exercised, VAT must be levied on all rents and charges made to tenants. The decision to opt to tax the property will, of course, be influenced by the VAT status of the tenants.

For example, financial institutions are generally exempt for VAT purposes (either partially or fully) and are therefore able to recover little, if any, of the VAT charged to them. So, if VAT is charged on rent, it may reduce the range of possible tenants or else require rent to be offered at a discount. Conversely, other sectors, such as retail, are fully taxable under VAT rules and there should not be a problem with VAT recovery on rent. In such cases, opting to charge VAT may prove advantageous to the investor.

If the option to tax is exercised, then VAT is charged on the proceeds arising from the eventual sale of the property. This may be an unattractive option if the purchaser is partially or fully exempt from VAT: an insurance company or financial institution, for example.

These rules may vary across the EU. For example, a 19% charge is levied on the acquisition price of Italian real estate if it is purchased from an Italian company or individual entrepreneur. The VAT is not then recoverable by a non-resident private investor.

Portugal does not charge VAT on real estate, but instead charges property transfer tax (see below). Belgium, France and the Netherlands charge VAT on new buildings at rates in excess of 17%. The extent to which this tax is recoverable depends on the individual circumstances of the investor.

TRANSFER TAX

The purchase of real estate in European countries generally attracts transfer tax based on its cost or market value at the time of transfer. This is usually paid by the purchaser and is therefore an additional cost of investment. In France, rates are in the range of 7% to 18.5% for direct investment, or 4.8% on shares in entities holding real estate. This tax only applies when VAT is not chargeable on the purchase.

Spain and Belgium do not charge transfer tax if the transfer attracted VAT. If it did not, the Spanish authorities will levy a 6% transfer tax. This tax is also charged on the transfer of a majority holding in a company more than 50% of whose assets consist of Spanish real estate.

Under Belgian law, the charge is typically 12.5%. In Portugal, transfer tax of between 0% and 10% will accrue, depending on the value of the real estate being acquired. A charge is also levied on the grant of a lease longer than 30 years and on the acquisition of 75% or more of a private company owning real estate.

Certain other European countries charge transfer tax on the value of the acquisition of real estate in excess of *de minimis* limits (for example, above £60,000 in the UK). The rates vary greatly between jurisdictions and according to the nature of the assets being transferred. There are common practices in each European jurisdiction for the minimisation of transfer tax liabilities, including the acquisition of shares in a real estate company as opposed to a direct investment in real estate. The fiscal authorities are, however, seeking to restrict such opportunities to minimise liabilities. A draft bill in the Netherlands, for example, proposes to charge a 6% tax on the economic, as well as the legal, transfer of real estate.

TAXATION ON DISPOSALS

The gain realised on the sale of real estate situated in European countries such as Germany, Italy, Portugal, Belgium, the Netherlands and the UK can, with appropriate structuring, be exempt from capital gains tax provided that it is clear that the gain accruing to the non-resident investor is in respect of real estate held as an investment. Should the local fiscal authorities successfully impute to the investor the motive that the real estate is acquired for the sole or main objective of realising a gain on the disposal, the gain may be treated as general income and may thus be taxed accordingly.

Furthermore, in certain jurisdictions the ability to realise a tax-free gain is gradually being eroded. In Germany, for example, this is now possible only where a non-resident individual invests directly in real estate. Non-resident corporates have been taxed on their disposal gains since January 1994. In the UK, there is general anti-avoidance legislation charging to tax any gain realised on the disposal of real estate, where it is held that it was originally purchased with the intention of realising a profit from its disposal.

In other European countries, should any gain accrue, it is possible to defer the gain by deducting it from the tax cost of new assets acquired. This is known as 'rollover' treatment. In Belgium and Portugal, this treatment applies, subject to certain conditions, if the investment is not structured appropriately to take the gain outside the charge to tax on gains.

Other countries charge regressive rates dependent upon the amount of time the asset is held; the longer the asset is held, the lower the effective rate of tax. This treatment applies, for example, in France, Spain and certain cantons of Switzerland.

Certain jurisdictions adjust the cost of the asset for the purpose of calculating the gain to take account of inflation during the period of ownership. This attempts to protect the investor from being taxed on inflation gains.

Another notable departure from the general rule is Ireland, where 40% tax is charged on capital gains, and the purchaser is entitled to deduct a withholding of 15% of the proceeds if the consideration is greater than I£100,000. This amount is paid over to the fiscal authorities on behalf of the vendor. No withholding tax is required when a clearance certificate granted by the Irish revenue authority is presented to the purchaser.

In Spain, non-residents are generally charged 35% on capital gains accruing. This tax may be mitigated by investing via a local company, although the ability to reduce the tax charged depends on the terms of the tax treaty between Spain and the investor's jurisdiction. In addition, purchasers of real estate from non-residents are required to withhold 10% of the proceeds on account of the vendor's tax liability on the gain realised.

Commonly, when tax depreciation has been allowed against rental income arising, and certain sale proceeds are attributed to the qualifying expenditure incurred, the tax depreciation given in prior years will be clawed back and taxed as income in the year of sale.

OTHER MATTERS

Certain European countries have an annual 'net assets' or 'wealth' tax, or charge tax on contributions to company capital. These taxes may apply to both resident and non-resident companies and individuals. The method of valuing assets for the purpose of calculating such taxes varies greatly between jurisdictions. Specific advice should therefore be taken in relation to any proposed investment.

Taxes such as rates and other local taxes are outside the scope of this commentary. However, these taxes may amount to a significant cost and should be taken into consideration in any contemplated investment. ❖

US REAL ESTATE INVESTMENT

Anthony E. Malkin, President, W & M Properties Inc., and Peter L. Malkin, Chairman, W & M Properties Inc. and Wien, Malkin & Bettex Esqs, New York

Peter L. Malkin was born in Brooklyn, New York in 1934. He graduated *summa cum laude* from Harvard College and *magna cum laude* from Harvard Law School. Since then, Mr Malkin has gone on to become a general partner in numerous prestigious buildings in New York, including the Empire State Building and the Lincoln Building. He also has an interest in apartments and retail properties in five US states.

Anthony E. Malkin received his BA degree from Harvard College in 1984. Before joining W & M Properties, he was with Chemical Venture Partners, LP. Mr Malkin is a member of the board of Lechters, a publicly traded kitchenware and houseware retailer. He is also a member of the board of Lincoln Center Theater and a member of the Overseers Committee to Visit the Loeb Theater at Harvard University.

Phone: (+1-212) 953 0888
Fax: (+1-212) 983 1385

US REAL ESTATE INVESTMENT

Anthony E. Malkin, President, W & M Properties Inc., and Peter L. Malkin, Chairman, W & M Properties Inc. and Wien, Malkin & Bettex Esqs, New York

A balanced investment portfolio in North America will inevitably include real estate. Real estate opportunities in the United States typically fall into two categories: income-producing property and individual residences. Occasionally, however, the two may overlap and individual residences may also serve as a source of income.

There are five main types of investment property:

(i) office;
(ii) residential;
(iii) retail;
(iv) hotel and hospitality; and
(v) industrial (including warehouse and distribution).

Each of these categories of investment property has its own risk and return ratio.

For the purposes of analysis, the United States can also be divided into seven geographical regions: the northeast, the mid-Atlantic, the southeast, Florida, the mid-west, Texas and the southwest, and the far west.

Ownership of real estate in the USA is widely dispersed. Pension funds and life assurance companies are by far the largest owners of real estate, but entrepreneurs and opportunity funds also hold substantial positions in this sector. Real investment trusts, a form of public shareholding, also experience popularity from time to time.

There are generally three main advantages to investing in US real estate. Firstly, there are the benefits associated with owning real assets in the United States, paid for in dollars and with fixed-rate leverage. Secondly, the value of US property is appreciating as a result of improvements in location, strengthening markets and superior management. Finally, US real estate offers protection against inflation thanks to the scope to increase rents and because the cost of land and buildings generally keeps pace with the rate of inflation.

There are, however, some disadvantages to owning property in the United States. The market itself is relatively illiquid and historically prone to significant price fluctuations, reflecting levels of development and the economic cycle. The availability of finance is also subject to variation, depending on government regulation of those institutions which invest in real estate and market sentiment. Furthermore, the intensive nature of the real estate management process requires a hands-on approach by skilled professionals.

However, the greatest difficulty which real estate poses for a foreign investor lies in identifying the correct party with whom to invest, structuring the investment in such a way that it meets the economic interests of both the managing party and the investor, and then maintaining close contact with the management of the assets purchased.

There are three main ownership structures under which real estate can be purchased, namely corporations, limited liability companies and partnerships, both general and limited. Occasionally an individual will own real estate outright. However, this is not the preferred method in the United States as the potential risks associated with personal liability are great. Foreign investors must also be sure to assess the tax implications of purchasing real estate carefully; certain investment structures may entail higher US withholding tax liability for foreign owners of real estate than others.

There are a number of options available in terms of the capitalisation of real estate acquisitions, equity on its own, equity combined with bank funding, pension funds, the public markets, and occasionally private sources of capital. It is important to note, however, that it is not always possible to arrange real estate financing in the United States, no matter how financially sound the individual borrower or how good a prospect the individual property may be. There have been periods when the regulatory framework and financial market upheaval have significantly restricted the availability of finance for real estate investment. When available, however, US leverage is often offered on advantageous terms: rates of interest are frequently fixed and repayment periods are longer than in most other countries.

INDIVIDUAL RESIDENCES

Americans typically regard individual residences as non-investment real estate. A foreign investor looking to acquire residential real estate in the United States should take into consideration the primary purpose of the property: will it be his or her main personal residence, an occasional personal residence, a residence for his or her children while they are attending university, or is it intended to serve as a source of rental income?

In major metropolitan areas, owned residences typically take the form of apartments either in a cooperative or within a condominium. Cooperative apartments are not actually owned outright: cooperative ownership entitles the purchaser to a permanence lease on an apartment through the purchase of shares in the corporation which owns the building in which that apartment is located. The majority of apartments in New York City are sold on this basis. Condominium apartments, by contrast, are purchased 100% outright and can be individually financed. They typically afford a much greater degree of flexibility in terms of purchase and sale than cooperative apartments.

Cooperative apartments have boards of directors made up of other share owners who determine whether or not financing may be placed on shares through acquisition. They even have the power to reject a prospective share purchaser if he or she is not considered suitable. The transfer of condominium apartments is not generally subject to such restrictions.

Any investor considering purchasing real estate with a view to letting the property should also be aware of the fact that cooperatives do not generally allow sub-letting. Furthermore, there are significant differences between cooperative apartments and condominiums with regard to real estate taxes, which may have a bearing on an investment decision: co-operatives will usually pay any real estate taxes on behalf of the shareholder, who then pays the cooperative in the form of maintenance fees; in condominium apartments, by contrast, real estate taxes are paid directly to the relevant tax authority.

THE IMPACT OF LIMITED LIQUIDITY

Substantial, cyclical fluctuations in the value of property can be caused by liquidity crises, the relatively long-term nature of real estate development, and the interaction of these factors with the underlying economic cycle. From a long-term perspective, real estate can generally said to be a sound investment. However, the property market is not characterised by steady growth, but by a series of peaks and plateaux, with an occasional valley.

The fate of property prices therefore depends, to a large extent, on the liquidity of the market and the timing of real estate development.

During the most recent real estate recession, liquidity was extremely limited. The government imposed new capital adequacy requirements in line with international agreements. Banks and savings and loan associations reduced their real estate exposure dramatically within a very short period of time. The fall in property values precipitated by this development in conjunction with the general economic slowdown, prompted many life insurance companies, pension funds and other lenders to withdraw from what they considered to be a deteriorating asset category.

Furthermore, many institutions sought to liquidate their real estate holdings in the face of an environment with little equity and no debt financing available. This lack of liquidity was the cause of a serious downturn in the property market.

Developers were particularly hard hit by this turn of events. From inception to completion, a real estate development project typically takes a number of years. The conditions which inspired the development in the first place therefore often bear little or no relation to the prevailing climate at the time of completion.

Developers, by their very nature, develop when they have access to the necessary funds and popular demand is strong. There is no limit on the number of developers who can develop, only on the liquidity available to them. Since real estate, in common with many other investment vehicles, is subject to the 'herd mentality' many developers and sources of funds will therefore make the decision to develop at the same time.

The property market débâcle of the 1980s and 90s provides clear evidence of the potential instability of a major building boom. During this period, projects were planned on the basis of their ultimate value upon completion. However, as prices fell below the equity invested, many properties were actually worth less than the debt on them.

Value redemptions of 25% to 30% in the recent real estate crises rapidly obliterated the 0% to 10% equity requirements maintained by lenders during the preceding construction boom. Even without the liquidity crisis which followed, these projects would have been in dire straits. Ironically, the experience of the last 10 years bears a close resemblance to the history of real estate between 1929 and 1931. Fortunately, however, the general economy has recovered on this occasion – in contrast to the sustained Depression of the 1930s, which held down real estate prices and stifled construction until after World War II.

THE CURRENT INVESTMENT ENVIRONMENT

Some analysts contend that the general recovery in the US economy, greater liquidity available for real estate, and the fact that properties are now available at less than their replacement cost due to the high level of repossessions, currently make real estate an attractive proposition. Indeed, massive amounts of capital have been flowing into real estate over the last four years. In the last 18 months in particular, there has been a marked increase in the amount of capital (both equity and debt) sunk into real estate.

However, caution is the watchword here. The foreign investor must bear in mind the volatile nature of the property market – characterised by peaks, plateaux and the occasional valley and the effect of the herd mentality, which can temporarily intensify downturns or upturns given the long product cycle for real estate development and acquisition. And in real estate terms, 'temporary' should be taken to mean a period of between 18 and 36 months.

The old adage that things are never what they seem is particularly applicable to the property market. The real estate investor must be conscious of the fact that perceptions define reality on the financial markets and the availability of cash in good times is often greater than is justified by the economics of the market.

In the context of this general overview of real estate investment opportunities in the United States, let us now look at each of the five main investment categories in turn and assess the risks and returns, our current position in the development cycle, pricing considerations, and the prospects for the future.

OFFICE PROPERTY

Office investment opportunities can be subdivided into three categories: city, suburban and single tenant.

FIGURE 1: OFFICE PROPERTY

Risk/return	*High*
Location in cycle	*Top*
Prospects	*Current prices have outpaced economic returns*

Based on the available economic and demographic data, the broad consensus of opinion is that, within this category, central business districts pose the greatest risk. As accommodation in these areas becomes ever more expensive and people place greater emphasis on quality of life, the population is shifting away from city centres and into suburban and exurban areas[1]. Cities without a vital core of occupied housing, cultural and educational facilities, night life, entertainment and services will fail to attract prospective tenants and will not flourish in the future. Every city is, of course, different and investors must be extremely discerning in this respect. The likely success stories include New York, Boston, Chicago and San Francisco. Washington DC, home to the federal government, also looks sure of a secure future.

Suburban office properties were the hardest hit by the recent economic downturn and real estate crisis, but are now seeing the strongest recovery. Analyses of demographic trends, tentative signs of economic recovery and speculation have pushed prices up close to replacement costs in many markets. Of course, suburban markets face unique issues in terms of transport and economic diversity.

In markets where land is readily available and there are currently few constraints on development, new development is more likely and consequently the scope for continuous price appreciation is severely limited. The 'herd' has selected suburban office property as the number one investment priority and investors must exercise caution if they are to avoid paying over the odds.

Some market analysts believe that single tenant office properties are an attractive way to invest in the credit of the tenant, as with a bond, but with the added bonus of protection against inflation, which a bond does not offer. However, such properties must incorporate user specific design features and require substantial alteration and thus additional investment when a tenant vacates on termination of the lease.

Investors considering this particular investment strategy might be better advised to investigate alternative avenues such as investment in industrial properties (see below).

[1]*Exurbia: the region outside the suburbs of a city, consisting of residential areas that are occupied predominantly by wealthy commuters.*

RESIDENTIAL PROPERTY

Inner city residential property is a high risk proposition. In some large cities, rent control regulations limit the value of existing rental property. The acquisition of rental property for conversion to cooperative or condominium ownership status offers interesting potential returns, but with the risk that failure to convert will leave the investor illiquid and with an inadequate return or, worse still, with actual operating losses. Development is an extremely high risk, but potentially lucrative, method of investing in residential property. However, we do not advise foreign investors to take this avenue.

FIGURE 2: RESIDENTIAL PROPERTY

Risk/return	*Moderate*
Location in cycle	*Middle*
Prospects	*Some new developments under way; nationwide vacancy rates are generally very low*

By far the most conservative and yet more promising investment category – and at the same time the one with which foreign investors are least familiar – is suburban or exurban apartments. More commonly referred to as 'garden apartments' these properties are generally two- or three-storey buildings located in more extensive grounds than would be available in an urban setting. Their location is determined by demographic trends.

Garden apartments offer significant inflation protection as the leases are usually only for one year and are not subject to rent controls, which means that rents can be increased on renewal. Although such properties are management intensive, they offer stable returns.

RETAIL PROPERTY

Retail properties are liable to suffer some of the most dramatic fluctuations in the next few years. The information superhighway, greater specialisation and changing retail space requirements may render many retail properties superfluous and others inadequate or functionally obsolete. Furthermore, the general retail environment in the United States at the current time is highly unsettled, with most retailers experiencing only small, if any, growth in inflation-adjusted sales, constantly rising operating costs, and significant financial instability.

FIGURE 3: RETAIL PROPERTY

Risk/return	*High*
Location in cycle	*Near top*
Prospects	*Danger of serious problems due to instability in the retail industry*

Regional shopping malls have historically been the post popular investment with major real estate owners. Such malls are used to offer a shopping experience encompassing every conceivable type of retailer, entertainment facility and food outlet. However, with the advent of television retailers in the United States – Walmart, Price Club and Home Depot, to name but a few – many shopping malls are losing their traditional customers and indeed their very future is in question.

Strip retail centres vary greatly in size and,

depending on location and design, can present a worthwhile investment opportunity at a lower cost than shopping malls, but in some cases, with as great a wealth of retail experience. Strip centres provide investors with a low-cost alternative to shopping malls and, in some instances, are even more attractive. This is particularly true of 'power centres' which include highly specialised 'category killers'.

Television retailers typically occupy stand-alone facilities. It is very tempting to view these properties in the same light as a corporate bond with inflation protection. However, like single-tenant office properties, single-tenant retail properties are specifically constructed to meet the requirements of an individual retailer and may prove to be of little value should that retailer vacate the premises upon termination of the lease or cease to operate as a result of insolvency or market saturation.

HOTEL AND HOSPITALITY PROPERTY

Hotel and hospitality real estate is the riskiest of the five categories. This sector is particularly susceptible to downturns in the economy and currency fluctuations. Recruiting and retaining staff is a constant problem. Property management contracts favour the management company, which gets paid regardless of whether or not the hotel makes any money. Hotels require constant reinvestment in the form of plant, furnishings and equipment and should be classified as businesses rather than real estate.

Urban hotels suffer acute labour problems, hefty operating costs, and are often subject to high taxation by the cities in which they are located. The fortunes of hotels which rely heavily on convention business can vary considerably, depending on the prevailing economic climate. Furthermore, development costs for new hotels are prohibitively high; all but a few major US hotels conceived during the 1980s have had transfers of ownership forced upon them by lending institutions at a fraction of their original cost. Strange as it may sound, the most successful investors in hotels have been the next owner.

Economy or budget hotels have become increasingly popular over the last five years. However, the quality of construction of these properties tends to be relatively poor, requiring high levels of maintenance. Furthermore, most such hotels are franchises of group reservation networks which are thinly capitalised and offer few barriers to entry.

FIGURE 4: HOTEL AND HOSPITALITY PROPERTY

Risk/return	*High*
Location in cycle	*Middle*
Prospects	*The economic resurgence and US currency devaluation are boosting occupancy and rates. However, when the downturn comes, it is both rapid and severe*

INDUSTRIAL PROPERTY

Industrial real estate such as warehouse and distribution property offers the best prospects for investors seeking an investment vehicle with the characteristics of an inflation-proof corporate bond. Industrial properties are typically leased in their entirety to one or a limited number of tenants. Many of these tenants will have credit ratings which can easily be assessed by the specialised agencies such as Standard & Poor's or Moody's.

There is generally a close correlation between the yield on these properties and the

FIGURE 5: INDUSTRIAL PROPERTY

Risk/return	*Low*
Location in cycle	*Middle*
Prospects	*Well-located properties offer good prospects for continued price appreciation*

yield on corporate bonds of a comparable term. If the lease is not renewed at the end of the specified period, industrial properties can quite easily be transferred to another user with very little capital investment and at a rate which reflects the inflation-linked cost of the fixed assets. The main risk associated with this category of real estate investment occurs when the lease expires or is terminated: if the economy is generally weak, it may prove difficult to find a new tenant.

On the whole, investment in industrial property offers a relatively low risk and a moderate return. However, it is imperative that prospective investors examine the environmental credentials of a property before committing themselves, and ensure that it has not been polluted or contaminated in any way by previous tenants.

REGIONS

Having analysed the five main real estate investment categories in some detail, let us now turn our attention to the relative merits of some of the different geographical regions within the United States.

The northeast

The northeast of the United States was the hardest hit during the recent recession. The region lost more manufacturing jobs than any other, suffered the greatest loss of corporate headquarters and proved to have one of the highest levels of excess capacity in the US property market.

Nevertheless, the financial services sector is continuing to flourish in Boston, Massachusetts and New York, three of the most important cities in the northeastern United States. Hi-tech and entertainment companies have also targeted the northeast for significant expansion and growth. Although it would be somewhat premature to claim that the region is in resurgence, the severe downturn which the northeast experienced is now over and the worst of its troubles are behind it.

The mid-Atlantic

The star of the mid-Atlantic property market is undoubtedly Washington DC. Cuts in the defence budget have led those companies that rely upon the federal government for major contracts to relocate closer to Washington in order to protect their interests. Northern Virginia and Fairfax County in particular, lying as it does directly between National Airport and Dulles Airport, have seen astounding growth and job creation in the last decade and look set to continue on this path for the foreseeable future.

The outlook for most other urban areas in the mid-Atlantic region is unfortunately not so bright, but the prospects for the suburban regions are more promising.

The southeast

The southeastern United States has been one of the main beneficiaries of the de-industrialisation that has blighted the northeast. The region has also attracted substantial foreign investment by automotive and other manufacturers. The creation of large numbers of industrial jobs in states such as Kentucky, Tennessee and Georgia can be attributed to greater harmony between unions and employers, fewer state employment and social welfare requirements, abundant land and utilities and excellent distribution networks.

The risks in these markets are twofold: barriers to entry are low and much of the development which took place was for single-user occupancy and therefore has little economic value for any potential future tenant.

Charlotte, North Carolina and the Raleigh/Durham research triangle as it is known stand out as urban or suburban centres that have witnessed strong growth and are poised to play an important role in the technology and financial sectors in the future. Atlanta, on the other hand, may well experience a downturn in its fortunes after the Olympics are over and therefore represents a significant risk.

Florida

Florida is something of a law unto itself. It is a marketplace which is stimulated by the travel and lodging industry, the retirement economy, entertainment centres and its proximity to Latin and South America. However, there are a number of problems with which Florida must contend; land use controls, limited water supplies, pockets of extreme urban decay and racial tension, and an inability to develop major, sustainable urban centres.

It is difficult to envisage many markets in this region achieving appreciable price gains. However, Jacksonville and Miami stand out as two financial centres with a realistic prospect of improvement: Jacksonville as the economic capital of the state and Miami as the economic capital of the Latin American and South American region.

The mid-west

The mid-western states sailed through the recent recession virtually unscathed. While there was slow growth, and indeed no growth, in certain marketplaces, the former 'rust belt' experienced its economic contraction in the 1970s and early 80s. Consequently, the mid-western states remained largely dormant until the recent economic recovery in the United States propelled them into significant growth and expansion.

There have been appreciable price gains in almost every asset category with the exception of major office blocks in downtown Chicago, Illinois, where over-development is a problem. Even there, however, newly developed space has been largely absorbed into the market, although to the detriment of older office buildings. Denver represents a major risk for investors, with property values quite high relative to economic return.

Texas and the southwest

Texas has never recovered from the collapse of the oil boom. Although some investors who purchased properties such as garden apartments when the market was at its lowest have seen the value of their investments rise substantially, commercial properties in Texas have not returned to anywhere near their levels during the peak years of the oil boom.

Texas is the perennial boom and bust market. With few restrictions on land use and plenty of available land, it is difficult to foresee any appreciation in this market. Commercial office rents have at best remained constant and at worst declined over the past five years.

The southwest as a whole has staged a tremendous recovery during the recent economic upturn. The urban centres of the southwest are typically more urban sprawl than cities, but hi-tech industries and the financial services sector have generated jobs and encouraged growth in the region. A word of warning to the prudent investor, however; although there is abundant land for development and few planning restrictions, much of the price gain associated with the general economic recovery has already occurred.

The far west

Dominated by California and its problems, the western United States is second only to the northeast in terms of the decline in real estate during the recent economic downturn. Large numbers of jobs are still being lost in the defence, aerospace and hi-tech industries. Wage costs in California are particularly high and the entire state has been overdeveloped.

The bankruptcy of Orange County, the main economic driving force for the Los Angeles area, has cast doubt on the government's commitment to supporting business in the county in the years ahead. The stringent development regulations in San Francisco will enable that market to recover the soonest.

The remainder of the west coast is home to a large concentration of hi-tech, aerospace and timber-related industries. In all probability, jobs will continue to disappear in the aerospace industry as ever more work is out-sourced to contractors and foreign companies. The timber and paper industry may also be adversely affected by federal restrictions on logging. However, the region will retain its status as a desirable location for hi-tech industries.

On balance, investors should exercise extreme caution when contemplating investing in the city centres of this region. However, other asset categories – particularly industrial – should fare better because of the region's proximity to the Pacific Rim.

REAL ESTATE AND THE CAPITAL MARKETS

Real estate investment trusts have existed for decades. They have been issued and purchased by institutions and individuals seeking to combine the benefits of real estate ownership with the liquidity of public markets. In practice, however, publicly held real estate companies tend to have limited capitalisation and only a small proportion of the shares actually rest with outside investors. Even a small exodus from the stock of one of these companies can cause significant price volatility.

Additional interest in real estate investment trusts has come from those who view the returns as more advantageous than bond yields, which are currently depressed. If one considers benefits which include cash distributions and price gains, there is something to be said for the argument that real estate investment trusts have outpaced bond returns.

However, if bond yields were to increase, they would exceed the cash distributed from real estate investment trusts. This would encourage people to sell their shares and thus jeopardise any capital appreciation to date. All things considered, therefore, investors should seek professional advice before opting for real estate investment trusts.

Mortgage-backed securities enable institutions to offer debt but to sell that debt in the public marketplace, therefore avoiding capital restrictions and regulatory risks. Investors are strongly advised against purchasing such securities as it is impossible to predict how they will perform during a real estate downturn when repossessions and property transfers take place. Furthermore, it is unclear how the interests of holders of mortgage-backed securities would be represented if they were the ultimate lender on a property whose value was less than the debt.

CONCLUSION

Investment in US real estate can be extremely lucrative: fortunes have been made and sustained over generations in this way. However, as with any investment, there is no guarantee of success.

The foreign investor must assess which type of property is best suited to his or her risk and return profile. Markets must be carefully selected according to their prospects for growth and fundamental economic development. Expert legal advice should be sought on ownership structures and taxation.

It is also important to consider the optimum ratio of equity to debt based upon the element of risk in a given investment. The value of an investment in well-located property is almost certain to rise. However, there will inevitably be peaks, plateaux and valleys along the way. An inadequately capitalised real estate acquisition or development becomes a purchasing opportunity for someone else during an economic downtown.

The prerequisites for successful real estate investment are information and experienced management. It is essential for a foreign investor in US real estate to recruit either an adviser or a partner to provide local knowledge and product expertise. We cannot stress too highly the importance of the adviser or partner in the investment process; he or she will be the investor's eyes, ears and voice. Indeed, the most frequent – and most serious – mistake on the part of foreign investors is the failure to consult an expert in the initial stages and arrange a professional acquisition analysis, and then not to utilise specialist property management services.

In the final analysis, it is virtually impossible for an individual foreign investor to acquire the enormous expertise and information necessary to minimise the risk associated with a particular transaction. The key to success for the discerning investor therefore lies in the careful selection of a suitably qualified and experienced adviser or partner. ❖

PLAYGROUNDS, PRIVATE ISLANDS AND LARGE ESTATES OF THE WORLD

Christopher Cornell, Partner and Head of Residential Division, Knight Frank, London

Christopher Cornell is a partner at Knight Frank, and head of its residential division in the UK and overseas.

Phone: (44-171) 629 8171
Fax: (44-171) 629 2639

PLAYGROUNDS, PRIVATE ISLANDS AND LARGE ESTATES OF THE WORLD

Christopher Cornell, Partner and Head of Residential Division, Knight Frank, London

Homeostasis is defined as the perfect balance – within the body itself, and between the body and its environment. It is the object and desire of human existence. Imagine the satisfaction of total harmony, being at peace with oneself and one's surroundings – there can be no greater luxury. However, harmony is not constant; the dynamic and perpetually swinging pendulum is what makes life on earth a challenge.

The acceptance and mastery of such challenge is demonstrated by success and wealth. Individuals show control over their own particular situation by the acquisition, retention and re-creation of wealth, through which they influence the lives of others, mobilise the economy and help maintain the status quo. All paths to wealth eventually lead to property, for people must have a home, a base, a hideaway, a sanctuary.

Today, satellites and air links make it possible to communicate and travel across time zones with ease, wherever one chooses to live. The technical revolution of the past 30 years has thus allowed wealthy individuals to select their ideal home, regardless of location, and maintain contact with their business bases.

The 'playgrounds of the world' are where the most wealthy settle; these include the great cities – London, Paris, New York – and the breathtaking beaches of Miami and Malibu. The wealthy buy the world's sequestered islands and properties in tropical paradises such as the Caribbean; and they look to the large sporting estates and historic houses of Scotland and Ireland, the Channel Islands and Europe. Elsewhere, beyond the western hemisphere, new properties are sought in locations such as Tanzania. A large-scale cattle ranch with ocean frontage may be of interest to the purchaser with a taste for the exotic – the Mkwaja Ranch, for instance, which boasts a private game-reserve with elephant, giraffe, buffalo, hippopotamus, lion and the nesting grounds of the green turtle.

Of course, the purchasing power of the ordinary householder and that of the most affluent are not comparable. While the former amounts to a few thousand pounds, the latter may reach many hundreds of thousands. Reasons for accessing the property market are quite diverse, and wealthy individuals seeking a second, third, or indeed any number of additional homes, need not concern themselves with the practical concerns or individual components of a property purchase. Wealth affords choice and a certain freedom: no need to consider the central heating or renovation requirements, an army of experts and designers will be on hand to deal with such matters.

Let us examine here where the wealthy individuals who buy island properties and estates in the playgrounds of the world come from, and how they make their choices. What attractions are sought and where are tax concessions to be found? How does one value such properties and, indeed, what does the future hold for buyers?

ISLAND LIFE

The epitome of exclusive living must surely be the ownership of a private island: the control of not just a property but an entire island 'empire' lies in the hands of the owner. There is also a certain romantic appeal in the image of a land of your own, adrift in the sea. Island owners can have all the creature comforts of their own home while enjoying space and seclusion in a natural environment. An island has a limited land boundary, but on clear days the owner can enjoy distant horizons.

What kind of person buys an island? Island life attracts free spirits who revel in privacy and seclusion, or those with a taste for the exotic. Living on an island can become addictive; it gets into the blood, like sailing some say, and those discerning individuals who have both the means and the desire to enjoy such an unconventional lifestyle, often go as far as buying a second, or even a third, island. Some private-island collectors remain in the market for many years in an attempt to buy what is, for them, the quintessential piece of land.

Private-island owner Richard Branson comments: 'When I first saw Necker Island I fell in love with it; it really is the most beautiful place in the world – a precious jewel!' He continues: 'Guests who visit Necker want to return; my aim was to make it a home to be enjoyed by everyone. All along I realised it would be ridiculous to use Necker as a private resort for it's an island to be shared.'

One island abode that offers an exciting opportunity is the Mamara Estate, a beautiful South Pacific plantation of 1,166 hectares in the Solomon Islands. The property, which is as big as the capital city, Honiara, offers an opportunity for creative development in a fabulous location. The national government of the Solomon Islands has recently approved, in principle, a unique land-use development plan for the area that will include tourist hotels, shopping, residential and sports facilities.

Although islands in tropical locations around the world have their own special appeal, many islands around the coasts of the United Kingdom are equipped with landing strips or helicopter pads. Scotland alone has 787 islands located off its coastline. Of course, not all islands are habitable; a piece of land jutting above the sea must be able to support at least one sheep and one human to be classified as an island!

Ireland, too, has its share of private islands. Waterford is a historic private island situated on Ireland's southeast coast with a regional airport and an international airport at nearby Cork. The island has a five-star hotel, an 18-hole championship golf course and a country club, as well as excellent development potential. Interested parties should expect to pay in the region of £5m.

THE BUYERS

In the world's largest cities buyers are, generally speaking, foreign governments and embassies, diplomats, corporate investors, wealthy industrialists and celebrities. Nine-tenths of this last category – if in show business – will own homes in Beverly Hills.

Rich Spanish, French and some Middle Eastern buyers are attracted to properties in Marrakesh. Properties in Tuscany appeal particularly to Germans and Americans, and Greek expatriates like to buy Greek islands. (A foreigner cannot buy an entire Greek island but island properties can be bought independently, without resorting to the traditional method of forming a company in conjunction with a Greek lawyer.) The ex-colonial Briton, who once owned huge farming estates in Kenya, is now more interested in the essentially stable European economy. Wealthy Chinese buy in Canada, the country with the largest Chinese community outside Asia; affluent Singaporeans tend to buy in neighbouring Malaysia; and Hong Kong Chinese will invest in Singapore property if business takes them there.

But most Middle Eastern and Asian purchasers continue to be lured by the leisure-property centres of Europe, with its air of sophistication, and the Caribbean, with its offer of winter sun. In a world in which the wealthy are obliged to keep several homes for business and pleasure, London, Paris and New York constitute the global 'golden triangle'.

In the early 1970s, buyers from the Middle East began buying London properties in large numbers because of the financial security offered by the city; Paris, the South of France, San Francisco and Los Angeles also ranked as favourites. At that time, US citizenship was made easier to obtain and the United States had become more accepting of the 'oil-rich' who hold much of the world's wealth in their wells.

The favourable rate of exchange and relatively low interest rates in the UK are currently attracting overseas buyers: in 1995 individuals and companies from Singapore, for example, targeted the London property market after their government issued a directive to invest overseas.

In France, celebrities such as Nick Faldo and Sophia Loren have properties in such exclusive ski resorts as Champéry and Villars, while the Comte de Paris favours France's most expensive ski resort of all, Méribel. Across the Irish sea, Ireland has become home to a host of celebrities: Tom Cruise, Julia Roberts, Andrew Lloyd-Webber, David Puttnam and John Hurt all own homes there.

REASONS FOR BUYING

Half of all the land in the world is uninhabitable, covered as it is with rock, desert, tundra, dense jungle, swamps and ice. Less than half the remainder is suitable for cultivation and nearly half the people in the world live on one-thirtieth of the total land area. Average population densities vary from one to 720 inhabitants per square kilometre. Small wonder, then, with the majority of the populace settled in a relatively small area, that the wealthy house purchaser is looking for a Utopia – somewhere secluded and with a unique character.

Investment criteria should always be borne in mind when making an important purchase. The wealthy, while they may have amassed their fortunes by controlled risk-taking, nevertheless maintain their status by holding on to that wealth. They look for political stability within a country, a safe currency and amenable local attitudes before they store their assets, which may include residential property. In their choice of residence they demand security and privacy, and enjoy being among sophisticated and like-minded people; they may or may not be seeking a tax haven but like to live in pleasant surroundings and, most important of all, close to an international airport.

Convenience

For the owner of several properties around the world, travel time is a key factor. Flights on Concorde, which shorten dramatically the flying time to the Caribbean and the US, the new rail service through the Channel Tunnel, and the advent of newer and faster motorways on the European continent, all influence the international buyer.

London is a very convenient place to live, with time zones in mind, since it is located roughly halfway between New York and Hong Kong, and conveniently close to Europe's major cities. With its excellent business services, communications and travel links, private educational and cultural facilities, shops, health clubs, private clinics and eminent doctors, it is high on the internationalist's list.

Language will also influence many people's choice of residence. Approximately 346 million people speak or understand English, making

FIGURE 1: TYPICAL FLYING TIMES BETWEEN MAJOR CITIES (HOURS)

	Durban	London	Los Angeles	New York	Paris	Rome	Sydney	Dar es Salaam
Durban		*13:15*	*23:20*	*15:00*	*14:20*	*13:00*	*17:40*	*4:15*
London	*13:15*		*11:00*	*8:00*	*1:00*	*2:25*	*21:45*	*11:40*
Los Angeles	*23:20*	*11:00*		*5:50*	*16:45*	*17:21*	*13:30*	*19:00*
*New York**	*15:00*	*8:00**	*5:50*		*7:45*	*8:30*	*20:00*	*13:20*
Paris	*14:20*	*1:00*	*16:45*	*8:45*		*2:00*	*20:40*	*10:00*
Rome	*14:30*	*2:25*	*17:21*	*9:25*	*2:05*		*19:30*	*10:15*
Sydney	*17:40*	*21:45*	*13:30*	*19:00*	*21:20*	*21:25*		*19:15*
Dar es Salaam	*4:15*	*11:40*	*20:30*	*17:10*	*10:30*	*11:40*	*19:15*	

**Flying by Concorde, time is 3:50*
Source: Country Trails Travel

it, after Mandarin Chinese, the most widely spoken language in the world. The business world demands English, making London a magnet, and buyers are currently arriving in London from all over the world.

London and Paris

Business travellers may be tempted to view London and Paris as virtually indistinguishable: both are elegant European capital cities which also offer functionality from a travel and business point of view. At the top end of the market, the two cities offer very different types of accommodation, however; London has its 19th-century houses in Belgravia, Mayfair and Knightsbridge, whereas Paris has incomparable period apartments in the 8th and 16th *arrondissements.*

While the performance of the Paris property market has in recent years been lacklustre by comparison with London, the shortage of high-quality houses and apartments for sale in Paris means that the best properties are always in demand from international buyers. Prices at this level are at least comparable to their equivalents in London. On the market at present are three superb properties in Avenue Foch, with prices ranging from US$18.6m for a superb first-floor apartment of some 1,100 square metres to more than US$77.5m for a magnificent private mansion of about 5,000 square metres.

In London, Middle Eastern property buyers traditionally have a penchant for Mayfair, Japanese buyers have a taste for St John's Wood, Greek purchasers favour Grosvenor Square, and the Chinese seek properties bordering Regent's Park. The main problem in London, as in all cities, is the lack of land available for new building. A developer's consummate skill lies in the art of reconstruction.

One property that has recently been sold in London is the Old Rectory, Chelsea, which had at least US$12m spent on renovation. Minimalist ideas were incorporated into the luxury design, combining the old with the new, with ochre Siena marble gracing the shower rooms, cupboards constructed from cedar of Lebanon and a swimming pool lined with exquisite black marble. Another such made-to-measure property in London is an old gothic warehouse in Wapping High Street in which the enormous open space has been redesigned into flats. The flats sell for approximately US$700,000 each.

Historical appeal

Established properties often appeal to buyers for the slice of history they represent. A 'novelty' historical item can increase the attraction of certain properties. For example, a kitchen floor made from Portland Stone taken from the Tower of London appears in Blackland Park, a classic Georgian house, listed Grade II, set in acres of beautiful Wiltshire parkland.

The work of famous architects and designers of yesteryear is also highly sought-after. Stobs Castle in Roxburghshire, Scotland, for example, was designed by Robert Adam and is also a rare example of a Georgian version of a castle. The original drawings of the building are in a London museum. The famous Victorian architect G.E. Street, who designed the Law Courts in London, built a convent in Lostwithiel, Cornwall, which is now Peregrine Hall. Containing eight holiday cottages and a private house with swimming pool, this property is currently on the market for US$697,500.

Some overseas buyers are drawn to Ireland's colourful history. The long struggle for survival in the 1800s meant that many ancient Irish families forfeited their homes and estates, emigrating in their thousands, largely to America. Now, with their fortunes secure, the descendants of these émigrés return – buying large estates for permanent residences or smaller properties as second homes for themselves and their heirs. Their roots take them back to a rural landscape without heavy industry and a place where the local people have a healthy respect for an individual's privacy – making Ireland particularly attractive to pop stars and celebrities who wish to preserve a semblance of seclusion in their lives.

Leisure appeal

In the UK, property outside the London area often appeals for its lifestyle and leisure qualities. Of large English country houses recently sold, 35% have gone to overseas buyers, and in Scotland foreign purchasers equal the numbers of Scots buying large houses, estates and farms. Some overseas purchasers may want to become agriculturists who can leave a working farm to their children, free of inheritance tax.

If a beautiful garden is of foremost importance, the buyer need look no further than the gentle hills and vales of Gloucestershire. The British gardener, who must contend with a varied climate, is unequalled in versatility and creativity. Just two hours by car from London, estates such as Standish Park offer beautiful gardens amidst stunning countryside. The gardens have been carefully planned with local-stone walls, terraces, a stone-pillared gazebo and beautiful shrub and flower borders. Gardens in general create enormous interest in Britain, contributing significantly to the economy and offering a relaxing and enjoyable hobby.

Sporting attractions

Sport figures highly in many people's lives today and properties with facilities for swimming, golf, grouse shooting, fishing, racing, sailing or scuba diving are a priority for many wealthy purchasers. Even if buyers are not sports enthusiasts themselves, such provisions may be *de rigueur* for the entertainment of friends and family. Whatever the chosen sport, the basic premise remains: the super-wealthy family, company, foreign embassy or government requires its own personal sporting privacy.

The leisure appeal of the Caribbean islands, enhanced by the weather, is almost without rival, and the retired magnate or holidaymaker need look no further than hiring a yacht and lazily cruising around to assess the market in property offering leisure facilities. In a region renowned for its enviably slow pace of life, such a pleasant pastime could take up to 18 months.

If land-based pursuits are more to your taste, the serious horse enthusiast could do well to consider purchasing property in South Africa. The Aldora Stud Farm in Natal, for example, is one of the top 10 studs in South Africa, and is currently up for sale. It is situated about 90 minutes by car from Durban and offers splendid surroundings and a successful stud operation.

Golf has become such a popular sport in recent years that it could be said to have been responsible for most luxury residential developments built in the last decade. The trend has been for the construction of new golf courses with exclusive residential properties and leisure facilities attached. The Mediterranean is now a very popular golfing centre, due to its excellent climate and infrastructure and improved communications. Spain, and in particular the Costa del Sol region, is recognised as the most popular holiday destination in the Mediterranean. Increase in demand over the past 10 years has caused an explosion of golf course developments in the Marbella area, many of them of championship standard and designed by well-known personalities such as Robert Trent Jones Jr, Seve Ballasteros and Gary Player. Regular players, either retired or semi-retired, may choose to purchase a substantial, detached property overlooking their favourite golf course, preferably with direct access to it.

Nearer to the centres of industry, however, a business executive may need more immediate fulfilment. This is where the large country estate, with easy transport links, is the winner. Looking for an estate with a particular form of sport or pastime in mind may well dictate your choice rather than the house itself.

For a keen hunter, the River Halladale presents a unique opportunity to acquire an entire salmon river in the highlands of Scotland in an area of outstanding natural beauty, famous for its salmon rivers, grouse shooting and deer stalking. The river rises to the south of Forsinard and flows north through moorland, native hardwoods, traditional farmland and the occasional rocky gorge to join the sea at Melvich Bay. There are spectacular sandy beaches at Melvich and Strathy and a number of golf courses within easy reach. Halladale enjoys a long season with spring salmon entering the river from March onwards and increasingly prolific runs of grilse and summer salmon appearing in June, July and August. Fresh fish continue to enter the river until the season's end. Fishing is by fly only and there are a variety of holding areas, from fast rocky streams favoured by the running grilse, to deeper pools where the spring salmon lie. Halladale provides exciting fishing from 12 January until 30 September. The property is easily manageable in size and has been recently renovated. The river flows beside the garden of the traditional estate lodge and sweep-netting rights in the river mouth are included in the sale.

Dunbeath Estate is a castle set dramatically on the North Sea shore in Caithness, Scotland. It offers a comprehensive range of sporting activity in its 30,000 acres. Red-deer stalking brings an average cull of 45 stags and 168 hinds. Also available is a 20,000-acre red-grouse moor, and exciting salmon and trout fishing on the Dunbeath River from source to sea. The nearby hills, cliffs and mixed woodland provide a natural habitat for woodcock, snipe, pheasant, pigeon and duck. Photographers, artists, wildlife enthusiasts and walkers can also enjoy roaming through natural deciduous woodlands with ancient birch and rowan, hazel and bird cherry trees, or strolling through summer meadows ablaze with wild flowers, mosses and ferns.

In southern England, within 40 miles of London, the country sportsman could investigate Westbrook House, a superbly located and secluded estate of about 88 hectares. Situated in the middle of the Wey Valley and currently for sale at US$2.32m, the grounds offer scope for a small rough shoot, equestrian or private polo activities and local hunting amenities.

The appealing image of the 'landed gentry' may also attract buyers to the fine and historic estate of Bagnor, approximately 38 miles from London's Heathrow Airport. Here, for a price of between US$10.8m and US$15.5m, country pursuits abound: fishing for wild brown trout in a river with fast-running water and natural pools; and an arable farm, yielding winter-grown wheat, oats and barley, with a small number of stock under management, offers the rustic lifestyle. Woodland allows for pheasant and partridge shooting, and stables and training facilities are available for horses and ponies.

Winter sports often attract families with children, especially teenagers, to mountain resorts for skiing, mountaineering, sledging and skating. A superb chalet with exceptional accommodation is on offer in Méribel in the heart of the French Alps. In the region of US$3.72m with direct access to lift systems and slopes, Chalet Brames has a landscaped garden and large balcony area with outside Jacuzzi. There are eight bedrooms, seven bathrooms, separate staff accommodation and several sitting and dining areas in this luxury home.

Although wine is not quite a sport, perhaps, the wine-lover will nevertheless appreciate the attraction of the vineyard in choosing a country estate, especially in continental Europe. One hour from Nice airport 52.5 hectares of Côte de Provence vines are for sale. The beautifully refurbished 18th-century Château Les Crostes comprises approximately 1,046 square metres and the property also includes the latest in automated wine-making equipment. A property of this sort could be purchased by a company, with the individual buyer as director and main shareholder. Careful advice, however, should be sought before entering the French property tax and inheritance system.

HOW PROPERTIES ARE VALUED

Islands are valued on the basis of supply and demand. Price is related to sales that have taken place in the preceding year or two. Opportunities to buy a suitable island are limited and purchasers must remain relatively flexible in their selections.

Valuation of a property in continental Europe is usually made on a price-per-square-metre basis for a house and a separate price for land, taking into account market conditions and actual market prices for properties that have recently been sold. In France, valuation does not take into account the additional 10% payable by the purchaser, which covers sales tax to the French government and relevant commissions; it must be paid when the final contracts are exchanged.

There is no set formula for valuing farms and large estates and the main guide is comparable sales prices. In the farm market, value is usually measured per acre or hectare, plus the price per square metre of the farmhouse, cottages and farm buildings.

In the estate market, the constituent parts of the estate are added together and then an element of 'marriage value' is added. Marriage value is a premium achieved over and above the sum of the constituent parts; it is not always obtained and it is only after a sale has taken place that the level of marriage value may be calculated. The quantum of marriage value varies considerably and reflects the marketability of the estate as a complete entity.

Large country houses are also valued comparatively, although one problem with houses valued at more than US$1.55m is that there are few on the market. The valuer takes into consideration the distance of the property from a business centre, its condition, its grounds and other positive or negative factors that may affect the price, such as the popularity of the type. Georgian houses, for instance, tend to be popular and will command a premium over Victorian houses.

Sporting estates are valued on their output, that is the catch from the beat, the cull from the hill and the bag from the moor. It is usual to look at the past five years' average output, and also to consider the potential of the sport. Beyond the sporting interest, the potential for afforestation, the quality of the house and location, and accessibility by loch or sea are taken into account. The experienced estate agent or valuer must know how to assess the ground itself. For example, to encourage grouse, heather must be burned at exactly the right time; young plants provide food for the birds, old plants provide shelter and, as the regeneration process can take up to 15 years, enough moorland for rotation purposes should be available.

The Scottish system of offering estates by sealed bids does not allow for flexibility and a purchaser must be encouraged to offer a final figure at the closing date for bids. It is not possible to go back after that date to change the bid price, so some purchasers will offer a premium over and above the value of the estate in order to increase their chances of success. In Scotland, offers are made after a survey has been undertaken and after finance has been arranged, and failure can therefore prove expensive.

CURRENT TRENDS

In today's complex market it is increasingly necessary to employ expert agents and advisers. A purchaser in France, for instance, is legally bound to a property within a very short time of agreeing a price. An international agent should therefore be sought, with the experience to guide purchasers through overseas business, using local branch offices, and the ability to maintain a management relationship with the purchaser on home territory.

A new type of buyer has emerged for large estates in Scotland in the past five years – the dedicated conservationist willing to spend money on the preservation of the natural habitat. These 1990s lairds aren't looking for sporting facilities or for their property to make money: they have bought for the pleasure the property can afford them and are interested in the quality of the lodge, the beauty of the countryside and in the local community.

The farm and estate market has also seen dramatic change during the past 10 years and the market is now led by agricultural and foreign buyers. As in other areas, properties above a certain price are in fairly short supply and a good medium-term outlook for the agricultural sector is keeping interest in agricultural properties high.

The South African market has enormous potential in both the commercial and residential sectors. The political situation is easing and the economy is becoming more open. Buyers must, however, register their introduction of sterling into the market with a local bank; failure to do so could result in complications upon selling. At present, vendors cannot take excess profits out of the country.

Values in residential and farming properties

in South Africa are extremely low by international standards. This is also true elsewhere in Africa, but South Africa is particularly appealing to overseas investors at the moment. Another country to consider for property investment is Zimbabwe, where US$69,750 can buy a three-bedroom house with a swimming pool, overlooking the Bulawayo golf course. However, investors should beware of residential agents who work on direct commission; obtaining independent professional advice is not necessarily easy, but assistance should be sought from a recognised name.

Ireland, which has traditionally been popular with Europeans and North Americans, is now experiencing a revival of interest from British buyers and returning Irish nationals. German buyers are also in evidence. The peace process in Northern Ireland has done much to increase the attractions of buying both there and in the Republic.

Confidence among foreign buyers and a marked increase in market activity in Ireland have resulted in a scarcity of prime properties located within one hour of the main centres, Dublin and Cork; this has caused prices to continue on their upward trajectory. Property viewing figures have shown an increase of 25% in 1995 and Ireland now looks attractive to the international buyer thanks to its recent investment in road networks, excellent accessibility through nine airports, the recovery in the economy and lower interest rates.

The property market reflects economic conditions, which affect confidence and currency fluctuations, and is also dependent on changes of goverment, presidential elections and world events.

FINANCIAL INCENTIVES

In Great Britain, if a house is historically important it may be graded by the Department of the Environment. The stately homes of England sometimes owe their very existence to the fact that they are open to the public and some, indeed, are run as profitable business ventures. The National Trust owns more than 200 dwellings, as well as gardens, historic monuments and religious buildings.

Tax benefits may be found in buying listed properties; an exemption from inheritance tax is available for properties listed Grade I and II, although this may be subject to strict conditions. It is generally the case that to qualify for exemption the property must be open to the public, maintained, kept within the family, and so on. Repair grants may be awarded by English Heritage if the property is of particular interest or historical value, but there are very strict rules as to what can be changed both inside and outside the building. All in all, owners with beautiful gardens, for example, are more likely to show them to the public for their intrinsic value than hope to profit financially from them.

The most fortunate buyer is the one who can afford to buy a home simply for the joy of living in it. But the wealthy make sure that they hold on to their money, and tax havens throughout the world, such as Monaco, the Channel Islands and Andorra, are deservedly popular. Monaco, an independent state since 1215 with a total area of 195 hectares, has long been regarded as an extremely stylish and glamorous location for the jet set. The official language is French and Monaco's laws are based primarily on French law, but the principality attracts people from all over the world. No income tax, capital gains tax or property tax is payable by foreign residents of Monaco, making it extremely attractive to foreign investors. It is not necessary to be a resident of Monaco to own or lease property in the principality and there are no significant restrictions on the purchase or rental of fine accommodation.

The Channel Islands also offer tax benefits and a high quality of life. Both Jersey and its sister island, Guernsey, have an income tax level of 20% and exemption from VAT, capital gains tax, and inheritance and corporate taxes. The islands impose restrictions on the sale of property to nonresidents, however. Guernsey's Open Market Properties Register has recently closed and potential purchasers in Jersey must offer proof of sufficient personal wealth to be able to assist the local economy.

Similarly qualified is Andorra, situated in the eastern Pyrenees between France and Spain. This ancient, safe and independent state offers spectacular scenery, a practically perfect climate and property in all price ranges.

Ireland offers certain personal tax exemptions, and artistic or literary types wishing to live in Ireland may follow the trail of blockbuster authors such as Frederick Forsyth or Leon Uris, or the geographer Anthony Summers. The Irish government is currently revising tax laws to gain favour with foreign investors. Ireland already allows foreign property owners to be considered resident but not domiciled in the island, so investors may live in Ireland for 183 days a year (280 days over two years) and pay tax only on Irish or British income, or income sent to Ireland from other jurisdictions, avoiding tax on worldwide income. Foreign managers of Irish companies may also qualify for tax benefits, subject to certain conditions. The Irish government now offers tax breaks to artists and film companies wishing to produce epics, such as the recently filmed *Braveheart*. Incentives work by reducing the high cost and up-front tax burdens that other countries impose.

AND FINALLY . . .

If you are looking for the perfect London property, with a combination of homeostasis, challenge and luxury, then the exciting Royal Naval College, Greenwich, may appeal. This magnificent group of buildings is of outstanding cultural and historical importance. The buildings, which adjoin the River Thames, are listed Grade I Baroque and were originally conceived by King William and Queen Mary, masterminded by Sir Christopher Wren and principally constructed between the years of 1690 and 1750. ❖

DUNBEATH CASTLE, CAITHNESS, SCOTLAND

A 30,000-acre estate with grouse moor, and salmon and trout fishing

Photo: Knight Frank

THE ROYAL NAVAL COLLEGE, GREENWICH

Grade-1-listed historic buildings

Photo: Knight Frank

WATERFORD CASTLE, IRELAND

Excellent development potential on the southeast coast of Ireland

Photo: Knight Frank

VAN GOGH'S 'SUNFLOWERS'

Its 1987 sale marked the start of the art market boom

Photo: Christie's

BLACK-GROUND GREEN-GLAZED DISH

Chinese, Qianlong period

Photo: Christie's

PAIR OF PAINTED POTTERY LOKAPALA GUARDIANS

Chinese, Tang Dynasty

Photo: Christie's

EARLY MING CARVED RED AND YELLOW CIRCULAR BOX AND COVER

Chinese, Yongle period

Photo: Christie's

SPINACH JADE CYLINDRICAL BRUSH-POT

Chinese, early 18th century

Photo: Christie's

CHRISTIE'S FINE ART SECURITY SERVICES WAREHOUSE

Photo: Christie's

'NATURE MORTE AVEC VASE D'ORFÈVRERIE, FLEURS ET FRUITS ET PARTITION DE MUSIQUE'

Pierre-Nicolas Huillot 1674-1751

Photo: Didier Aaron & Cie

'LE REPOS PENDANT LA FUITE EN EGYPTE'

Jean-Baptiste Marie Pierre 1714-1798

Photo: Didier Aaron & Cie

'ECHO'

Alison Wilding, 1989.
Stainless steel and polished brass,
104cm x 296cm x 135cm

Photo: Karsten Schubert Ltd

'SET FAIR'

Bridget Riley, 1989.
Oil on linen, 165.5cm x 227.2cm

Photo: Karsten Schubert Ltd

SILK KESI IMPERIAL HANGING

Woven in silks, peacock feather filaments and gold-wrapped thread. Chinese, Wanli period, 1573-1619

Photo: Spink & Son

SNUFF BOTTLES

Selected highlights from the collection of Lionel Copley. Dates between 1750 and 1850

Photo: Robert Hall

PORCELAIN MOONFLASK

Chinese, Qianlong period, 1736-1795

Photo: Spink & Son

CLOISONNÉ VASE AND COVER

By Namikawa Yasuyuki. Japanese, Meiji period, 19th century

Photo: Spink & Son

'THE COVENANT OF ABRAHAM'

Edouard Moyse, c.1860

Photo: Sotheby's

JUDAICA SILVER

Torah finials, silver Torah shield and parcel-gilt silver Torah crown. Italian, 18th to 19th centuries

Photo: Sotheby's

OUR LADY OF EVSIMANSK

Moscow, 16th century. Gilded silver frame with silver and niello appliqués, 44.5" x 32.25"

Photo: A La Vieille Russie Inc.

ST NICHOLAS THE WONDERWORKER

Novgorod, late 15th century 25.5" x 16.5"

Photo: A La Vieille Russie Inc.

ST DEMETRIUS SLAYING THE PAGAN

Early 17th-century icon in silver-gilt frame

Photo: A La Vieille Russie Inc.

THE VERONICA VEIL

By Ovchinnikov, Moscow 1889.
Gilded silver and enamel, 16" x 13"

Photo: A La Vieille Russie Inc.

ST MACARIUS OF ALEXANDRIA
AND ST MACARIUS OF EGYPT

17th-century silver-gilt-framed wooden icon, 12.36" x 9.12". Part of the imperial treasure of Tsar Nicholas I

Photo: A La Vieille Russsie Inc.

THE MARIE HALL STRADIVARI

By Antonio Stradivari, Cremona, 1709. Sold for £473,000 in March 1978

Photo: Sotheby's

A TWO-MANUAL HARPSICHORD

By Jacob & Abraham Kirckman, London, 1776. Sold for £66,000 in November 1992

Photo: Sotheby's

JIMI HENDRIX'S GUITAR

Fender Stratocaster, c.1968. Sold for £198,000 in April 1990

Photo: Sotheby's

GOLD AND SILVER
STAR-SAPPHIRE BROOCH

With diamond mount

Photo: A La Vieille Russie Inc.

FABERGÉ STYLE

The smaller pieces are what Fabergé was renowned for

Photo: A La Vielle Russie Inc.

'OBJECTS OF FANTASY'

These pieces utilise a wide range of materials and techniques

Photo: A La Vielle Russie Inc.

DIAMONDS

A pride to own,
a pleasure to wear

Photo: Dehres International

DIAMONDS

Concentrated capital

Photos: Dehres International

VAUXHALL 30/98
VELOX FAST TOURER

Photo: Brooks

2.3-LITRE ALFA ROMEO
8C-2300 BERLINA
SPORT, 1933-35

Photo: Brooks

BRISTOL CHANNEL
PILOT CUTTER

Built 1904, rebuilt in the UK 1995.
60' on deck, 15' beam, 10' draft

Photo: Wooden Ships

'WHITE WINGS'

Interior

Photo: Kos Picture Source Ltd

RIDING AT ANCHOR

Photo: Kos Picture Source Ltd

'WHITE WINGS'

Under sail

Photo: Kos Picture Source Ltd

CHAPTER 4

ALTERNATIVE INVESTMENTS

INVESTING IN ART

Colin D. Sheaf, Senior Director of Oriental Art, Christie's Europe

Colin Sheaf is senior director of oriental art for Christie's Europe, and a director of Hong Kong-based Christie's Swire. He regularly lectures, writes and advises clients around the world on Chinese art at auction and the international art market.

Phone: (+44-171) 839 9060
Fax: (+44-171) 839 1611

INVESTING IN ART

Colin D. Sheaf, Senior Director of Oriental Art, Christie's Europe

The 1980s witnessed the most astonishing boom in prices that the world art market has ever seen. Nothing in the past has remotely approached that decade's spectacular increases in price levels and market turnover, or in public awareness that art was available, attractive and affordable and conferred a certain social cachet.

This increased awareness changed attitudes towards the acquisition and aesthetic appreciation of every kind of art: both the traditional, classic kinds such as old masters, fine silver and European and American furniture, and the newer, less familiar categories, including contemporary paintings and sculpture, as well as a whole range of more ephemeral categories of art generically known as collectibles.

One of the most extraordinary developments to accompany this boom was a change in philosophy: it became acceptable to consider the purchase and holding of art purely for investment. But has this altered attitude survived the market slump of the early 1990s? Is the art market fundamentally flawed by the manifest inefficiency of the art market compared with the markets in most other income-bearing or capital investments? Are the underlying variables too eccentric to allow any real understanding of price trends?

These are all questions that continue to perplex those art-market professionals, fund managers, statisticians and supply-side economists who try to deduce long-term trends and predict market strengths and weaknesses from the evidence of the 1980s.

ART AS WEALTH

Analysing a work of art by the same criteria applied to more conventional investments was a pretty radical innovation in the 1970s and 1980s. It offended huge numbers of collectors and academic purists, to whom it was unthinkable that the finest tangible products of the human imagination should be equated, as investment vehicles, with pork bellies and frozen concentrated orange juice.

As a new form of asset play, art-holding briefly intrigued money-market professionals. After mature consideration, however, most apparently decided that art was an area of such volatility and illiquidity that art objects should make up no more than a tiny section of an investment portfolio. In any case, most funds lacked the expertise to run an art portfolio, although a few committed corporations such as British Rail did hire specialist fund managers to cope with the administration.

Meanwhile, art-world professionals, unfamiliar with the ability of financial traders to view absolutely anything with a market value as a possible vehicle for profit, almost always gave a frosty reception (at least in public!) to the concept of investing in art. Many times during and since the 1980s the question has been asked, 'Is art an investment?' And, almost always, the same pious but irrelevant answer comes back: 'Even if it is, only buy what you like: art by its very nature is not an appropriate vehicle for achieving capital growth.'

This has always struck me as an unhelpful answer to a straightforward question. There is no apparent logic in relying on your untutored, or even tutored, personal taste to assemble a portfolio of art objects for investment purposes: any more than there would be in buying sugar futures because you have a sweet tooth, or gold-mining shares because yellow happens to be your favourite colour.

Without either condoning or criticising the general principle, therefore, there are clearly a number of basic considerations that must underpin a realistic assessment of art as an 'alternative investment'. Very little has been published on the subject, and mine is only a personal view. Art is an intriguing asset vehicle and one that, with lucky timing, can offer unusually high returns and splendid opportunities to add value to the basic commodity. Yet there are such a bewildering number of irrational and incalculable variables that buying, holding and selling art as a formal investment is not the obvious option at all for the fainthearted fund manager of conservative inclinations!

Changing attitudes

It is self-evident that in any inflationary period it is quite possible to make money out of buying, holding and reselling art. Retail art dealers do so all the time, and my fund-management advice includes many of the guidelines a dealer follows when building his stock. To a great extent it is inflation that has focused attention on buying art for investment. One side-effect of living in a permanently or potentially inflationary economy, like the UK during the early 1970s, is that holding art, just as much as any other tangible asset, has come to be regarded as a way to outpace other financial indicators.

In the past, art has been used as a simple hedge against currency depreciation; apparently this was established practice among the financial elite of pre-Revolutionary France and, much more recently, it had some appeal in postwar Europe. But the use of statistics as a guide to decision-making and to predict and measure all the highs and lows of this fickle market, is a leap forward. The concept of art as investment will not, to my mind, fade away, and it will change art dealing and the way many people decide what to buy.

When an object is highly priced, I hear the question more and more, 'Is it a good investment?' Buyers are frequently preoccupied with the likely value, years ahead, of a purchase to which they are primarily attracted as established collectors. The question deserves to be squarely addressed.

Rather than addressing the question, those who offer the 'only buy what you like' school of investment advice leave you with nothing but a feeble consolation: in the state of penury brought on by your inability to master the intricacies of handling your personal finances, you will at least be surrounded by beautiful and rare, albeit unsaleable, works of art. This is not an obvious encouragement to a fund manager! Buyers who want to understand art-portfolio performance deserve better advice than this.

The right response is to try to understand what causes the rises and falls in supply and demand; to attempt to chart, empirically or statistically, the mercurial price movements of this (relatively minor) market; and to see what is the best combination of fine objects, good timing and quantifiable variables to make the possession and management of an art portfolio as profitable (and hopefully twice as enjoyable) as the management of an equivalent fund of capital employed elsewhere.

Calculated investment or arrant gamble?

Is holding 'art assets' really an investment at all? After all, art differs from almost all other investment areas in that it offers no income or dividend: the profit resides entirely in any capital gain, and that can be hard to extract even when perceived to be available! It is only the experience of the boom years of the 1980s that has changed attitudes towards the purchase and holding of art by private or corporate entities outside the traditional groups of art dealers managing their inventory.

In the UK the Inland Revenue has treated art holdings as an acceptable formal investment (with different tax implications from income tax) only since a court judgement in 1986. The traditional view was always that an investment must yield some kind of regular income. Revenue commissioners were unmoved by the emotional claim that the regular income was the non-pecuniary, aesthetic enjoyment of a work of art hanging on the drawing-room wall! Revenue practice has now caught up with the money markets. New approaches to investment include seeking capital profits without any concomitant income yield from an asset. Art is now no exception in this: trading short-dated stocks to give an income-less capital yield, or establishing split-level investment trusts, can have the same effect.

The evidence of the late 1980s is that art can indeed be a very profitable form of investment. Although the early 1990s saw a savage reaction to the excessive prices paid so recently, art is now too well established as a commodity vehicle for its investment potential ever to be ignored again. Art is still, in many ways, one of the riskiest speculations on offer in the markets, but there is no doubt that in cautious hands it is indeed a calculated investment.

MARKET INEFFICIENCY

Technology and competition are helping to reduce what has always been one of the art market's greatest benefits for professional players: its inefficiency. (No market that can offer an overnight profit margin of several hundred percent can possibly be considered efficient.)

The competition to find important pieces is very strong among the top auction houses and retail art dealers, all seeking 'fresh' consignments from untapped sources. This has led to pressure on traditional trading margins levied from vendors, representing an increase in efficiency on the supply side. But the fundamental inefficiency of the art market is its lack of consistent, chartable, globally similar price levels. This remains the distinguishing characteristic, and the supreme variable of the whole market.

Many other variables can – and usually do – play a role in establishing what is rarely an inflexible price:

- Dealers often quote widely varying prices to different clients, in proportion to the strength or weakness of the buyer's bargaining skills.
- Auctions are notoriously unpredictable: pre-sale estimates, themselves very widely spread, are routinely either surpassed or not reached.
- The physical condition of an object is very important, and is complicated by the use of new conservation methods to conceal damage.
- The relative strength of regional economies, and relative movements of currencies, can alter market prices dramatically in the long term.
- The arrival in the market place of a rich new collector temporarily fascinated by one small sector of the market can lead to a surge in price levels, since supply is, after all, very limited in the less well-known art sectors. Price levels, once raised, are nevertheless easily depressed (equally whimsically): the disappearance from the

market, for whatever reason (death, bankruptcy, boredom?), of this same collector; the fact that he missed a plane to an important sale or, the ultimate unpredictability, that he didn't happen to like an example of his speciality because the colour gave his wife a headache.

These are all considerations that I have had to cope with when appraising the value of objects to be included in an auction at Christie's.

Towards an international market

Technology is now having some effect on this inherent inefficiency. For example some 1980s innovations are already increasing price stability and potential buyers can now bid on the telephone from almost anywhere to well publicised auctions around the world. This helps to maintain local prices at something at least approaching perceived international levels, since it ensures a bigger active pool of private and 'trade' bidders, who are unlikely to let things sell absurdly cheaply wherever they are auctioned.

Arbitrage is less easy nowadays, too. In the past, because the market was less global, overnight profits could be made by shifting property between auction centres like London, Paris, New York and Hong Kong. Often it was only art dealers who appreciated the widely divergent price levels for a particular object at different selling centres around the world.

Easier access to statistics about prices actually achieved, usually at public auctions, has also made a considerable contribution to market efficiency. Art-price indices can provide an unprecedented perspective on the way sectors of the art market have moved, although none but the most fervent disciple of actuarial science would base a major investment decision on their projections.

Auction-house sale catalogues, and especially lists of prices actually achieved, provide a basic pricing guide for vendors of comparable objects – however remote the comparison between 'similar' objects may actually be! In my own experience, everyone with a serious interest in art or antiques, from the backstreet Asian-art dealer in Saigon, to the antique-paperweight vendor in the suburbs of Mexico City, regularly receives the important auction-house catalogues for his specialist area and revalues his own stock accordingly.

There is no more immediate guide to the vitality of an art-market sector, than to monitor the prices paid for a group of pieces put under the hammer in a big public auction. This has injected considerably more consistency into worldwide pricing – at the same time, sadly, lessening the possibility of finding unregarded treasures in remote places.

Art and antiques fairs also play an important role. They encourage dealers to travel more widely and exhibit new types of stock, they draw in new buyers, they familiarise (sometimes astonished) buyers with international price levels and promote art knowledge and active collecting. Above all, they build consumer confidence by inviting independent experts to 'vet' the exhibits, by offering specialist lectures to disseminate knowledge and by promoting openness in art trading. This sheds some light on what can, in many countries, be a somewhat murky profession!

ART FOR INVESTMENT: THE OPPORTUNITIES

As a capital asset, art (like real estate) has one considerable advantage over most other forms of financial instrument: the owner can personally increase its capital value. There is wonderful potential for 'adding value' to a piece of art in a number of ways: for example, by establishing a distinguished provenance, by further research and by restoration.

Provenance

The previous ownership (provenance) of an object can have a crucial effect on current market value. Ten pictures by the same artist, even if purchased separately and from different sources over a long period, gain cachet as a collection. Rather than hanging for sale in 10 dealers' galleries, the pictures will have greater appeal to likely buyers because as a collection they are perceived to have a homogeneity created by the collector's taste.

Art gains a 'fresh' quality if it has not been offered in the market for five or 10 years. In the art world, it is hard to sell any object that is perceived to be 'tired', that is to say it is unsold in a gallery or an auction, too visibly over-restored by a recent owner, or rejected by buyers for whatever reason.

An owner can easily enhance the cachet of a collection by having his paintings (or other pieces of art) exhibited in a museum, lent to a retrospective show or – best of all – published in the artist's definitive *corpus raisonné* credited to the owner. This does not always work, however: a buyer can run too far ahead of a speciality market, which then fails to catch up either because he has simply overpaid or because the objects are basically undesirable or of dubious authenticity.

All this activity does, nevertheless, establish provenance. The term implies a certain value judgement for art buyers: the persona of the previous owner casts lustre on the object or collection, its authenticity is unquestionable and it is of sufficiently high quality to have appealed to this refined or knowledgeable collector. Here is 'added value' indeed!

Many journalists defined the most spectacular boom years of art the market in the 1980s by the Christie's auction of two masterpieces, whose record-breaking prices owed a great deal to the lustre of provenance: Van Gogh's *Sunflowers* from the Chester Beatty estate in June 1987, and his *Portrait of Dr Gachet* in June 1990, from the trust under the will of Siegfried Kramarsky.

Even the most nakedly investment-oriented of all 1980s art collections, the wide-ranging holdings of the British Rail Pension Fund, were (and still are) seen to be something greater than just an assemblage of objects for investment. They were skilfully gathered together to represent the top end of each category in which the fund invested, and in many cases the prices achieved on dispersal of the portfolio reflected the public belief that this was really a collection rather than a fund. The objects had been carefully selected, were in good condition and had been widely exhibited, and each piece gained value from the other objects around it in the category.

Research and restoration

Establishing provenance requires very little expense for a profitable effect; research and restoration require a bit more. Effective research can add markedly to the value of a work of art, and every sale contains one or more relative bargains whose value will be enhanced by further information. In the same way, a good restoration, whether cosmetic or conservative, can bring out understated decorative qualities, or accentuate the inner power of a genuine masterpiece – the sympathetic cleaning of darkly varnished old masters or broken porcelain, for example.

With a relatively small outlay a collector can reap rich profits. The danger is that new, sophisticated techniques of restoration may cause the hidden work of a previous owner to reappear, to haunt the unwary owner who seeks to sell his unknowingly over-restored object.

Price tracking

To assess the advantage of adding value in these ways, it is now possible to track a general movement of prices, by reference to regularly revalued indices of art-auction sales. The UK's Daily Telegraph produces the best-regarded index. Indices can indeed be helpful but, insofar as they reflect an individual's perception of current prices for an arbitrary 'basket of goods', they will always be far more subjective than stock-market indices, commodity tables or even property prices, where the precise value of a square foot of land can be established daily on the basis of actual transactions.

Art indices need to be handled with enormous care. Even when extreme elements and fluke results are discounted, an index cannot possibly reflect all the variables that affect the actual price achieved. No art professional refers to an index when deciding how much to buy or sell an object for.

Price appreciation

Art can, and between 1984 and 1990 did, offer exceptional rates of capital growth if one gets the timing right. During the 1980s prices for most art doubled or trebled and some even quadrupled. There was a freak combination of strong demand (fuelled by cheap money) and a ready supply (boosted by the high prices publicly achieved at auction sales year after year, which encouraged more top-quality consignments such as the Van Goghs at Christie's). It was even possible by borrowing from banks, to obtain the very high gearing available in other financial markets to fund speculative asset purchases, since banks for the first time showed a willingness to lend money using art as collateral, and some auction houses began to offer extended credit on purchases.

New buyers came into the market in droves, enthused by continuous press publicity that almost invariably, and uncritically, stressed the successful sales, world-record prices, unprecedented public interest and so on. The traditional centres of art dealing were joined by new ones: buyers flocked not only to London, Paris and New York, but now to art auctions and retail galleries in Geneva, Monaco, Hong Kong, Amsterdam and, above all, Tokyo.

In some ways, it is a tragedy for the market that in certain areas (notably contemporary art and collectors' cars), prices have plunged so precipitously. The late 1980s reminded us how very limited the supply of art really is when actively traded and collected. The big drops of the early 1990s, although a necessary corrective bringing prices back to a more realistic and consistent long-term (but still rising) trend, have driven a wide range of buyers out of the market. This has made people forget just how rare a pristine Rembrandt print, an early Ming blue-and-white dish or an important Jackson Pollock drip painting really is.

ART AS INVESTMENT: THE DANGERS

During the past five years, we have seen all the disadvantages of investing in art as a short-term play, and a large number of investment buyers will go into the late 1990s chastened but wiser.

An illiquid market

Many of the disadvantages stem from the illiquidity of the art market. The inefficiency of market pricing and wide buy-sell spreads mean that desperate sellers may find it difficult to recoup a large proportion of their cost when the market is stable. It will be even harder to achieve the purported current value in a depressed market. Price indices mean nothing then, for there is no inherent value in anything except bullion and coins (and, if one accepts the labour theory of value, the current cost of manufacture).

A nervous buyer, in a narrow and weakened market, may simply not want the object for his stock or collection at any price. What was a great rarity six months ago becomes an unsaleable and rapidly depreciating asset: less saleable, probably, than even the other great illiquid asset, property, where at least land price and construction costs underpin some kind of basic value.

The narrowness of speciality markets, although probably an advantage in boom times, is a disadvantage when the market turns. Limited marketability is the most serious problem in bad times – which is when the pressure to trade is really on. We have seen this time and time again, notably in new areas that stress the narrow market, such as modern limited editions – cars, stamps, commemorative coins and ceramics, and prints by contemporary masters. All of these have suffered very badly from the recession.

A narrow market is a very volatile market. Consequently a few bullish trades affect prices significantly. There is nothing unusual about this in terms of financial markets; what is of note is the unpredictability of the outcome after two or three serious new collectors enter a market and markedly change its vitality. The resultant press coverage compounds the effect, the whirlpool sucks in extra willing buyers and the cycle is maintained, and accelerates. Anyone considering investing in art should be acutely aware of this. Apparent growth may be very artificial, and an art investor without sound market knowledge is very vulnerable indeed when the music stops!

Not only is the property illiquid at the best of times, but there will almost always be a delay in selling even during a buoyant period in the market. Even a top auction house will need about three months before your consignment can be put under the hammer. It is even harder to predict the time it will take to sell through a dealer, but most will be pleased if they sell an important painting or object within six months, unless the market is extremely active.

The ideal market for an art investor, therefore, is one with:

(i) plenty of trades;
(ii) a steady increase in price on a relatively consistent curve; and
(iii) a wide base of private-collector support.

If any combination of circumstances can give the investor a modicum of confidence over the longer term, this seems to be it; the most obvious disadvantages will, at worst, be contained.

Costs

The physical nature of an art portfolio puts it at a clear 'disadvantage' to other assets. Maintenance costs, restoration and conservation charges, insurance and security charges will all add to the capital cost, without any compensating income. Loaning pieces to a museum will help: the provenance is enhanced and pieces receive proper care in secure conditions.

Trading costs when selling (and sometimes when buying) make another hole in the profit. No small fractions of a percent for broking services in the art world: a more substantial 10% or 20% is the norm.

Proving authenticity

Is your purchase genuine? Another disadvantage of long-term investment in art is that much 'authenticity' is relative. The scholars examining the entire *œuvre* of Rembrandt have reminded us again of this truth: some familiar paintings, regarded as genuine for a century, are now being reconsidered and demoted, but by the next generation may quite possibly have been reinstated as works by the master.

The art market is faced by twin threats (or opportunities?) with regard to authenticity. On the one hand, academics are constantly re-evaluating works of art, identifying new masters and reconsidering traditional attributions. Scholarship within the commercial art market has never been so important, and this can be a big problem if your own art portfolio, bought in good faith and relying on the scholarship of the time, comes under critical review.

On the other hand, deliberate fakes and forgeries (meant to deceive) are becoming more and more convincing. Science helps to identify many of these: spectrographic study, X-ray photos, thermoluminescence, carbon-14 testing and pigment analysis are standard tools for detecting the new generation of fakes.

The art-buyer must take advantage of every scientific test available to ensure the authenticity of acquisitions because, in the event of any controversy, the hostile critic will certainly seek to do so. Really skilful art faking is a very old practice, but a decade of rising prices has made the rewards much greater. The relaxed standards of the late 1980s, when demand was outstripping supply, meant that a new crop of fakes found their way on to the market. These are surfacing only now, as owners seek to realise their assets.

Political risk

Political change can affect art prices badly but, in this respect, the art market is no different from others. Any international crisis that sucks liquidity from financial markets, or a local revolution that disturbs trade flows or exchange rates or causes loss of public confidence in assets of any kind, is bound to be bad for a sector as dependent on confidence as art. Where the reaction differs is in the speed of recovery. Recent experience suggests that conventional markets recover more quickly and consistently, while art trading and collecting depend much more upon individual circumstances.

As an example, the communist uprising in Portugal in 1974 had one side-effect that markedly influenced the art market for several years. In the early 1970s, a small group of Portuguese bankers, assisted by one or two dealers, began to spend heavily on Chinese porcelain. Prices predictably shot up (Hong Kong had not yet come to play its later role in establishing world price levels for Chinese art), credit lines were extended, supply expanded to fill a demand that widened to Asia and across Europe and prices continued to rise.

Suddenly, the 1974 revolution disrupted everything, after the oil crisis had threatened all trading the year before. In six months prices halved – if a buyer could be found. Dealers, holding overpriced stock, made little effort to expand inventory or support the market; auctioneers scaled down their pre-sale estimates drastically, further depressing market sentiment; and the Portuguese collectors, their assets frozen, owed money to London agents, who owed the trade suppliers.

The earlier enthusiastic press coverage of successful auctions disappeared. Collectors fled from Lisbon to Brazil, Geneva or London, but now they had other priorities than buying *famille rose* dinner services or Ming Dynasty blue-and-white dishes.

It took several years for this sector of the market to regain any kind of buoyancy, but general art-market indices largely missed this big slump in a minor area. It takes an international oil-price crisis for the general indices to reflect whatever may really be happening to individual sectors.

No income

There is no income from art investment, a circumstance that has also traditionally been seen as a disadvantage. Given a new sophistication in assessing real investment returns, even with no income yield the main disadvantage here lies in the possibility of missing the right moment to sell. If profit accrues solely from capital growth, timing the moment to sell is all that matters.

There is nothing surprising about the conclusion that, with no regular income and considerable volatility in inefficiently priced markets, accurately timing the moment to sell means the investor must follow the market very closely indeed.

In theory it may be better to plan to sell when a specific return on capital is likely to be attained and the market is confident, even if this is ahead of a further peak in prices. Market confidence can evaporate months before real falls in value begin to work through the system. The gap between apparent value and attainable sale price can be widened if there is nobody in the market who actually wants to buy, at however low a price. This disadvantage – the total collapse of marketability – is shared primarily by the property market.

A matter of taste

The most severe structural disadvantage in the art market must be the fact that, in the long run, it relies on little more than personal whim to generate turnover and price movement. There can be no financial market more unpredictable than one in which personal taste determines what is bought and sold.

Over the decades, this has led to colossal shifts in what the market perceives as the most rock-solid blue-chip art investments. The famous New York dealer Joseph Duveen was selling ready-made social status on the English-country-house model to rich American collectors in the early decades of this century. Nothing was more suitable for the drawing room walls of East Coast millionaires than a group of (perfectly nice) 18th-century English portraits. By aggressive buying, and even more aggressive selling, Duveen pushed prices up to a level that in nominal, much less in real, terms they never reached again until the mid 1980s, since American taste soon turned away from this unsubtle imitation.

Tastes in contemporary art have proved equally fickle. Sir Edwin Landseer's heroic wildlife compositions were more expensive sold hot off the easel in the 1880s than they were to become again for a century. Even schools of old masters have shown considerable shifts in popularity from one generation to another: Sienese primitives, Dutch marine scenes and still lifes, Italian baroque and French academic art have all boomed (and bust) with some regularity since the 18th century, as art-auction records reveal.

For the investor, it is hard to determine the likely direction of rising schools. The changes here are probably led by academic, rather than commercial, opinion. A scholarly reassessment of a forgotten artist or movement can stimulate a market quite dramatically. In England, both Pre-Raphaelite and Camden Town School paintings were rescued from complete obscurity by the endeavours of a few scholars, and are now the objects of mainstream collecting. Renewed public interest brought enough pieces on to the market to offer a wide range and variety, and several rich collectors (with the assistance of their essential advisers) positioned themselves to assemble important groups of paintings, at price levels not approached in real terms since they were first painted.

When patrons feel rich enough to remodel their houses, even decorators can affect the direction of the art market: a flurry of big 'traditional' commissions, and 18th-century English furniture is hot property; a few lifestyle articles puffing country life, and oak and pewter are back in fashion.

When supply is restricted, which is very often the case, it takes only a small demand to move prices, even without long-term rationale. Much of the 1980s was spent trying to find new areas of art to collect. A number of these fringe markets were considerably manipulated, with private buyers tacitly accepting the rigging of price levels for everyone's perceived short-term benefit. Many of these new-found 'collectibles' are still traded, but the speculative element largely evaporated during the recession. It is, once again, genuine collectors who buy early toys or corkscrews, for instance.

Oversupply

An unexpected event occasionally unleashes a large group of objects into a market believed to be finite, inevitably disrupting prices. During the 1980s archaeological treasures were looted in large quantities from newly discovered sites: pre-Columbian and early Chinese pottery plummeted in price, since collectors and dealers lost confidence when hitherto rare objects came up for sale in unprecedented quantities.

New technology made sea-bed treasures more accessible in the 1980s. Old shipwrecks yielded huge cargoes of antique porcelain; in 1986 the Nanking Cargo poured 150,000 pieces of standard 18th-century Chinese export porcelain into a narrow market. Unexpectedly, these auctions have mostly followed the pattern of country-house sales, in which middle-range objects often wildly exceed conventional price ranges. Sale prices were astonishingly high and re-offered pieces have, predictably, rarely reached the same heights at auction.

These blue chips with a famous provenance began at a price level too high to be sustained. As with contemporary art, the initial trading price is of crucial importance if long-term investment is being considered.

CONCLUSION

There are definitely profits to be made by treating art as an investment commodity. Ignore the platitudes about following your personal taste, however, and analyse everybody else's – that is what makes the market move.

This requires serious study of the commodity itself, of the individuals who buy it, their motivation, longevity and range of collection. Who, if anybody, is controlling and promoting the sector? A dealer, an auction house, a museum academic? If there is any one key to building a successful portfolio, it lies in knowledge and understanding of the underlying mechanisms.

In assembling an art portfolio, several options are available. The long-term outlook suggests concentrating on what have become established as the blue chips of the art market: French furniture, early English silver, classic Chinese porcelain, Impressionist paintings. A wide spread seems the safest approach for an investor who relies on a group of external advisers to analyse different market sectors; this was the British Rail approach. But the British Rail fund showed clearly that although many blue chips were profitable a number were not, even over 15 years. The investor running his own fund with acute art-market understanding would probably do better to focus on individual specialist areas that he is very familiar with.

For the conservative who takes the trouble to delve deep into long-term market performance, it is quite possible to identify blue chips that have performed consistently well. For the more aggressive, indices will suggest the under-performers as a speculative buy for future growth. At the moment, the market is 'soft' – the perfect opportunity for the 'recovery-stock' enthusiast to take a profitable position?

The only really buoyant economies in the world today are those of the 'four dragons' and the emerging nations in Asia. Should an investor follow evolving political stability and economic dynamism and buy Chinese, Japanese, Korean, or Vietnamese art? Or should one buy German art, for the same reason but on a more parochial level? A cold-blooded analysis of the global art market and the prospects for Asian economies strongly suggests that Chinese and Japanese paintings and porcelain are, in reality, much more of an international artistic currency.

It seems appropriate that my conclusions are essentially simply a series of questions. We are dealing with a market almost entirely arbitrary in fundamental pricing, where the agents of growth and decline are diverse, whimsical and inconsistent. Prices in the art market – unlike other markets – are a function, ultimately, not of the inherent value of the commodities that it comprises, but of the personal predictions, obsessions, and self-interest of the individuals who are the active players. This is what gives the market its enormous public appeal, its volatility, its potential for 'added value' and its traditional profitability.

For the investor the only realistic conclusion must be: the only thing that one can predict with complete confidence for the art market is its unpredictability! ❖

ART INDICES

Robin Duthy, Art Market Research, London

Robin Duthy, the creator of the *Daily Telegraph* Art Index, trained as an investment analyst in the City of London. He has published two books on the art market – *Alternative Investment* and *The Successful Investor* – and has written art market surveys for the *Financial Times*.

Phone: (+44-171) 249 8071
Fax: (+44-171) 254 5619

ART INDICES

Robin Duthy, Art Market Research, London

The following indices demonstrate the movements charted in the art markets since 1974/75. They represent a comparison of the average market value for paintings and sculptures in a particular group, or for an individual artist, relative to their value in the base period, which is 1974/75.

These indices serve to provide a comparison between groups at any one point in time and also to show the trends in art values over time.

For each artist included in an index prices are recorded for every one of the works that are sold in the period from January 1974 to November 1995, on a monthly basis. The indices are technically 14-month weighted moving averages. The data is originally recorded in sterling and then converted into US dollars and weighted as an index.

The figures for 1995 are provisional, and may be subject to revision after late-reported results are included.

DAILY TELEGRAPH ART 100 INDEX

* *Refer to the note at the end of this article for a list of the artists included in this index*

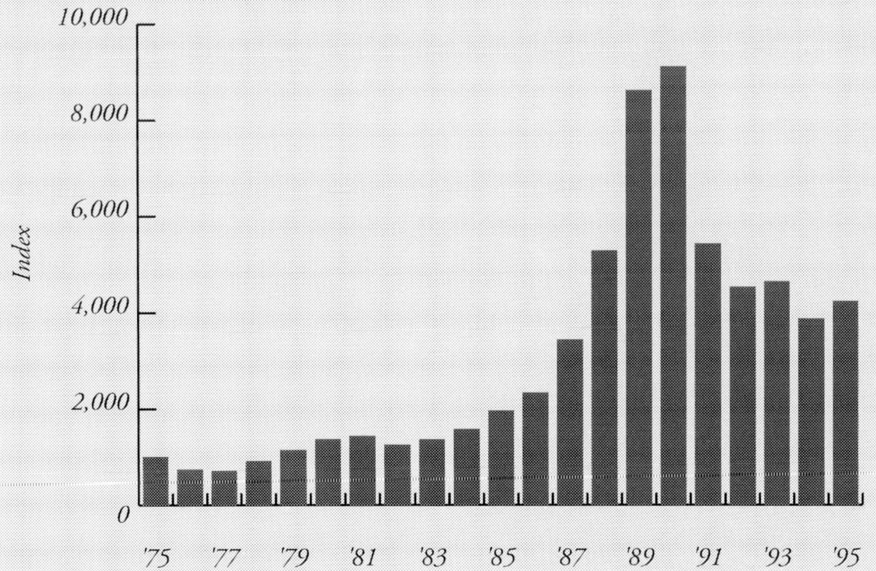

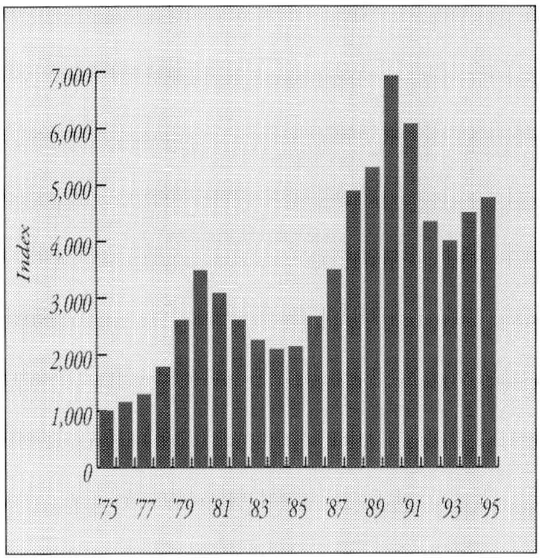

CHINESE CERAMICS: EXPORT PORCELAIN

(Christie's – Art Market Research Index)

Christie's – Art Market Research indices are updated twice yearly. The Chinese Ceramics: Export Porcelain Index is based on a representative group of 20 objects from the period 1550-1850.

CONTEMPORARY MASTERS

(*Daily Telegraph* Art Index)

Component Artists: Jean-Charles Blais; George Baselitz; Jean Michel Basquiat; Sandro Chia; Francesco Clemente; Keith Haring; Mimmo Paladino; A.R. Penck; Sigmar Polke; Antonio Tapies.

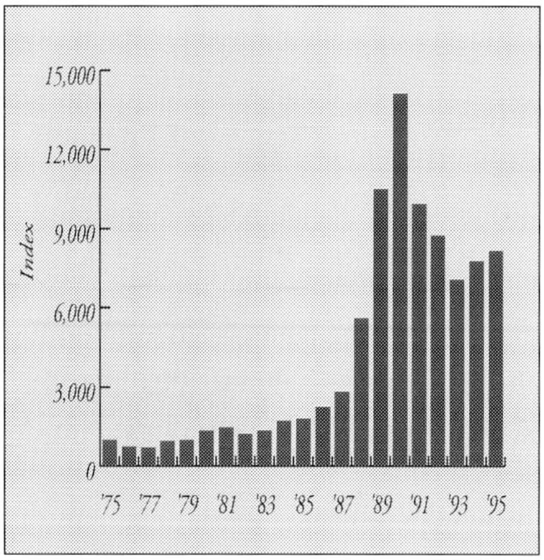

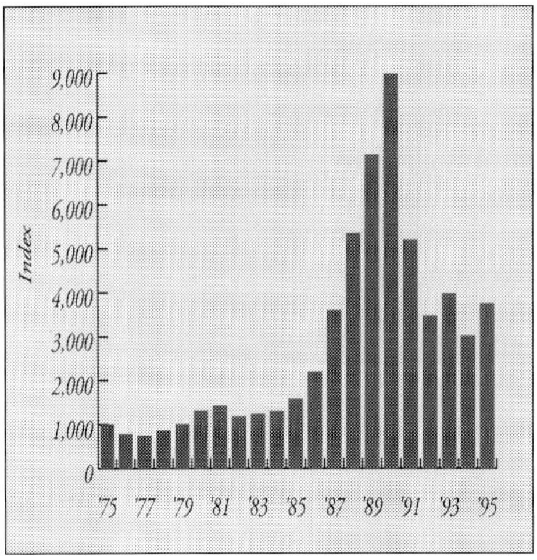

MODERN EUROPEAN PAINTING

(*Daily Telegraph* Art Index)

Component Artists: Pierre Bonnard; Georges Braque; Marc Chagall; Kees van Dogen; Moise Kisling; Marie Laurencin; Pablo Picasso; Georges Rouault; Maurice Utrillo; Edouard Vuillard.

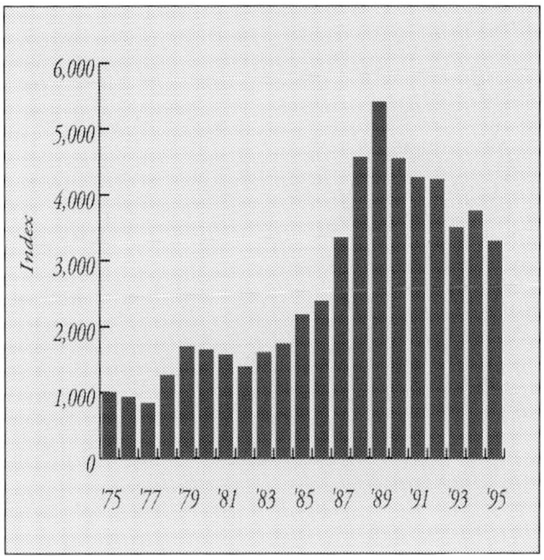

OLD MASTERS

(*Daily Telegraph* Art Index)

Component Artists: Francois Boucher; Canaletto; Pieter Claesz; Jean-Honoré Fragonard; Thomas Gainsborough; Jan van Goyen; Francesco Guardi; Jean-Baptiste Greuze; Jan van Kessel; Nicolas de Largillière; Sir Peter Lely; Nicolas Maes; Michele Marieschi; Ben Marshall; Sir Joshua Reynolds; Hubert Robert; Salomon van Ruysdael; Abraham Storck; Giovanni Battista Tiepolo; David Teniers the Younger.

EUROPEAN 19TH AND 20TH CENTURY SCULPTURE

(*Daily Telegraph* Art Index)

Component Artists: Alexander Archipenko; Fernandez Arman; Jean Arp; Antoine Bourdelle; Lynn Chadwick; Jules-Aimé Dalou; Sir Jacob Epstein; Dame Elizabeth Frink; Dame Barbara Hepworth; Henri Laurens; Aristide Maillot; Carl Milles; Henry Moore; Louise Nevelson; Auguste Rodin.

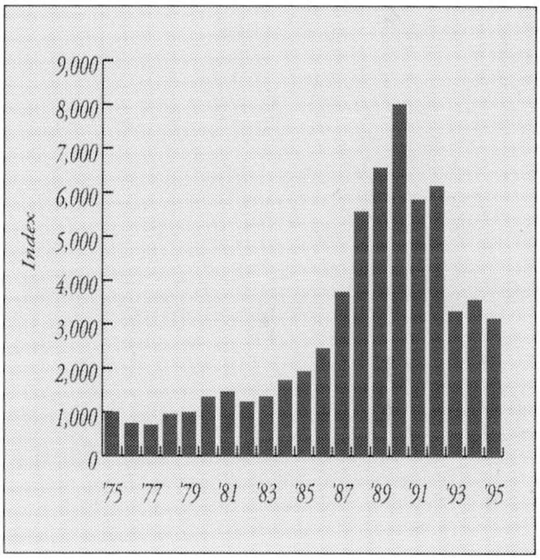

* DAILY TELEGRAPH ART 100 INDEX

Based on 78 individual artists and two groups. These are:

Pierre Alechinsky; Helen Allingham; Michael Ancher; Karel Appel; Georg Baselitz; Jean-Michel Basquiat; Albert Bierstadt; Pierre Bonnard; Fernando Botero; Eugène Boudin; Arthur Merric Bloomfield Boyd; Georges Braque; Bernard Buffet; Marc Chagall; Sandro Chia; Giorgio de Chirico; Jean-Baptiste Camille Corot; Gustave Courbet; Salvador Dali; Montague Dawson; Otto Dix; Jean Dubuffet; Max Ernst; Henri Fantin-Latour; Lyonel Feininger; Lucio Fontana; Myles Birket Foster; Sam Francis; Atkinson Grimshaw; Keith Haring; Henri Harpignies; Childe Hassam; Paul-César Helleu; John Frederick Herring (Snr); Ferdinand Hodler; Antonio Jacobsen; Johan-Laurents Jensen; Johan Barthold Jongkind; Asger Jorn; Ernst Ludwig Kirchner; Moise Kisling; Paul Klee; Gustav Klimt;

Willem Koekkoek; Oscar Kokoschka; Willem de Kooning; Carl Larsson; Marie Laurencin; Fernand Léger; Bruno Lilgefors; René Magritte; Henri Matisse; Joan Miro; Claude Monet; Giorgio Morandi; Sir Alfred Munnings; Emil Nolde; A.R. Penck; Pablo Picasso; Serge Poliakoff; Pierre Auguste Renoir; Jean Paul Riopelle; Diego Rivera; Gino Severini; Dorothea Sharp; Leon Spilliaert; Carl Spitzweg; Alfred Stevens; Antonio Tapies; Fritz Thaulow; Archibald Thorburn; Maurice Utrillo; Louis Valtat; Edouard Vuillard; Andy Warhol; Tom Wesselman; Jack Butler Yeats; Anders Zorn.

Old Master Group (weight of 20): François Boucher; Canaletto; Pieter Claesz; Jean-Honoré Fragonard; Thomas Gainsborough; Jan van Goyen; Francesco Guardi; Jean-Baptiste Greuze; Jan van Kessel; Nicolas de Largilliere; Sir Peter Lely; Nicolas Maes; Michele Marieschi; Ben Marshall; Sir Joshua Reynolds; Hubert Robert; Salomon van Ruysdael; Abraham Storck; Giovanni-Battista Tiepolo; David Teniers the Younger.

Victorian Group (weight of 2): Sir Lawrence Alma-Tadema; Sir Edward Coley Burne-Jones; John William Godward; Frederick Lord Leighton; Sir John Everett Millais; Dante Gabriel Rossetti; Marcus Stone; James Jacques Joseph Tissot. ❖

OLD MASTERS

Olivier Aaron, Didier Aaron & Cie, Paris

Olivier Aaron studied art history in Paris. He worked for two years at New York's Metropolitan Museum of Art in the European paintings department and then at the Victoria & Albert Museum in London. For the last five years Mr Aaron has been concentrating on the work of Jean-Baptiste Marie Pierre, and is preparing the *catalogue raisonné* of this 18th-century French artist.

Phone: (+33-1) 4742 4734
Fax: (+33-1) 4266 2417

OLD MASTERS

Olivier Aaron, Didier Aaron & Cie, Paris

The art market, although it has recovered significantly, remains fragile in the wake of the crash in art prices five years ago. High quality, that rare property, is once again in demand from institutions around the world. This demand has contributed to some high prices in New York, London and Paris auction houses, and in sales between private dealers and collectors. Against this backdrop of burgeoning confidence in the marketplace and the underlying fragility caused by the still-recent crash, Michael Hall's admonitions in the Asian edition of this book should remain the cornerstone of any collector's philosophy: 'Fine art and antiques do not make good investments . . . art pieces should be purchased primarily for the pleasure they give, and . . . the longer they are kept the greater the reward, which may take several decades.' This advice remains sound for collectors of all art forms, old masters included.

PAST MASTERS

The artists now commonly referred to as 'old masters' have always been collected. Throughout the 15th to 18th centuries their works were commissioned by the dominant art patrons of the day, mainly the Church and the royal courts. Well-known artists were the recipients of monumental commissions all over Europe: in Italy, Carpaccio, Giovanni Bellini, Sano di Pietro and Titian, amongst others; and, in France, such artists as Fouquet, Vouet, Chardin, Hubert Robert and Jacques-Louis David. In Flanders, quality was considered the essential element of a work, and paintings, usually in a small format, by artists such as Jan van Eyck, Memling, Dirk Bouts and van der Weyden were commissioned for private chapels. In Spain the Church commissioned paintings by artists including those of the Catalan School, then Murillo and Velasquez.

The paintings themselves graphically illustrate the importance of the Church and royal patronage. From the 17th century, for example, there are several Flemish paintings showing an Austrian archduke visiting private collections. In the 18th century in Rome, Panini depicted some important contemporary collections which had been assembled by cardinals, and some of which included the most important monuments of contemporary and ancient Rome.

CONDITION, AUTHENTICITY AND PROVENANCE

Collecting old masters requires a keen eye and attention to detail. Apart from a collector's love for a particular work of art, he should follow certain guidelines to make certain of its intrinsic value. A painting's condition, authenticity and provenance all affect how much it is worth.

With respect to the quality and condition of a painting, a collector should rely on the guarantee of an expert or a reputable gallery. Although restoration is an accepted practice, repainting or adding a signature is certainly not!

Authenticity is an essential quality in a work of art, and old masters are rarely signed. A private collector should consult the expert at the sale if the painting is being auctioned, or another expert who specialises in the art of the period. The certificates of the master experts of the beginning of this century, such as Bernard Berenson and Friedlander, are often contested. This is because these scholars expressed opinions based purely on their own judgements; they did not have the benefit of the wealth of photographic documentation, textual publications, X-ray and other modern technological devices which today contribute to more refined and focused analyses. This is obviously less of a concern with modern paintings, where the artist or the artist's family is often still alive. One has only to recall (perhaps with a shudder) that an art historian recently decided all of the Rembrandts in Queen Elizabeth II's collection were workshop paintings, and that his best known painting, *L'homme au casque d'or,* was not by him at all. Perhaps in another 30 years or so another expert will come to contradict those judgements.

Henry-Clay Frick, in the United States, provides an excellent example of the importance of the relationship between collector and expert. Between 1889 and 1918 Frick established a collection with the guidance and assistance of renowned art critic and connoisseur Bernard Berenson and art dealer Joseph Duveen. His collection, though relatively small, is composed of irreplaceable masterpieces. It is a testament to the success of collecting done by Frick, Berenson and Duveen that, when the administration of the collection called for a revaluation of the collection's insurance, the unanimous response of the renowned experts who were consulted was that the collection was so magnificent that no price could be put on it, for purposes of insurance or otherwise.

Aside from the crucial questions of authenticity and condition, it is desirable that a work should be part of a coherent collection, with a good provenance (past ownership or lineage). A good provenance will often contribute to the saleability of a painting. A case in point is the recent sale of part of the collection of the New York Historical Society, one of the oldest museums in the New World. Despite a lack of quality in the pieces auctioned, most of the 1,834 pieces sold, with a few fetching high prices. The total sale exceeded US$10m. The buyers, connoisseurs rather than spectators, were attracted not by the provenance alone, but also by the 'virginity' of the pieces, which had not been seen on the market for a long time.

For love or money?

It is often difficult to distinguish between pure speculation on the expected increase in value of a piece of art, and the buyer's aesthetic appreciation of a particular work. Maurice Rheims, the famous French auctioneer and academic, recounts in his book *Les Fortunes d'Apollon* the recommendation of Louis XIV to one of his courtesans: 'Buy paintings by Le Brun: he has just died, his paintings will go up in value.' (Le Brun became painter to the king in 1662.)

A MARKET FOR QUALITY

A survey of the world's main art markets for the years 1994 and 1995 reveals the importance of collecting quality works of art. In Paris for example, most of the recorded auction sales have been around FFr100,000 although some good bids of more than FFr400,000 have been made. The highest price of 1995 was FFr2.2m, for a still life by Balthasar van der Ast at Maître Tajan in June. Despite the success of this particular work, only one-third of the entire collection from which it came was sold. Privately sold pieces fared much better by comparison; a pair of paintings by Hubert

Robert, *Cascade de Saint-Cloud* and *Pont de Mereville*, were sold for FFr1.9m. In the sale of Maître Picard on 11 April a very nice still life by Desportes sold for FFr1.2m. At Maître Kohn, a small painting by Pieter Brueghel the Elder, *La trappe aux oiseaux*, sold for FFr1.45m. This genre of paintings, although in great demand, did not reach the price levels set in the years 1985-1988.

A review of French auction records for 1994 demonstrates that the prices of old masters are extremely sensitive to the quality of the pieces. By way of example, Andrea Solario's *Christ au roseau* was sold for FFr4.93m in June 1994 and Brueghel the Younger's *L'été ou la maison* reached FFr3.83m at Drouot George V in December that year. A beautiful still life by Flores van Dyck, a painting of superb quality in both composition and execution, fetched the exact same price in March 1994. The masterpiece of the season, *Le retour du Bucentaure le Jour de l'Ascension* by Canaletto, prompted pre-auction jousting between French and British experts over the final sale price. British experts estimated that the painting would fetch FFr40m and in Paris the owner was promised more than FFr60m. The actual price, FFr72.27m astonished everybody. This painting had all the desired attributes: outstanding quality, a sought-after subject and a good size of 86cm x 135 cm.

In both London and New York the market for Dutch and Flemish masters has stabilised after several chaotic years. British and American demand is strong, as illustrated by Sotheby's sale of a Ruysdael to an art dealer for £1.5m on 5 July 1995. At that sale, 73.2% of the works auctioned were sold for a total of £14.4m. By comparison, the sale one year earlier made only £9.6m. Furthermore, two records were set at Christie's in London in December 1994; a painting by Giulio Romano, *Spinola Holy Family*, brought £1.541m and, at the Houghton Hall sale, *La lecture de Molière* by Jean-Francois de Troy sold for £3.96m. This is a delightful painting representing French society in the 18th century. In contrast, the Christie's and Sotheby's sales in Monaco were much less successful due to an absence of quality pieces.

This review of recent successes and failures in the art market proves that when quality and rarity are present, high prices can be achieved whether in Paris, London or New York. The only true hedge against the turmoil and volatility of the art markets is to buy what one loves. Because, while stocks and bonds cannot be framed and exhibited, owning a masterpiece brings an immeasurable amount of pleasure. The purchase of a quality work of art, if it is held for a long time, can provide financial pleasure, too. ❖

SCULPTURE: 'THE MOST VALUABLE ACQUISITION A MAN OF REFINED TASTE CAN MAKE . . .'

Joanna Barnes, Joanna Barnes Fine Arts, London

After graduating from the Courtauld Institute of Art, Joanna Barnes worked as director of sculpture at the Heim Gallery in London. In 1986 she founded Joanna Barnes Fine Arts, a gallery specialising in sculpture, particularly from the 18th and 19th centuries. Joanna Barnes has organised several sculpture exhibitions, and has contributed articles on 18th-century European sculpture and painting to various art-historical publications.

Phone: (+44-171) 930 4215
Fax: (+44-171) 839 8307

SCULPTURE: 'THE MOST VALUABLE ACQUISITION A MAN OF REFINED TASTE CAN MAKE . . .'

Joanna Barnes, Joanna Barnes Fine Arts, London

For many people the word 'sculpture' conjures up images of deceased dignitaries in civic squares, tedious glass cabinets and ranks of busts in museum displays. But there is much more to sculpture than this.

To be thoroughly appreciated and enjoyed, sculpture needs to be carefully observed. Sculpture is not confined to museums, architectural design and public monuments, but over the centuries has also played a key role in the decorative arts – in works executed in silver or ceramics and as mounts on furniture.

THE FIRST COLLECTORS

Sculpture is not widely accepted on a domestic scale today, although many pieces were actually intended for this setting, from the small bronzes collected by the Renaissance humanists to the statuettes of Alfred Gilbert and the 'New Sculpture' of the 19th century. Small bronzes have been collected since the mid-15th century, when many ancient bronzes were excavated, and during this period antique and contemporary sculptures were displayed side by side. Renaissance humanists collected these bronzes not only for their classical associations, but also for their fine craftsmanship. Statuettes were the currency of prestigious diplomatic gifts throughout the 16th and 17th centuries.

Throughout the 17th and 18th centuries and even into the 19th, sculpture received major ecclesiastical and secular commissions, and in certain instances was even more highly prized than painting. Sculpture was not only commissioned as spectacular decoration, but was also sought after on a smaller scale for the private enjoyment of the rich – the bronzes of the *kunstkammer* were the 'toys' of princes.

In the 17th and 18th centuries, terracotta *bozzetti*, the preparatory fired clay models for larger projects, came to be prized by the leading Italian collectors of the day, such as Cardinal Flavio Chigi (1641-1693), the nephew of Alexander VII. At his death, Chigi owned 25 terracottas, among them a group of important works by the greatest exponent of Baroque sculpture, Gianlorenzo Bernini. Abbot Filippo Farsetti was also an avid collector in this medium and his collection has recently been the subject of an entire exhibition: *Alle Origini di Canova*, held at the Palazzo Ruspoli in Rome and the Ca d'Oro in Venice. During this period the intimate scale of the terracotta model found favour in France, too, with celebrated French collectors such as Jean de Julienne and Pierre-Jean Mariette.

This was also the century when collections of sculpture were formed by gentlemen on the grand tour, and galleries were set up in English country houses devoted to their display. The dealer Thomas Jenkins was instrumental in the formation of many of these collections. The idea of a 'sculpture gallery' was pioneered by Lord Burlington at Chiswick House and continued by Matthew Brettingham at Holkham and Petworth and by Jeffrey Wyatville at Chatsworth. Sadly, many of these famous holdings have subsequently been dispersed at auction. The contents of Ince Bundell Hall, for example, became part of the Liverpool Museum. But fortunately, and instructively, some collections have remained in the settings for which they were designed, such as Petworth House in Sussex and Chatsworth House, Derbyshire.

Towards the end of the 18th century, Scottish painter and dealer Gavin Hamilton wrote to the collector Charles Townley: 'Never forget that the most valuable acquisition a man of refined taste can make is a piece of fine Greek sculpture.' This outlook reflects the passion for classical art that was at the centre of the grand tour ethos. Englishmen on the grand tour acquired antique sculptures, coins and cameos. One of the most notorious of these was the Earl of Elgin; the repercussions of his purchases are still being felt today. The Italian sculptor Bartolomeo Cavaceppi was a famed restorer of antique pieces and specialised in producing faithful copies after the antique. Similarly, contemporary copies of bronzes were made for the grand tour market by Giovanni and Giacomo Zoffoli, Francesco Righetti and Massimiliano Soldani-Benzi. In 1710 Soldani-Benzi made casts of the famous group of classical marbles in the Uffizi, Florence – *The Dancing Faun*, the *Medici Venus*, *The Knife-Sharpener* (*Arrotino*) and *The Wrestler* – commissioned by John Churchill, first Duke of Marlborough, for Blenheim Palace.

Throughout the 19th century and until around 1920, the market for sculpture was elitist and almost on a par with that for painting. Dealers like Seligman, Wildenstein and Joseph Duveen formed collections for wealthy clients such as Andrew Mellon in Washington and William Randolph Hearst.

REVIVING INTEREST

After 1920 some of the bronzes from the princely collections of Vienna and Berlin were dispersed, and great scholars of this medium began to emerge. The work of Wilhelm von Bode has had a lasting influence in this field, as has that of Hans Weihrauch. In more recent times John Pope-Hennessy, Jennifer Montagu and Anthony Radcliffe have made major contributions to the subject.

Although interest in Renaissance items declined during the 1930s, it was during this decade that the dealer Julius Goldschmidt, a protégé of von Bode, arrived in London, where he centred his activities. A member of a celebrated family of German dealers, Goldschmidt had a particular interest in sculpture, especially Renaissance bronze statuettes. Earlier in the century he had played an important part in the formation of the Pierpoint Morgan, Wiedner and Bache collections.

Alfred Spiro was also active in London at this time, as an important dealer in Renaissance bronzes. In the USA there was a slow revival of interest in sculpture after the Second World War and prices for sculpture both there and in Europe were ridiculously low. The activities of Goldschmidt and Spiro and the exhibitions of David Peel focused attention on sculpture.

In 1963 the renowned dealer Andrew Ciechanowiecki arrived on the international scene, again based in London – first at Malletts of Bourdon House and then at the Heim Gallery. He began a series of pioneering sculpture exhibitions such as *Les Animaliers* and *Jules Dalou* aimed directly at the small collector. Cyril Humphris also appeared at this time, and influenced many collectors. The Robert Smith collection of bronzes for example, owes much to his advice and expertise.

Other recent dealers of note have been the late Anthony Roth, who combined connoisseurship with scholarship, and the larger-than-life figure Danny Katz, who specialises in Renaissance bronzes and is famed for his discerning eye.

Commercial galleries have had a pivotal role in heightening awareness and focusing on specific areas of academic research. Knoedler & Co.'s 1968 exhibition *The French Bronze*

1500-1800, held in New York, for example, was a true trailblazer.

Great collectors of our day have included Arthur Sackler, whose collection was celebrated in the catalogue *European Terracotta Sculpture from the Arthur M. Sackler Collections*. Dr Sackler's collection revealed the breadth of his interest, which ranged from the arts of the eastern Mediterranean and East Asia to the splendid collection of European terracottas from the 15th to 20th centuries. The collection was first shown at the National Gallery of Art in Washington and later at New York's Metropolitan Museum. In a lecture in 1986 Dr Sackler declared: 'I have found myself moved when I feel closest to the impulse which I believe impelled the artist to communicate feelings, as in terracottas . . .' Other important collectors include the Renaissance bronze collector Robert Smith, whose bronzes have recently been published in *The Robert Smith Collection 1500-1650*, and Margaret and Ian Ross in Canada, who assembled a fine collection of baroque works.

FORMING A COLLECTION

The best private collections are often the most personal ones, reflecting the particular taste and discernment of the collector. This was illustrated by the 1988 Royal Academy exhibition *In pursuit of the Absolute: Art of the Ancient World from the Collection of George Ortiz*. Critics lambasted the collection for lacking archetypal examples from each chosen sphere, and confusing the visitor with the collector's quirky choices; and yet the collection's unique character rests on this very individuality. However, by displaying the collection as an 'exhibition', almost a museum status is conferred – hence the misguided premise of the critics that the collection should have a specific art-historical 'agenda' and a certain responsibility to the viewer, like a public collection in a museum.

The collector should not necessarily rely on buying signed pieces or commonly collected names. Although a signature may impart a feeling of security, this can often be illusory. A 'signed' work may be a workshop piece or, in the case of a bronze, a later cast; an unsigned work of quality may be a masterpiece.

Be cautious, be observant, look at as much comparative material as you can, and then trust your instinct. Buy at auction, and the experts will answer your questions, buy from specialist dealers, antique shops and even market stalls.

Develop your confidence slowly. If you can, find a specialist dealer who will take a friendly interest in your collection and offer breadth of experience and connoisseurship. A piece's provenance, authenticated by relevant documentation, can not only increase its value, but also provide reassurance for the purchaser.

Items that warrant particular attention as potential areas for the collector are unattributed terracotta models, Italo-Flemish bronzes, English 19th-century works, particularly the Pre-Raphaelites, and European 19th-century pieces. Medallions and medals also have much to offer: often small pieces of sculpture in their own right, these works can be obtained relatively inexpensively and are a fascinating area of research.

The most important aspect for the collector is to look for quality, and to observe and 'train the eye'. Visits to museums such as the Victoria and Albert or the Wallace Collection can teach much about bronzes, and Liverpool's Walker Art Gallery and the Musée d'Orsay in Paris are excellent for 19th-century works. At the Musée d'Orsay, although not allowed to touch the pieces, the visitor can examine the surface closely and fully appreciate a sculpture's three-dimensional qualities without the barrier of a square glass case. Intimacy with the surface of a sculpture and the various media, such as terracotta, marble and bronze, on display in these museums can lead to highly informative comparisons of interpretation, technique and finish. The Louvre also has a magnificent collection of sculpture, which is displayed in an accessible way. Several of the bronzes, for example, are placed on glass shelves that allow the visitor to see the interior of the casts.

The profitability of sculpture as an investment tends to depend on the vagaries of fashion and market forces. For example, a medal by Lynn Chadwick which had been issued to members of the British Art Medal Society for £25 in 1983 sold for around £400 at auction at Christie's five years later. But an indication of how fashion and the market can change is provided by the following example: at the height of his career Bernini was paid 6,000 *scudi* (the equivalent of several thousand pounds at 1920s prices) for a single portrait bust. This was an exceptional sum even for Bernini. The same bust, which represented Englishman Thomas Baker (the man who was entrusted with the responsibility of accompanying the Van Dyck painting of three views of Charles I to Italy) was bought at auction by Messrs Durlacher on behalf of the Victoria and Albert Museum in 1921 for only £1,563 10s, less in real terms than the original purchase price.

CONCLUSION

In conclusion, collecting sculpture offers rich rewards, apart from any financial investment. Sculpture in the round has much to offer and will 'come alive' when viewed from different angles. This was beautifully illustrated in the recent exhibition of bronzes, a tribute to Wilhelm von Bode, *Von Allen Seiten schön* (beautiful from all angles) at the National Galerie in Berlin. ❖

HOW TO COLLECT CONTEMPORARY ART

Karsten Schubert, Karsten Schubert Gallery, London

Karsten Schubert, born in Berlin in 1961, set up Karsten Schubert Gallery in 1987, specialising in contemporary art. In 1995 Mr Schubert co-founded Ridinghouse Editions, with Charles Asprey and Thomas Dane, to publish multiples and editions by UK and international artists.

Phone: (+44-171) 631 0031
Fax: (+44-171) 436 9255

HOW TO COLLECT CONTEMPORARY ART

Karsten Schubert, Karsten Schubert Gallery, London

When the editors of *The Credit Suisse Guide to Managing Your Personal Wealth* asked me to write an article on contemporary art as an alternative investment, my first impulse was to decline. I felt it would be misleading to stress the investment potential of contemporary art so soon after the collapse of the speculative art market of the 1980s – in which contemporary art played such a conspicuous part.

While financial gain is of course possible, it should not be the main reason for collecting contemporary art. People are drawn to contemporary art because, like all other aspects of contemporary culture, it is a way of understanding ourselves and our time. It is history in the making: unpredictable, challenging and exciting. Collecting contemporary art is a pleasurable activity and financial considerations should be a secondary concern.

AN INTELLECTUAL CHALLENGE

Contemporary art is probably the most intellectually demanding field of collecting. If you collect stamps or coins, you just need one of each kind; and a complete collection is quite possible. If you collect antiquities, period furniture or old masters you can make choices with the benefit of other people's hindsight. It is a fairly safe bet to assume that painters who have been appreciated for a century or more will retain their appeal, although even then there is no guarantee of sustained popularity – look at the ups and downs in esteem suffered by artists like Rembrandt over the centuries. English 18th-century portrait paintings offer another good example; it is hard to appreciate their popularity amongst American collectors in the 1920s and even harder to believe that they will ever again achieve such status. Works of art from the past may go in and out of fashion but one can be reasonably sure about their status in the future. There are enough collectors around to sustain even the most obscure niche market.

Contemporary art is an altogether different ball-game. There is no hindsight and no precedent. To collect contemporary art means getting involved in a current debate. By buying contemporary works, collectors (if they make the right decision) not only become part of this debate: they may also to some degree influence the outcome.

MAKING THE RIGHT CHOICE

How then do you make the right choice? And anyway, what are the 'right choices'? Unfortunately, if you talk to the people who have made the right choices, people whose collections contain a lot of important and desirable works of art, you will get a lot of mumbo-jumbo about instinct and hunches. 'I just knew it' is an oft-heard and infuriating refrain.

'Instinct' is a misnomer as far as judgement of intellectual matters – like books, films or works of art – is concerned. What people are really talking about is some sort of subconscious knowledge. It is like driving a car. You cannot be an instinctive driver: you have to learn how to drive first. But the knowledge can become second nature, and you can apply it without being consciously aware of it most of the time.

This is good news for would-be collectors, because it means that they can acquire the requisite knowledge. It must be said, though, that just as there are good and bad drivers so there are good and bad collectors. You may know all the right things but be unable to apply your knowledge properly.

Works of art from the past to the present are part of a historical continuum. No work of art is created outside a historical context and this context determines the way we look at any piece of art. If, for example, you took a stroll through the Louvre, you would realise very quickly that all the works of art we admire the most are the ones that push back the boundaries of what was considered the artistic norm at the time. This rule applies to works of art of the past and of the present; they defer to tradition and at the same time subvert it.

Thus, to make a proper judgement you need to know a little about art history. You could not really grasp, for example, Cézanne's achievement without a rough knowledge of what preceded him and of what consequences his work had for the generations of artists who came after him (which included Picasso and Braque).

How then do you become a collector of contemporary art? Here are some suggestions:

(i) To get a rough overview of this generational interplay over the centuries, I suggest you read E.H. Gombrich's *The Story of Art* (Phaidon Press). This volume, first published in 1950, is now in its 16th edition. It will give you a very good outline of art from cave paintings to abstract expressionism and beyond. Gombrich's elegantly written volume has probably introduced more people to the visual arts than any other book. For a more detailed look at events since Impressionism, I suggest H.H. Arnason's *A History of Modern Art* (Abrams, New York and Thames & Hudson, London). It was first published in 1969 and is now in its third edition. It is still the most comprehensive book on the subject, with more than 500 pages and 600 illustrations.

(ii) Acquire knowledge. Try to see as much contemporary art as possible. Visit the main modern art museums of the world and see as many temporary and permanent exhibitions as you can. There is a museum devoted to current art in virtually every Western city. You will see certain things over and over again and patterns will emerge.

(iii) The next best thing to looking at the actual works is to have a comprehensive library of books and catalogues on contemporary art. Read as many as you can. Subscribe to the main contemporary art magazines. I also suggest that you subscribe to the contemporary art catalogues of the main auction houses. They will give you a very good overview of what, as far as the market place is concerned, is 'hot'.

(iv) After a while you should develop personal preferences. You might discover that you prefer photographic or video works, that you favour minimalist works or that you are really 'a painting person'. Follow your instincts and concentrate on the area you are drawn to. There are a lot of things you could collect and it is important that you limit yourself from the outset. Otherwise there is a great danger that you will end up with a mixed bag of ill-matched acquisitions rather than a coherent collection. Your aim is to assemble a group of works that adds up to more than just the sum of its parts.

(v) Visit contemporary art fairs, especially the main events in Chicago, Basle and Cologne. Professionals might moan about how awful these events are but they are a very useful way to get an overview of what is interesting and in vogue. Within a few hours you will gain a comprehensive overview of what many of the main international contemporary art galleries exhibit.

(vi) Get to know artists whose work you admire. Artists are often the best and most insightful judges of what their colleagues are doing.

(vii) Your local art museum probably has a group of 'friends'. For a very modest fee you can join them. In return you will receive invitations to private views and special events (studio visits, talks, etc). It is a good opportunity to meet like-minded people, talk to them and see their collections. These groups often raise funds for their museums to purchase contemporary art. Try to get on to the acquisitions committee. This will give you a very good insight into how museum curators make purchase decisions.

(viii) Visit commercial galleries. Now you know what you like it will be quite easy to identify the galleries that cater to your taste. Make a habit of visiting these galleries whenever you can. Ask to be put on their mailing lists so that you can be kept informed of their activities and forthcoming exhibitions. (Do not be put off by the dragons at reception. They usually turn out to be quite amicable once they have worked out that you are serious and mean business.)

(ix) When you finally feel that you have acquired enough knowledge and you feel secure in your dealings with the professional world it is time to think about your first purchase. To make matters easier and to deal with the overwhelming number of works on offer, I suggest that you set up a few basic rules for your own guidance. Decide what your collection will be about and stick to it, and set yourself a budget (an overall figure or a limit on the number of works you wish to buy over a certain period). If you do not do this, you may become a slave to your own collection. You will want everything and be satisfied with nothing.

(x) Be patient. Instant results are very unlikely. It will take many years before your collection will really take shape and grow together. Great collections are like great claret: the older they get, the more complex, subtle and rewarding they become.

(xi) Do not be too dogmatic: allow yourself the occasional transgression (you might want to try out a few things). Corrections can always be made at a later date.

If you follow these tips, you may end up with a good, maybe even a great collection. It will take years. Nothing is guaranteed. Along the way you will participate in a great and rewarding intellectual adventure. You will be entertained. You will have fun. And maybe in the process, without even thinking about it, you may make a number of good investments. ❖

ORIENTAL COLLECTIBLES

Jacqueline Simcox, Departmental Director, Oriental Department,
Spink & Son Limited, London

Jacqueline Simcox studied European painting, sculpture and architecture at university before joining Spink & Son in the early 1970s. She specialises in Chinese works of art. Ms Simcox writes on Far Eastern art and lectures on Chinese art all over the world.

Phone: (+44-171) 930 7888
Fax: (+44-171) 839 4853

ORIENTAL COLLECTIBLES

Jacqueline Simcox, Departmental Director, Oriental Department, Spink & Son Limited, London

The appreciation of Far Eastern art has a venerable history. The patronage of eastern emperors set the standard for taste and craftsmanship and influenced demand amongst the nobility. Less refined examples circulated amongst the merchant classes, or were made to satisfy an export market. For more than 2,000 years Imperial gifts, especially from China, were sent to vassal states or presented to visiting dignitaries, and these, together with trade goods, made Chinese art and workmanship a byword for excellence. Similarly, Japanese works of art reaching Europe from the 17th century onwards stimulated interest in the country and visitors to Japan brought to the West examples of its fascinating and unusual arts.

It is the finest of these works of art, whether made in China or Japan, that can truly be called 'oriental collectibles'.

CHINA

Great collections were formed in China from early times: there are paintings exhibiting the seals of collectors in unbroken succession from the 13th to the 20th centuries. In the cultured society of the court, the appreciation, study and collection of art were considered a mark of a true scholar and were highly regarded.

After the Cultural Revolution, archaeology was resumed in China, with spectacular results: the awe-inspiring Xian pottery army was unearthed in 1974, the magnificent tombs of Mawangdui in 1971. Increased excavation and access to the countryside have unfortunately resulted in smuggled goods coming on to the market, mostly through Hong Kong, and forgeries have followed in their wake. Rarities previously unknown to the West have been looted from royal tombs: remarkable bronzes inlaid with gold and silver from the Warring States period (475-227 BC), later pottery tomb figurines, animals and vessels of the Han, Wei, Sui and Tang periods (2nd-10th centuries), and even stoneware from the Song dynasty (960-1279). The same is true of sculpture, bronzes, jade and lacquer, whether from burial sites or ancient monasteries. This influx of illegally acquired items has created a minefield for the collector, and has had a catastrophic effect on the price of common examples from the period. Green-glazed Han vessels sold at auction for £22,000 around 1980 would today realise only £1,500; unglazed Tang tomb figures abound and can be bought in Hong Kong's antique shops for HK$8,000 to HK$20,000, or less.

Ceramics

Ceramics can be divided into three main categories:

- Early stoneware, pottery and funerary wares
- Imperial porcelain made for the court
- Export porcelain made for the West

Early stoneware, pottery and funerary wares

It is the rare items with exquisite decoration that are now the most sought-after. Pottery figures and animals made for the tomb are still popular amongst Western collectors who, unlike many in East Asia, are not averse to having grave goods in their homes. Horses with elaborate trappings with amber, green and cream glazes can command prices in excess of £100,000, as can rare figures of court ladies or foreigners. On all these pieces, the details of the moulded and carved features should be exceptional and, whether small or large, the object should have a distinct 'presence'.

Imperial porcelain

These are the porcelains from the Yuan dynasty (1279-1368) and those with Imperial reign marks, made between the early Ming dynasty (1368-1644) and the fall of the Qing dynasty in 1912. This category of ceramics has shown the most consistent rise in demand and so in prices. Outstanding examples can realise more than £1m, and in general the finest pieces range from £100,000 to £500,000. However, only a few pieces of this calibre come on to the market each year and more generally the price range is £20,000 to £80,000.

Early Ming dynasty (1368-1644) blue and white decorated porcelain remains a favourite with collectors, as are the Imperial wares of the 18th century, particularly those of the Yongzheng (1723-1736) and Qianlong (1736-1795) periods. The contrast could not be greater: the Ming pieces with big, bold designs, in rich blues on a sparkling white ground where dragons, phoenix or flowers are bordered by waves or by more flowers; the 18th-century pieces with exquisite polychrome enamels, also featuring the classic designs of dragons, phoenix and flowers but more controlled, in smaller-scale displays of supreme elegance.

Shapes that display well, such as vases and plates, have more general appeal than shallow bowls with white interiors and a narrow band of decoration on the outside. In this category, condition is all-important, as is the colour of the decoration. Ceramics with a good provenance, whether derived from inclusion in public exhibitions, from their illustration in books, or from a well-known collection, command a premium.

Export porcelain

Collections of export porcelain are more naturally to be found in the western countries which had trade links with China.

There is a buoyant market for armorial porcelain bearing great family names or interesting surrounding decoration. The 18th century was the prime period of production for these items, and collectors prefer such pieces, favouring particularly those made for their own domestic markets.

Other popular export wares include porcelain copies of silver or pewter originals, such as candlesticks or tankards; and the more spectacular pairs of figurines, birds or animals also have a wide appeal. A pair of birds realised £50,000 at auction in London in autumn 1995, and attractive pairs in good condition and more than 25cm high, whether of animals, birds or figures, are unlikely to sell for less than £20,000. The rarest forms, and those in fine condition, are usually found in the galleries of specialist dealers in New York, Paris, London, Amsterdam, Stockholm or Lisbon.

Snuff bottles

This is an attractive, specialist area of the Chinese market. Snuff-taking was introduced from Europe to China and became highly popular at court during the reign of Kangxi (1662-1722). The bottles to hold snuff were initially made of glass or painted enamels.

Leading artists were brought to the Imperial Palace workshop in Beijing to decorate the enamel bottles, including a French enameller, Gravereau, in 1717. By the time of the reigns of Yongzheng and Qianlong, bottles were being made in a wide range of shapes and in many materials, including jade and other hardstones, bamboo, lacquer and porcelain. The fashion for interior-painted glass bottles came in the late 19th century, reaching a climax in the reign of Guangxu (1875-1908). Although a small number of snuff bottles are still made today, they are ignored by the major collectors as they are purely decorative and not for actual use.

The market for snuff bottles peaked around 1980 and declined somewhat until 1986, since when values have almost trebled. As in all areas of Chinese art, the market is selective; a jade bottle holds the world-record auction price of £51,000. Most bottles have no reign mark and collectors look for interesting shapes and finely executed, well-conceived designs in this miniature art form. Good examples can be acquired for between £1,500 and £5,000, but the upper end of the market for the serious collector starts at around £10,000.

Jade

White jade is extremely popular, especially in East Asia, and there is a strong and rising market for 18th-century and earlier examples. Prices vary according to the size, the purity of the colour and the quality of the carving. Prices range from £3,000 to £5,000 for small animals and handling pieces, to £50,000 for larger, exceptional pieces in classic shapes.

Textiles

Textiles are a new market but one which already attracts dedicated private collectors. Museums, too, have been adding to their collections of silk textiles as earlier pieces – some dating from the third century BC – have come on to the market in the last 10 years. A successful Hong Kong museum exhibition in the summer of 1995 generated new interest in this area. Prices vary according to the age of the piece and whether it is fragmentary or complete. Pre-1800 pieces are available from specialist dealers in Hong Kong and London, rather than at auction.

Scholars' items

These are the accoutrements of calligraphy and painting: brushes and brush-pots, water pots and brush-washers, wrist-rests, seals and seal boxes, etc. Many different materials were used, including bamboo, jade, various hardstones, ivory and porcelain, to make these highly finished works of art. It is possible to find some examples of such pieces, particularly wood carvings, that have been signed by the artist. These pieces were always popular in China, and the opening up of that country has created a new demand for them in Asia: prices have been rising steadily. Nevertheless, it is still possible to find quite attractive examples at prices less than £5,000.

JAPAN

Ceramics

In the West, the market for Japanese ceramics is almost exclusively in wares originally made for export to Europe from the 17th century onwards:

- Blue and white decorated Arita porcelain

- Predominantly red and blue decorated Imari porcelain
- Porcelain made by the Kakiemon family

The serious collectors' market is confined to ceramics from the 17th and early 18th centuries, including vases, bowls and dishes of all types, and especially the much rarer figurines or animals. The value of all these pieces fell in the late 1980s, and only the rarest pieces should be considered. Of the three types, the Kakiemon pieces are generally the most desirable because of the elegance of their designs and colouring.

The products of late 19th-century ceramic factories, such as Satsuma, have shown a dramatic increase in price and signed examples by the artists Sozan, Yabu Meizan, Kinkoran and others can command prices in excess of £20,000. Unsigned pieces are usually of a lower quality and realise only modest prices.

Netsuke and inro

These are the small toggles (netsuke) attached by a silk cord to a series of small interlocking boxes (inro) which hang from the belt of a kimono in formal Japanese dress. Inro and netsuke were frequently signed by the artist and in both cases the work of a select handful of artists is the most desirable. The history of a piece will also add to its value, particularly if it has come from a private collection of some distinction.

Netsuke are carved from wood, ivory or antler into animal, human or mythical forms and especially those from Japanese mythology. Most desirable at the moment are strong, bold examples of figurative netsuke, £97,000 having recently been achieved at auction for a dancing islander by Ryukei.

Inro are small works of art, about 10cm long, usually decorated with gold and colours portraying scenes of endless variety and imagination. Whatever the subject, the detailing is invariably superb. Lacquer is built up in layers on a wood surface and decorated with sprinkled gold or squares of gold leaf and tints of colours sandwiched between further layers of lacquer. Striking subjects are more desirable than dull landscapes and condition is all-important. Chips to the lacquer, scratches and fading are not acceptable to discerning collectors. Good examples can be bought for £5,000 to £6,000; top-quality examples frequently fetch between £40,000 and £50,000 or more.

Lacquer

Besides inro, many other items were lacquered: chests, writing-boxes, tables and other artefacts. These were often embellished with inlays of mother-of-pearl and given gilt metal mounts. Examples from the 17th century onwards are most common, and it is an interesting area to collect, with wide varieties of pattern and form. Especially attractive are small boxes dating from the 18th and 19th centuries, which many consider to be undervalued.

Bronze and enamels

Japanese bronze artefacts and enamels of the 19th century show outstanding workmanship. Names to look for in cloisonné include Namikawa and Ando, who won medals at exhibitions of Japanese art held in Europe and America in the late 19th century. Values have climbed steadily and fine examples range in price from £15,000 to £50,000, depending on their size. A recent exhibition at the British Museum in London illustrated the variety and quality that dedicated collectors of such objects can achieve.

GUIDING PRINCIPLES FOR A COLLECTOR

Condition

A work of art should be as close to its original state as possible. The following defects should be avoided:

- Cracks and breaks
- Fading
- Missing parts, e.g. a vase without a lid
- Removal of damaged areas, especially on jade or hardstone carvings or glass
- Over-restoration or new parts
- Alteration, e.g. a table cut down to make a coffee table

Quality

Fine workmanship, rarity and balance of design all contribute to quality, and clear, luminous colours are equally important. Avoid muddy colours, poor workmanship and misshapen or warped pieces.

Guarantees

Ensure that each piece you acquire carries a guarantee of age. If there is a scientific test that can be carried out, try to obtain it, e.g. thermoluminescence for pottery, carbon-14 for organic substances, spectrographic analysis for metalwork.

Works of art must date from before 1600 for these tests to be useful and, as the tests are expensive and time-consuming, they are not to be expected with every piece. The exception is early Chinese pottery, which all reputable dealers will sell with a thermoluminescence certificate. The tests should be carried out by a reputable independent laboratory.

Sources of information and expertise

(i) Reviews of auctions and exhibitions appear regularly in magazines such as *Orientations* and *Arts of Asia*, published monthly and bimonthly respectively. These two journals carry scholarly articles, advertisements by leading dealers or book suppliers, and notices of forthcoming exhibitions.

(ii) Dealers all over the world hold exhibitions, publish accompanying catalogues and supply prices.

(iii) Auction houses hold regular sales which are advertised in art magazines and the national press. The public can handle the works of art on the 'viewing' days and details of the prices achieved are available after each sale.

(iv) Museums hold specialist loan exhibitions drawn from other museums and private collections, as well as displaying their own holdings. Good museums will provide a reading list of books within a specialised field.

(v) There are many specialist societies throughout the world. They hold regular meetings, and organise conventions and visits to private collections. They are useful for meeting fellow collectors, curators and dealers, and as general sources of help and advice. Some examples are the Oriental Ceramic Society in Britain, USA or Hong Kong; the Snuffbottle Society; the Textile Society of Hong Kong; and the Netsuke Society. Details can be obtained from museums, leading dealers or large auction houses.

(vi) The major international dealers exhibit

their wares for sale at regular fairs or shows. Some such fairs are:

- **New York, The Armory**
  January: The Winter Antiques Show
  October: The International Fine Art and Antiques Fair
  March (from 1996): a new fair devoted to Asian Art
- **London**
  June: The Grosvenor House Art and Antiques Fair
  June: The Olympia Fine Art Fair
  June: The Ceramics and Glass Fair, Park Lane Hotel
- **Paris**
  September/October: The Biennale, Grand Palais
- **Maastricht, Netherlands**
  March: The European Fine Art and Antiques Fair
- **Singapore**
  May (from 1996): Tresors

Museums with outstanding collections

- **Asia**
  The National Palace Museum (Chinese only), Taipei, Taiwan.
  T.T. Tsui Museum (Chinese only), Hong Kong.
  There are important collections in many Japanese cities including the Tokyo National Museum (Idemitsu Collection) and the Kurita Museum, Tochigi Prefecture.
  The Palace Museum, Beijing, China.
  Shanghai Museum, Shanghai, China.
  The National Museum of Korea, Seoul.
- **Europe**
  Most European capital cities have fine collections in their national museums, including the Topkapi Museum in Istanbul.
  Other outstanding collections can be found at the Sir Percival David Foundation of Chinese Art in London, the Musée Guimet in Paris and the Baur Collection in Geneva.
- **North America**
  The Royal Ontario Museum, Toronto
  The Sackler Collection, Washington DC.
  Other cities with outstanding collections include New York, Boston, Cleveland, Chicago, Kansas City, Seattle, San Francisco and Los Angeles.

CONCLUSION

There is a wide range of oriental artefacts that are worthy of collection – far more than it is possible to mention within the confines of this article. The range includes some very expensive categories in which only one or two pieces come on to the market each year – such as Chinese paintings from the 10th and 11th centuries or early Japanese Buddhist wooden sculptures – all of which are snapped up by museums or one or two 'mega-collectors'.

This article has concentrated on the more accessible areas of the market that are particularly active at the moment. Whichever part of the market you choose to concentrate on, the key to success is to seek specialist advice at the outset and to consider both the aesthetic and financial merits of any potential purchase. ❖

JUDAICA: COLLECTING JEWISH ART AND HISTORY

David Breuer-Weil, Fine Art Expert, Sotheby's London

David Breuer-Weil is a Sotheby's expert in Impressionist and modern art and Judaica, a general antiques expert and an artist; he exhibits his work regularly in London and Tel Aviv. He studied at Central St Martin's School of Art and graduated from Clare College, Cambridge in 1989, since when he has worked at Sotheby's. Sotheby's Tel Aviv holds twice-yearly sales of Judaica and Jewish art in April and September or October. For information, confidential advice and catalogues, contact Sotheby's Tel Aviv on *(+972-3) 522 3822* or Sotheby's London on *(+44-171) 408 5334*.

JUDAICA: COLLECTING JEWISH ART AND HISTORY

David Breuer-Weil, Fine Art Expert, Sotheby's London

The term 'Judaica' encompasses everything that was made specifically for Jewish ritual use. It has also been used as a generic term for any work of art made by a Jew or for any object or document relevant to Jewish history.

Judaica embraces a number of distinct disciplines, including paintings, silver, manuscripts, printed books, graphics, ceramics, glassware and furniture. In this respect it is similar to Americana. As an investment field it therefore has the advantage of offering the collector a well-balanced portfolio. In the last few years many people forming Judaica collections across the different disciplines have been surprised by the security of their investment during unpredictable times.

There are a number of reasons that people choose to collect Judaica. For Jews, perhaps the main reason is their desire to maintain some tangible link with their heritage. Judaica appeals to the religious and secular alike, but for different reasons. The religious delight in being surrounded by items that can be used today for ritual purposes, while also providing a constant reminder of their religious ethos. For the secular, Judaica often provides their only link with their ancestors and with a way of life that is often thought of with great nostalgia and longing.

One incentive for collecting Judaica is its rarity. Demand for important pieces far outstrips supply and experience shows that fine pieces, particularly those from private sources, almost always perform well at auction, in many cases far exceeding pre-sale estimates – always a sign of a healthy market. The dependable nature of demand for Judaica has been reinforced by the recent growth of Jewish museums in Germany and throughout Europe, and in the United States. A number of these museums are specifically intended, either overtly or in an understated manner, as Holocaust memorials. This tendency is likely to increase over the next two decades in Eastern Europe, where a number of the great cities that boasted vital Jewish communities before the Holocaust have inadequate memorials or less than completely satisfactory Jewish museums. Museums, furthermore, frequently have a longer, if perhaps slightly more cautious, collecting life than many private individuals.

People wishing to start collecting Judaica should begin by viewing the outstanding collections of material in the Jewish museums in London, New York and Amsterdam. The Musée de Cluny in Paris holds one of the most significant collections extant, but it is rarely on view. One of the first complete Judaica collections, it was assembled by Isaac Strauss and was generously presented to France by the Rothschilds. The Israel Museum in Jerusalem has a collection with outstanding examples from the main classes of ritual objects. The Haaretz Museum in Tel Aviv houses a well-presented group of items, many concerned with Jewish folklore.

JUDAICA PAINTING

Judaica painting, which has seen an extraordinary investment growth over the last six years, is worth examining in some detail. What is a Judaica painting? In a sense a collector sets his or her own definitions. A number of collectors who do not regard themselves as specialists in Impressionism or modern art nevertheless assemble works by Camille Pissarro, Amedeo Modigliani, Jules Pascin, Chaim Soutine, Max Lieberman, Lesser-Ury, Marc Chagall and Louis Marcoussis, to name but a few, because the artists were Jewish – not for their great innovations. Since these artists are respected figures in the international art market, their art is not usually referred to as Judaica, although in the terms of our definition it certainly qualifies.

Modern art

Some critics have noted that artists of the New York school, such as Mark Rothko, have captured in their brooding abstract paintings the essence of Jewish monotheism and its Biblical prohibition of the graven image.

Modern Israeli art has mostly been abstract, and much influenced by the vision of the abstract expressionists. Israeli art has proved an outstanding investment over the last few years. Mordecai Ardon's works now often sell for between US$100,000 and US$200,000 whereas only a few years ago an important abstract oil by him could be secured for as little as US$30,000. Other Israeli artists such as Reuven Rubin, Moshe Castel, Yaacov Zaritsky, Yehezkel Streichmann and Avigdor Arikha are collected by both Judaica and contemporary-art collectors. Israel continues to inspire a number of artists and the wealth of output in this area should be seriously explored by those wishing to make discoveries for the future.

Biblical paintings

Biblical paintings have not traditionally been considered as Judaica for the obvious reason that the Old Testament, although the ultimate statement of Judaism, became, with Christianity, a possession of world culture. Michelangelo's Sistine Chapel cannot in any sense be considered Judaica – or can it?

Topographical art

Views of the Holy Land by artists of both Jewish and Christian descent have now come to the forefront of Judaica collecting. A case in point is *View of Jerusalem* by the little-known British artist S. Lawson Booth, which sold for US$18,000 in Sotheby's sale of paintings in Tel Aviv in October 1995. A canvas of similar size and quality by the same artist depicting a Scottish view struggled recently in a British saleroom to reach £1,100. If views of the Holy Land – by no means the sole preserve of Judaica collectors in the past – have become desirable, it is at least possible that paintings depicting Moses, the Children of Israel and Queen Esther, for example, by 16th- to 19th-century European hands could climb considerably in value.

MANUSCRIPTS AND HEBRAICA

The Jews entered the field of visual arts at a fairly late stage. Illuminated manuscripts are perhaps the only evidence of Jewish artistic activity during the Middle Ages and there are probably no Jewish panel paintings whatsoever from this period. Hebrew manuscripts have for a number of centuries formed part of major collections of Christian manuscripts and have been highly prized by Jews and non-Jews alike. The British Rail Pension Fund collection included a group of Hebrew manuscripts, which were sold by Sotheby's in the mid-1980s having proved an excellent investment.

Perhaps the most important modern collection of Hebrew manuscripts was that formed by the late David Solomon Sassoon. The sale of part of the collection at Sotheby's in New York in 1984 included such treasures as the Shem Tov Ben Abraham Gaon Bible from Soria in Spain dating from 1312, which sold for US$750,000, and the Great Legal Code of Maimonides, illuminated in Germany in 1356, which fetched US$410,000. Hebrew manuscripts continue to be seen as the princely items of the field and Sotheby's auction rooms in Tel Aviv, now virtually the only international auction house to hold regular (twice-yearly) sales devoted exclusively to Judaica, has continually sold manuscripts with prices ranging from US$20,000 to almost US$1m. Hebrew incunables (early printed books) have also been highly successful at auction over the last few years. An auction record was achieved at Sotheby's New York on 1 November 1995 when the early-16th-century Prague Hagaddah became the most expensive Hebrew printed book, selling for more than US$250,000.

JEWISH ARTISTS PAINTING JEWISH SUBJECTS

The first professional easel painters of Jewish origin to achieve distinction reached maturity between 1840 and 1860. There are now a steadily increasing number of collectors who focus on the history of Jewish painting, and individuals who had the foresight to invest in this branch of Judaica as little as six years ago have enjoyed an investment triumph. This market has witnessed greater stability than, say, Scandinavian art for the simple reason that Jewish collectors live all over the world. If the economy of one country suffers, collectors in other countries may not be affected.

First generation

The most significant painters of the 'first generation' of Judaica painters were members of important aesthetic movements in their countries of birth. Moritz Daniel Oppenheim (1799-1882) was a member of the Nazarenes in Germany. Part of his *œuvre* depicts scenes from Jewish daily life in Germany. These works are exceedingly rare and no major piece with a Jewish subject has appeared at auction within the last decade. If one were to appear for sale it could achieve as much as US$200,000 to US$400,000, despite the fact that Oppenheim is by no means the most important Nazarene. Oppenheim's secular works, although of far less value than his Judaica pieces, are considerably prized by Judaica collectors. They are worth many times as much as similar pieces by his contemporaries with no Jewish following. Maurycy Gottlieb (Polish, 1856-79), the subject of a recent exhibition in the Tel Aviv Museum of Art, is a legendary figure to scholars of both Jewish painting and Polish art. He was a pre-eminent pupil of one of the leading Polish history painters and his death at the age of 23 robbed both nations of one of their most promising talents. His greatest work, and possibly the most important picture in Judaica painting as a whole, is the celebrated *Yom Kippur* now in the Tel Aviv Museum of Art.

In England, Simeon Solomon, a member of the Pre-Raphaelites, was at 18 the youngest artist to have exhibited at the Royal Academy. Although he was extremely well known to collectors of Victorian art, and despite the fact that he produced a larger group of Judaica works than Oppenheim, including many of great quality and strength, he has not yet been adopted by the Judaica market. His near contemporary, Solomon Alexander Hart, likewise produced a number of memorable works. Important Judaica canvases by this English master of the first generation of Judaica painting can now be seen in the Jewish museums of London and New York.

In France, Edouard Moyse, Edouard Brandon and Alphonse Levy were the masters of the first generation. Although he is now almost entirely forgotten outside a small circle of Jewish collectors, Moyse's monumental depiction of a circumcision ceremony, a subject that many salerooms might rightly or wrongly deem off-putting to potential bidders, more than doubled its already generous saleroom estimate, reaching US$81,000 in Sotheby's Tel Aviv sale in October 1995. It is certainly a reflection on the strength of this highly specialised market that only 10 years ago such a work would have sold for barely US$5,000. Similarly, a year ago a painting of a Jewish school interior by Brandon brought nearly US$60,000; it had been bought at public sale five years earlier for one-sixth of the price. These figures are especially impressive in view of the change in the market for Impressionist works and modern art during the last six years.

Second generation

The most prominent artist of the 'second generation' of Judaica painting is Isidor Kaufmann. He painted both Judaica and secular pictures. Both genres are bought by Judaica collectors, but Judaica panels sell for at least three times the price of the secular pictures. Major works by the artist have for the last few years been selling for between US$40,000 and US$300,000. Kaufmann's works have always been highly prized and have held their value. One of his first important patrons was Emperor Franz Josef and there is still some limited interest in Kaufmann from non-Jewish collectors.

In general, however, it is probably true that most Judaica collectors (with the exception of those who collect manuscripts) are Jews. But as in other fields, such as Chinese or Korean art, it is highly likely that the collecting base will expand to encompass collectors of diverse backgrounds in coming years, thus strengthening the market still further.

This century a number of artists have followed the example of the first generation and of Kaufmann. Indeed, Kaufmann spawned a whole school of followers, who were active in Austria, Germany and Hungary before World War II. Major British artists, such as Mark Gertler, David Bomberg, Alfred Wolmark and Jacob Kramer produced works with Jewish subject matter and the Ben Uri Gallery in London has a fine collection of paintings by many of these artists. Modern British Jewish masters are generally moderately priced and afford excellent investment potential.

In America, the cubist Max Weber produced a number of wonderful religious works, some of which can be seen in the Jewish museum in New York. The school of Paris boasts a number of painters who fall into our category: the best-known are, of course, Chagall and Mane-Katz.

JUDAICA SILVER

An enormous quantity of Judaica silver was lost or destroyed during World War II. In the 1960s there were few experts capable of distinguishing between the more common pieces and those with unique virtues, so that common pieces were often sold for prices far in excess of what they might be expected to reach today. This was largely because some collectors assumed that almost nothing had survived the tragedy of the Holocaust. At that time Judaica silver was often bought as a poignant reminder of the wealth of Jewish culture destroyed by the Nazis. We now know that the more commonplace 19th-century Polish and Austro-Hungarian items of ceremonial silver were preserved in American, English and Israeli collections in quite large numbers, and it is possible to assemble a collection of the main ceremonial items quite inexpensively.

Individual 19th-century spice towers still sell for less than US$400, whilst silver Hanukkah lamps regularly fetch between US$2,000 and US$3,000. Good Kiddush cups from celebrated centres such as Augsburg and Nuremberg are available at auction for prices ranging from US$4,000 to US$20,000. Torah ornaments are generally popular, but Torah shields in particular seem to be undervalued, with rare pieces from Poland and the Ukraine still selling for between US$1,000 and US$5,000. Many of these pieces are bought by people wishing to use them for the festivals but, as the 17th- and 18th-century pieces become scarcer, the later pieces will inevitably become more desirable to collectors.

It should also be noted that almost no pieces of Judaica silver from earlier than the 17th century have survived, although there is little doubt that many pieces of ceremonial art must have been produced. In general silver terms, an 18th-century piece is fairly late; for Judaica silver it is quite early. In October 1995 in Tel Aviv, a silver crown for a Torah made in Venice around 1750 sold for almost US$110,000. Similarly, a pair of Torah finials made by Peter von Lohe in Hamburg around 1650 astonished a crowded room in November 1993 at Sotheby's general silver sale in Geneva, selling at more than 10 times their pre-sale estimate and establishing a world record for Judaica silver at almost US$400,000. Before that, many felt that there was a psychological price barrier beyond which no Judaica silver buyer would proceed; this sale proved that hypothesis to be a myth.

TRAGEDY AND HOPE

The 20th century has been one of the most momentous in Jewish history, in terms of both tragedy and hope. The years of the Holocaust (1933-45) were the most devastating in a long and often tortured passage through time, while the foundation of the state of Israel only three years later was the triumph of belief and long-cherished hope over practical near-impossibilities. The recent interest in works of art produced by survivors and victims of the Holocaust, and in Israeli art and artefacts of the early days of the state, heralds what might well turn out to be a most rewarding period for collectors of Judaica. ❖

RUSSIAN ICONS

Peter L. Schaffer, Vice President, A La Vieille Russie Inc., New York

Peter L. Schaffer, a graduate of the University of Columbia, is a former member and president of the board of directors of the National Antique and Art Dealers Association of America. He has lectured and taught at the Smithsonian Institution, the Cooper-Hewitt Museum, and the Metropolitan Museum of Art, amongst other notable institutions. He is an authority on the art of the goldsmith and jeweller, specialising in Fabergé, antique jewellery and the Russian arts.

Phone: (+1-212) 752 1727
Fax: (+1-212) 223 6454

RUSSIAN ICONS

Peter L. Schaffer, Vice President, A La Vieille Russie Inc., New York

As our knowledge of iconography increases, Russian icons have ceased to be looked upon purely as an expression of the spiritual devotion of the Russian Orthodox Church, and have come to be recognised as the brilliant works of art that they are.

THE BYZANTINE ICON

The word 'icon' is derived from the Greek *'eikon'* which means a venerated image. It is thought that icons developed from the funeral portraits found on coffins in many Egyptian and Syrian tombs. In Hellenistic Greek culture, people worshipped portraits of their rulers, a practice taken up by the Romans, who deified their emperors. The images also served a judicial purpose: in remote areas they stood for the emperor's person.

Very little is known about the appearance of icons in the first centuries of the Christian era, but it is believed that they began to be produced in the time of Christ himself. The disciple and evangelist St Luke is said to have painted portraits of the Virgin Mary, and a 4th-century Greek historian, Eusebius of Cheserea, refers to portraits of Christ and Saints Peter and Paul. We do know of paintings on the walls of the catacombs, alongside symbolic representations of anchor, fish or lamb.

In 330 AD the Byzantine Emperor Constantine rebuilt the city of Byzantium to establish his new capital, Constantinople (modern-day Istanbul). Before the battle at the Milvian Bridge near Rome in 312, he is said to have seen a vision of the image of Christ with a flaming cross and the words: 'In this sign, conquer.' He adopted the sign for his banner and was victorious. This brought about his conversion to Christianity.

In both Rome and Constantinople, artists worked for the new faith. Monasticism flourished and images of Christ, the Blessed Mother, the Apostles and prophets, as well as Church festivals and events from the Bible, were painted for veneration. (The veneration was not of the icon as a physical object, but of what it represented. St Basil the Great said: 'The veneration given to the effigy is transferred to the original image.')

The icons' imagery was drawn from daily life and was originally Hellenistic in its realism. Soon, however, all references to nature were eliminated, producing abstract depictions of 'the other world'.

Rules for the portrayal of the human image were set by the Trullian Council, 691-692 AD. Both imagery and style were strictly controlled by the Church. The early icon painters, monks who were believed to be inspired by visions, were obliged to follow a certain format and to adhere to traditional iconography. Free expression or interpretation was forbidden. This control by the Orthodox Church imposed a uniformity that is still followed today.

The unnatural, stylised appearance that was required was achieved by a 'flat' treatment of the figures, distorted proportions, abstract lines and the effect of light on colour. This is known as inverse perspective. It was not until later, in about the 17th century, that the influence of Western art and the technique of linear perspective began to be seen in icons.

The teaching of the Orthodox Church defines the icon as an expression of the basic doctrine of Christianity: God's physical incarnation. Icons depict the life of Christ, as described in the Gospels. Each saint is an image of Christ, a human being in whom 'Christ is found', the source of holiness. This relationship with Christ gives man his harmony: peace, silence, and calm. The whole world – the earth, the sky, the animal and vegetable kingdoms – is shown in its relationship to man. This world is one of light, where there is no shadow. Everything is penetrated by light – and the background of the icon is called 'light', whatever its colour.

ICONS IN RUSSIA

When Grand Duke Vladimir of Kiev converted to Christianity in 988 AD, he accepted the Greek-Byzantine rite. He ordered all Russians to be baptised and brought Greek priests and monks to Russia to help in the teaching of his new religion. Although icons had by this time undergone several changes, they continued to be produced in line with the traditions of Constantinople. As the spiritual life and art of the Russian people developed, however, local character gave shape to the many schools of icon painting that grew up in the different principalities of Russia. (With the expansion of the Russian Empire under Ivan IV in the 16th century, the local schools gradually lost their individual originality and identity.)

Kiev

The icon schools of 10th-century Kiev developed a simpler and less perfect technique than Constantinople's. In 1051 the priest Iliarian established the first Russian monastery, Kieva Percherskaya Lavra. The icon school there was made up of monk-artists, the first-known truly Russian iconographers.

In the century before the Mongol invasion of Russia of 1240, several schools of icon painting grew out of Kiev, in the towns and villages of Yaroslavl, Pskov, Novgorod, Tver and Vladimir-Suzdal. They became known as the Northern School, developing different techniques and styles from those of Constantinople. They used brighter colours and more decorative, rhythmic patterns with simplified shapes. In Tver pale, delicate colours were worked, with an emphasis on various shades of blue.

With the Mongol invasion and the advance of the Tentonic Knights, much of Kiev's greatness was destroyed. The few icons that remain depict one or two dark, earth-coloured figures with features of an oriental cast and large, heavily outlined eyes with severe and sorrowful expressions, against a background of gold or silver. This shows that the Byzantine influence lived on.

Novgorod

The city-state of Novgorod, having escaped the Mongol invasion, became an important commercial and religious centre, with an icon school which developed its own style. The figures were bold and strong with various highlights, and faces a natural oval, less stern in expression. Bodies were stockier and less elegant with a flatter look, owing to the lack of modelling, and their outlines were more sharply defined. Novgorod colours were pure and unblended, with red, green and yellow predominating.

Theophanes the Greek was working in Novgorod by the mid-1370s and in Moscow by the 1390s. His style was large and monumental, with elongated lines and simple colour schemes. He used sure, free brush strokes with white-lead highlights. The Novgorod school reached the height of its brilliance by the 14th and 15th centuries, with the disappearance of much of the Byzantine style. Andrei Rublev, a Russian monk-iconographer and student of Theophanes, combined the iconography of Constantinople with a youthful, fresh approach to composition, harmony of colours and rhythm of line. His work dominated the 15th century and established a 'Russian' style.

Moscow

Moscow's importance as a political and cultural centre grew following the fall of Constantinople to the Turks in 1453. As it developed, its sphere of influence absorbed Yaroslavl in 1465 and Novgorod in 1478. In 1485 peace was declared with Tver. Moscow remained the focus of the Russian court and for the next century many native artists were attracted to the tsars' workshops. Dionysius (1440-1508?) was a great icon painter of the second half of the 15th century. He established his own workshop, and developed a style of fine proportions (elongated figures with smallish heads), grace of movement and a joyful use of colour.

In the second half of the 16th century, the doctrinal meaning of the icon began to lose its importance, with narrative and moralistic content taking its place. Allegorical subjects, where the interest passes from the person to the idea, became common. The majestic simplicity disappeared and there was a tendency to complicate the composition and overload it with detail. Theological and spiritual content gave way to intellectualism and technique. The hues became darker and duller, and icons started to be adorned with halos and frames of silver and gold. A little later, a metal cover known as a 'riza' was worked into the design, leaving the flesh parts (face, hands and feet) uncovered so that the painted wooden section would show through. At times, the riza was encrusted with precious stones or enamel work.

The Stroganov School

Around the end of the 16th century the Stroganov School was founded in northeast Russia under the influence of the Stroganovs, a wealthy, mercantile family who emigrated from Novgorod to Soloychegodsk. They owned vast estates in the Urals and defended the eastern frontier for the Tsar. Being great art patrons, they established an icon school in the town of Perm. Here, the Stroganov masters painted icons in the style of 15th-century Novgorod: long bodies with small heads and a mannered grace of gesture that had an Asian quality. They were embellished with gold and a rich use of colour, giving the impression of precious jewels on a miniature scale. Procopieur Tchirine worked in this school from 1631 to 1642. Most of these icon painters were secular artists of the commercial world, for we now see some of the works signed.

Western influence

In the reign of Tsar Michael Romanov, between 1615 and 1645, the Russian court employed a great many Polish and German artists. With so many different painters working in the Tsar's workshops, icons began to acquire a Western look and a mixture of styles became common. One of the more famous icon painters of this period was Simon Ushakov (1620-1686).

This was a period of great change. In 1652 Patriarch Nikon insisted on reforms in the Church, which caused a schism that was not resolved until 1658. In addition, he wanted icon painters to return to the Byzantine tradition and to free themselves from the influences of Western art. Those who broke away from the Church were called 'old believers'. They wandered throughout Russia, some settling in small colonies and villages such as Palakh, Mitera, Nikolokoi and Kholuy.

After the Nikon Reforms, the Western influence, with its naturalism and perspective, continued alongside the monastic and Orthodox schools. A high spiritual and artistic level was maintained, but the link with tradition had been broken. The 18th and 19th centuries saw the rise of secular, realistic art, and metal icons of brass and tin began to be mass-produced.

COLLECTING ICONS

Icons are collected by people with both secular and religious interests. Since the collapse of the Soviet Union and the birth of free-market capitalism in Russia, many newly wealthy Russians have swelled the ranks of collectors. Buyers also include many people with an eye for great art, who do not have deep enough pockets to satisfy a desire for, say, the Italian primitives (Giotto, etc.): icons, while in the same league artistically, are much less expensive. Prices for late-18th and 19th-century pieces range from US$700 to US$4,000, making them quite affordable gifts. The earlier icons, those from the 15th and 16th centuries, can reach as much as US$300,000 and the purchase of these rare items is confined to the serious collector.

What to look for

Icons from all periods and in all price ranges and degrees of quality are available to buy. In general, the earlier and therefore rarer pieces are the most valuable, although icons decorated with gold, silver, enamel and jewels, most of which date from the 19th century, can be very expensive indeed. The quality of icons is particularly hard to define, partly because the same subjects tend to be repeated over and over again. The most common are the Madonna, Christ and St Nicholas. Only by studying icons for many years will you be able eventually to determine their quality.

Naturally, the condition of a piece also has a bearing on its value but, because of the unstable nature of the media, it is virtually impossible to find an icon that has not been restored. Painted in tempera and gesso on green wood, even the greatest of icons have had some work done on them over the years. But it is the degree of restoration that matters. That said, condition is less crucial to the value of an important icon than of a poor one, which must be as near perfect as possible to have any value at all. If you buy an icon, you should bear this in mind and keep it in a controlled environment, away from draughts, at a constant temperature and in normal humidity.

The final factor affecting value is, as always, provenance – although little is known about most of the early collectors. Still, an icon from amongst the possessions of a collector whose name is remembered today will certainly fetch a higher price.

Reputable dealers will be able to guarantee the authenticity of an icon. They should make a careful study of style, paint, technique and subject matter. A knowledgeable dealer will be familiar with books on the subject, particularly early ones, and keep up to date with current scholarship.

Where to look

The earliest icons, dating from the 6th to 8th centuries can be found in St Catherine's

Monastery on Mount Sinai in Egypt, and depict four scenes from the life of Christ: *The Nativity*, *Christ Meeting the Two Marys*, *The Crucifixion* and *The Ascension*. By the end of the 19th century, field research on Orthodox art by foreign schools working in Greece and the Ottoman lands had resulted in the establishment of valuable archives such as the Russian Archaeological Institute in Constantinople and the British School of Archaeology at Athens, where these collections are still held. ❖

FROM STRADIVARI TO JIMI HENDRIX'S GUITAR

Graham Wells, Director, Sotheby's London

Graham Wells was educated at the King's School, Canterbury, where his interest in organology (the study of musical instruments) was first kindled, and he later studied the oboe at the Guildhall School of Music and Drama. He joined Sotheby's in 1969 and was appointed a director in 1977.

Phone: (+44-171) 493 8080
Fax: (+44-171) 408 5942

FROM STRADIVARI TO JIMI HENDRIX'S GUITAR

Graham Wells, Director, Sotheby's London

There can be few occupations whose principal tool or essential equipment can, if chosen wisely, provide you with a pension on retirement. But that is one of the many advantages of being a professional violinist – or viola player, cellist or bassist, come to that. Of course, there is nothing to stop anyone benefiting from the ownership of a fine musical instrument: you do not even have to play it, although that could be seen as wasted opportunity. In practice it is most unusual for individuals to invest in a musical instrument unless they play it to some sort of standard. It is becoming far more common, however, for companies or financial institutions to purchase expensive instruments for loan to highly promising young players. In doing this, these buyers inevitably have an eye on the advertising and public relations potential of making such a loan.

Instruments of the violin family by eminent makers, particularly those by Antonio Stradivari, have long been considered the pinnacle of man's craftsmanship and have been valued accordingly. However, not all musical instruments gain value with the passage of time: a wind instrument, for example, will generally lose value from the time it is bought. This situation may be changing; indeed, in the last 20 to 30 years the value of historic wind instruments has risen dramatically.

Each category of instrument should be examined separately for its investment value, and we shall now look in some detail at instruments of the violin family, bows, woodwind, brasswind, keyboard and stringed instruments.

THE VIOLIN FAMILY

The popularity of the violin since its appearance in the 16th century is demonstrated by the sheer numbers made. Violins range in value far more widely than any other instrument: it is possible to buy a modest student instrument for less than £100, while a top Stradivari could set you back several million pounds. Nearly all the top violin makers have been Italian, and it was the demand for these fine instruments, together with the later makers' desire to emulate their tone, that quickly led to the production of copies. It is but a short step from a copy to a fake, and it is as well to know from the start that this is a field rife with forgery and has been since the 17th century.

Authenticity

The problem of authenticity is aggravated by a lack of independent academic expertise. High profit margins quickly woo anyone with the necessary knowledge and a good eye into becoming a dealer – with a consequent loss of objectivity. Despite this, certificates of authenticity issued by prominent dealers carry great weight. Traditionally these dealers were long-established firms, often family businesses going back several generations, such as W. E. Hill & Sons, Hart & Sons and J. & A. Beare in London, Wurlitzers and Hermann in New York, Hamma & Co. in Stuttgart and Möller & Zohn in Amsterdam. Of the firms quoted, only Beare's still exists, but many instruments still rely on certificates from now-defunct companies to support their authenticity. There is a tendency to question the opinions of authorities who are no longer around to defend their views and, while one must applaud the advancement of scholarship, one must be wary of those who denigrate past scholarship merely in the hope of enhancing their own prestige.

It is best to buy a violin from a reliable source, such as a reputable dealer or auction house. The main advantage of buying from dealers is that they usually allow a potential buyer to borrow an instrument for a short time in order to test its tone. Some people take advantage of this opportunity to get a second opinion on the instrument's authenticity. The disadvantage is that dealers' prices are usually much higher than auction prices. Whichever course you take, if you do not have at least some knowledge of the field it is best to consult a friend or adviser who does.

Condition

Having cleared the authenticity hurdle, the next problem is that of condition. Violin restorers go to great lengths to ensure that any work they carry out is as near invisible as possible. A fresh crack in the straight-grained pine of the table, if it is glued immediately, can be very hard to detect. It is more difficult to hide a crack in the maple of the back, although techniques to disguise these are becoming more sophisticated all the time. If there is a crack in the area of the sound-post, the instrument can lose up to half its value.

Some readers will probably be puzzled that so little mention has been made of tone. This is because it is a highly subjective area. Considerable differences between playing styles mean that one player may extol the tonal qualities of an instrument that another may find very ordinary. Furthermore, a high-value violin with beautiful tone will never make a great player out of a mediocre one, but a virtuoso can often make a mediocre instrument sound quite presentable.

Provenance

Provenance is another indefinable influence on value. It can be a psychological boost to know you are playing a violin owned and played by a great virtuoso. Owning and playing such instruments can also help professional musicians to market themselves to concert promoters and record companies. It should be borne in mind, however, that some of the very greatest players, such as Joachim and Kreisler, owned numerous instruments during their careers. It is important to know how long they kept the instrument in question and whether, like Joachim, they had a reputation for being particularly hard on their violins.

History

The violin first appeared in Italy in the 16th century and instruments from that country are still the most sought-after. The high prices of 17th- and 18th-century violins, as well as the fact that many are beginning to show their age, have led to an upsurge of interest in 19th- and 20th-century instruments with, in many cases, a consequent substantial increase in value. Good examples of this trend are the instruments of two 19th-century Turin makers, Pressenda and Rocca. In the early 1980s their average price stood at around £32,000; it has now risen to around £100,000. A 20th-century example is Fagnola, also from Turin, whose violins have risen in value from an average of £5,000 to around £20,000 over the same period.

After the Italian instruments, the most

desirable are those made in France in the 19th century, particularly those by J.B. Vuillaume or Lupot. The influential and successful maker Vuillaume was probably the first maker to deliberately 'age' his new instruments by rubbing down the varnish to imitate the patterns of natural wear. The practice was adopted by later makers, in particular commercial manufacturers, making it even more difficult for the uninformed to distinguish between a genuine instrument and a copy.

In the 19th century Paris was the centre of the violin trade, probably because of the presence of Vuillaume. In the 20th century London took up this mantle. For at least half the century this was due to the influence of W.E. Hill & Sons, the highly respected violin dealers, who are sadly no longer trading. In the second half of this century, the presence of three major auction houses, Christie's, Sotheby's and Phillip's, recently joined by a fourth, Bonhams, has undoubtedly contributed to keeping the market centre in London.

BOWS

Many people understandably do not realise that bows have a value in their own right. The auction record for a violin bow is £72,000, for a bow by the leading maker, Francis Tourte. Although the most sought-after violins are Italian-made, when it comes to bows it is those made in France that are the most coveted. The bow as it is known today developed only in the late 18th century and, although earlier examples have a certain value to a few specialist collectors, the market focuses on useable bows by 19th- and 20th-century makers.

Most quality bows are stamped with their maker's name. Unfortunately this does not stop the production of forgeries but they are at least much less numerous than counterfeit violins. The exception to the rule is Tourte; he never stamped his bows and his name was used indiscriminately by German commercial makers.

Bows are quite inexpensive, and a good-quality French bow by a respected maker such as F.N. Voirin costs between £3,000 and £5,000. They are easy to store, need little maintenance and are therefore popular items to collect, particularly with professional players.

It used to be the practice for makers to mount bows which turned out particularly well with exotic materials such as ivory or tortoiseshell. International legislation on the trade in endangered species has reduced the value of these bows, contrary to what might have been expected.

WOODWIND INSTRUMENTS

Interest in collecting historic woodwind instruments is a relatively new phenomenon. Until some 30 years ago it was quite usual to pay just a few pounds for most items. Early specialist collectors, mostly academics, have had a windfall. Prices began to rise in the late 1960s as the practice of performing early music on original instruments developed, and as Japanese interest in Western instruments was awakened. Buying was very unselective at first, certainly insofar as condition was concerned. Over the last 10 years, however, the market has changed in several ways. The most obvious is that very few professional players are buying original woodwind instruments these days, preferring to play on accurate copies. And a new breed of collectors has appeared who accept that fine examples are inevitably going to be expensive. Museums are now prepared to raise the necessary funds if the right piece appears on the market.

The important thing is that the instrument must be as near perfect as possible, as any cracks will have a substantial effect on value.

BRASSWIND

Brasswind from the 18th century or earlier is very rare and thus expensive. Until the invention of the valve, around 1825, most brass instruments had musical limitations or were at best very difficult to play. Most of the comments made about the woodwind market also apply to brasswind, and in the early days these instruments appealed to an even smaller group of collectors. However, there is now a fast-growing interest in 19th century brasswind. This was a time of great innovation as makers vied with each other to produce instruments that would eliminate the various acoustical problems that beset brasswind instruments. This gives great scope to collectors who enjoy seeking out rare and unusual designs, even if these instruments did come to dead ends in terms of development. Given the great strength of the brass and military band movements in both Europe and America, many unrecognised treasures are, undoubtedly, still to emerge on the market.

KEYBOARD INSTRUMENTS

For anyone thinking of investing in keyboard instruments, the limiting factor is undoubtedly space. This does not seem to have deterred some of the more determined collectors, who will happily fill almost every room in the house with instruments. This field can certainly be most rewarding, both financially and aesthetically. In the latter case this is the result of such instruments being built to fit in with contemporary furniture and, at least until the second half of this century, the casework of the instrument can have a considerable bearing on the value.

The earliest keyboard instruments, apart from the organ, were the harpsichord, its domestic versions the spinet and virginals, and the much rarer clavichord. All of these had become obsolete toward the end of the 18th century, when they were replaced by the pianoforte. The early instruments were much collected even in the 19th century, but pianos are a new field of interest, which has developed within the last 30 years – with a consequent rapid rise in prices.

There are a number of caveats to consider with any keyboard instrument. Harpsichords have attracted the attention of forgers, particularly in the second half of the 19th century. A favourite trick was to use an authentic instrument to make two or three instruments, each incorporating some of the genuine parts. At a time when scholarship in this field was in its infancy, this device was all too often successful, and it has left a troublesome legacy to the collector. This is compounded by the fact that it was usual practice in the 18th century to update the now very desirable 17th-century Flemish harpsichords by a process known as 'revailment'. This involved re-casing the instrument to match the current taste in furniture, extending the compass of the keyboard to fit the new musical demands and, often, fitting the latest style of keyboard and action. Sometimes little was left of the original instrument but the all-important sound-board, but an interesting precedent had been set.

When it comes to historic pianos, it is only grands and what now seem rather eccentric varieties of the domestic piano, such as giraffes and lyres, that are of interest. Square pianos, which were the most common domestic instrument until overtaken by the upright around 1840, have sadly to be avoided except as inexpensive pieces of Georgian or Regency furniture. They have survived in large numbers, are hideously expensive to restore, need constant tuning and are more likely to go down in value than up.

STRINGED INSTRUMENTS

This category contains enormous diversity and there is space here to touch on only its most common instrument: the guitar. First appearing as a distinct instrument at the end of the 15th century, the guitar was at first somewhat overshadowed by the lute. As the lute receded into obsolescence in the early 18th century, however, the guitar took its place and has never really looked back.

Guitars from the 18th century or earlier are very rare. Most of those that have survived were not the working instruments of professional players, but those built for wealthy amateurs. They were made with exotic materials and their value lies as much in their being works of art as in their quality as musical instruments. During the first 60 years of the 19th century, however, the guitar became available to a broader spectrum of society, and examples have survived in some numbers. Only in the past few years have these instruments started to realise anything more than a few hundred pounds. The rise in value has been substantial and may or may not be maintained.

The electric guitar appeared in the 1930s. There is a small but growing group of enthusiastic buyers for early examples. There is also a much more unpredictable market for guitars – and indeed other instruments – that once belonged to famous figures in popular music. Given the right star, such instruments can be worth dramatically more than could be expected without the provenance. At a Sotheby's auction in April 1990, a 1968 Fender Stratocaster owned by rock legend Jimi Hendrix and played by him at the Woodstock festival in 1969 fetched £198,000, the highest price ever paid at auction for a rock 'n' roll guitar. ❖

STORAGE OF WORKS OF ART

Gordon Brennan-Jesson, Director, Christie's Fine Art Security Services, London

Gordon Brennan-Jesson joined Christie's on leaving school in 1960 and has held various administrative and accounting positions within the group. He has been involved with the group's UK property holding company and is at present director and secretary of this company and of Christie's Fine Art Security Services Limited.

Phone: (+44-171) 622 0609
Fax: (+44-171) 978 2073

STORAGE OF WORKS OF ART

Gordon Brennan-Jesson, Director, Christie's Fine Art Security Services, London

So, you have found somewhere to store those treasures that you hope will appreciate so rapidly over the next few years; there may be any number of reasons that you might wish to put them into storage, but can you be sure that the store is safe?

An art depository is by its very nature a target for thieves: an Aladdin's cave of fine paintings, gleaming silverware, glowing silken carpets, items that their owners treasure sufficiently to fear storing in their own homes. But thieves are only one danger; there are a number of other threats to your precious possessions.

HOW SAFE IS IT?

Before choosing a storage facility you should ask the following questions:

- What fire protection does the warehouse have?
- How secure is it?
- Will your property be covered by insurance?
- Is the store conveniently located?
- Will there be any customs duty or taxes incurred in placing property in such a warehouse?
- Is it a suitable environment for your property?
- How much will it cost?
- Is there a minimum storage period?

A whole host of other questions may spring to mind, but these are the most important points to address when choosing a safe resting place for your treasures.

It would be quite disastrous if the store you chose were to be destroyed by fire or a bomb, or stripped by thieves. However well insured you are, mere money cannot compensate for the loss of something you love. The warehouse must have intruder and fire-alarm systems, and the plans of the building should not be open to public scrutiny. You should satisfy yourself by a visit to the premises that these matters have been addressed and security is maintained without compromise.

Many of the older depositories are built on a number of floors with large open areas, and are equipped with sprinkler systems which can cause irreparable damage to works of art. The most effective means of preventing damage by fire is containment. This can be achieved by dividing the space into separate areas, all with suitable fire-resistance. These areas are then further divided into storage cells or units of varying size, which are leased to clients on a periodic basis, and clients hold one of the access keys.

To ensure that only people authorised by the client can gain access to his or her property, a log must be maintained, preferably computer-linked to the unit door, which automatically records any access. These records can be compared with a log book showing the client's, or his or her agent's, signature and the date and time of the visit. If all is well, the two records will match. It would be unfortunate indeed if a client came to view property that had been put in storage some time before, only to find that a crooked warehouse worker had been systematically pilfering valuable pieces over a number of years.

Insurers are notoriously difficult to please, especially when they have a financial interest in both the building and the contents. A reputable, secure warehouse will be able to obtain favourable rates for its clients through its brokers. By the same token, a client's own broker, too, will be willing to offer better terms once a suitable warehouse has been chosen.

It is important that the warehouse should be in a convenient location (remember that you and your property will have to travel to it) and not in an area that suffers from a high crime rate. There should be secure parking and unloading facilities; you should not have to run the risk that someone might snatch from your hands the work of art you are about to put in storage. If the depository is well organised, you will be asked not to visit without having made a prior appointment, in order that a number of people do not all turn up at the same time; it may be important to you to remain incognito.

VAT

In the United Kingdom there has been a change in relation to the treatment of imported works of art, antiques and certain collectors' items. Value-added tax (VAT), and possibly duty, is payable on property imported from outside the European Union (EU).

Provided the correct procedures are carried out, works of art and antiques can enter a bonded or exempt warehouse without the need to pay either duty or VAT for as long as the property remains in store. In this case, tax becomes payable only when the property leaves the store and is delivered within the EU. No tax is payable if delivery is outside the EU. Christie's Fine Art Security Services was the first security store in the UK to be granted exempt status by the customs and excise department following the change in the VAT regulations on 1 June 1995.

The rate of any tax payable will be the rate applicable at the time of the property's removal from the warehouse. The rate payable on works of art and antiques, which before 1 June 1995 would have entered the UK free of charge, is at present 2.5%. This is a concessionary rate and the rate will increase by the end of the century. If you propose to use this scheme for the import of works of art, and their ultimate destination is within the EU, you must consider whether it is to your advantage to pay the concessionary rate now or pay a potentially much higher rate at some time in the future.

Certain collectors' items are included in this scheme. These tend to be items of zoological, botanical, mineralogical, anatomical, palaeontological, archaeological, ethnographic or historical interest. There is also provision for products designed by famous craftsmen, including De Morgan or William Morris tiles and ceramics; Peter Waals desks and tables; Voysey, McIntosh and William Morris home furnishings; and decorative glassware produced by Lalique, Galle, Tiffany and Clarice Cliff.

HANDLE WITH CARE

Obviously the physical handling of your property is of prime importance. It will hardly benefit from being dropped or knocked against doors or walls. Your main reason for leaving it in storage may be to guarantee its protection – and not to have it fashionably 'distressed'! If an item needs to be wrapped it should be with acid-free paper. Polythene should never be used as it does not allow the item to 'breathe' and encourages 'sweating', which will result in the growth of mould.

Ideal storage conditions include a good circulation of atmosphere-controlled air. Valuable and delicate works of art must always be kept in the right atmosphere. Japanese lacquer work, for example, is extremely sensitive to changes in temperature and humidity, and both must be kept constant. If the air is not humid enough, the lacquer-work will spring off its fixing and form an untidy heap on the floor. The resulting jigsaw puzzle would prove a near-impossible challenge for even the most experienced restorer.

Old books and manuscripts, too, require a regulated combination of temperature and humidity. If the air is too dry, the vellum will crack and fall apart, and if it is too damp mildew will discolour the pages and the covers. Violins and wooden musical instruments are particularly vulnerable to changes in atmosphere; the wooden cases are likely to split if badly stored. Furniture, too, will split and warp if climatic conditions are not maintained at constant levels.

Old master drawings are particularly sensitive to light and ultraviolet rays and therefore have to be kept away from windows and in dim lighting. They also benefit from high humidity. Ceramics and porcelain can be stored on shelves but are much safer if placed in a customised box fitted with foam or polystyrene supports. They are not sensitive to humidity but should not be subjected to extremes of temperature as, like timber, they will expand and contract.

The old-style depositories with their many floors were well able to deal with a variety of items; they had the space to store them according to temperature and humidity requirements simply by placing them on different floors. The most sensitive, and those requiring maximum humidity, would be put in the basement, while items requiring drier conditions would be stored on upper floors. But they were still at the mercy of sudden changes in the weather, and the vagaries of old-fashioned heating systems. Specially built, modern security stores offer the benefits of mechanical aids, used in conjunction with computer controls, to maintain optimum levels of both temperature and humidity.

Costs vary considerably but, as with most things in life, you get what you pay for. You must assess the cost of storage in relation to the value you put on your property. Some warehouses will have a minimum storage period of up to one month while others will accept property for any period, however short. Invoicing will always be in advance.

At one time the only form of security store was a safe-deposit box or bank vault. Thankfully, things have improved since then, and we can now provide environmental, as well as physical, security to our treasured works of art. Thus, we can rest assured that they will not deteriorate physically, but we can only hope that their value is protected. ❖

ANTIQUITIES AND INTERNATIONAL LEGAL CONSIDERATIONS

Lyndel V. Prott, Chief International Standards Section, Division of Cultural Heritage, UNESCO

Lyndel Prott, international and comparative lawyer and expert in jurisprudence, has specialised in cultural heritage law since 1976. Since 1991 she has been Professor of Cultural Heritage Law at the University of Sydney, from where she is on leave to serve in her position at UNESCO. She was closely involved with the preparation of the UNIDROIT convention, and is the author or co-author of 150 publications.

Phone: (+33-1) 4568 1000
Fax: (+33-1) 4567 1690

ANTIQUITIES AND INTERNATIONAL LEGAL CONSIDERATIONS

Lyndel V. Prott, Chief International Standards Section, Division of Cultural Heritage, UNESCO

The collection of antiquities for investment has many attractions. However, recent developments in the law relating to the international trade in antiquities suggest that buyers should be extremely discerning in their acquisitions or they may find that their investment is not a sound one.

Much of the material on the market was removed from its country of origin a long time ago: European museums and private collections hold many items collected during the 18th and 19th centuries which already have a long chain of title. These items can be bought and sold without difficulty.

If an object has been held in a collection for some time, it should be possible to trace its prior title. While claims are sometimes made for the return of such items to their country of origin, there is so far no legal mechanism that would displace the law of the country where the objects have long been held. Within UNESCO, there is the Intergovernmental Committee for Promoting the Return of Cultural Property to its Countries of Origin or its Restitution in case of Illicit Appropriation, established in 1978, but this body mediates rather than making legal rulings. Furthermore, it has jurisdiction only in claims between states (such as Greece's call for the UK to return the Elgin Marbles) and has no authority to investigate private transactions between individuals or non-governmental organisations.

More-recently excavated objects are a different matter. For many years archaeologists all over the globe have been very concerned about the damage being done to the world's cultural heritage by the clandestine excavations carried out by treasure-hunters, who pick over important historic sites – often tombs (such as the Etruscan mausoleums in Italy) or great monumental complexes, such as those in Guatemala or Mexico. When archaeologists excavate ancient tombs or monuments they painstakingly map and photograph every inch of the site, and every particle of soil is sifted; even the smallest traces of remains will help construct a picture of the site. Thus plant remains can tell us about human diet, health and climate; animal remains provide an insight into diet as well as patterns of husbandry; the careful recording of levels enables structures and artefacts to be dated. Archaeologists may work with a small brush, proceeding centimetre by centimetre, so as to lose nothing from the precious, and irreplaceable, record.

In contrast, unauthorised excavations are often carried out by bulldozers, cranes, hacksaws and explosives in the hands of professional raiders or pick and shovel in the hands of an impoverished local population. They damage incredible amounts of cultural material by loss, breakage and abrasion as well as causing the total destruction of commercially valueless, but scientifically very important, remains such as stratigraphy, small organic traces and humble human artefacts. This situation has led to a worldwide effort to stamp out the illicit trade in antiquities.

Market forces ensure that objects of archaeological importance flow from the poorer countries into the hands of collectors in the industrialised countries who are prepared to pay high prices for them. The trade has for many years operated on the basis that goods could be auctioned or transferred through dealers without the seller's identity being revealed; it was only too easy for dealers in illicitly gained goods, whether stolen or obtained by unauthorised excavation, to sell them into the licit trade and be amply rewarded. UNESCO, on the advice of cultural experts, has made numerous attempts to tackle this problem.

INTERNATIONAL CONVENTIONS

UNESCO

The *Protocol to the Convention for the Protection of Cultural Property in the Event of Armed Conflict* (the Hague Convention) 1954, to which 73 states are parties, requires signatory countries to prevent the exportation from occupied territory of the cultural property of that territory. Every signatory undertakes, at the close of hostilities, to return any illegally exported cultural property that is in its territory to the competent authorities of the territory previously occupied. There is a great deal of such property in circulation as a result of recent conflicts: Khmer antiquities from Cambodia, moveables from Afghanistan, Cyprus, Kuwait and Iraq, and artefacts from the former Yugoslavia. Many industrialised countries are party to this protocol: Austria, Germany, the Netherlands and Switzerland, for instance. Even many states not bound by the protocol are now unwilling to allow transactions in this kind of cultural property, as illustrated by the decision of a US court in Indiana that an American buyer should return to the Greek Orthodox Church certain mosaics looted from a church in occupied Northern Cyprus.

In 1970 UNESCO adopted the *Convention on the Means of Prohibiting and Preventing the Illicit Import, Export and Transfer of Ownership of Cultural Property*. This treaty requires the return of stolen cultural property to it country of origin as well as special action against the traffic in illegally exported cultural property and possible emergency measures to protect endangered archaeological resources. Although there are 82 parties to this convention, few of them are industrialised nations. Argentina, Australia, Canada and the United States, which have large collecting communities, are signatories, but the European states have so far remained aloof. The exceptions are France and Switzerland, both of which have recently announced their intention to accede to the convention.

The United States has introduced import bans on specific categories of archaeological materials from Bolivia, Ecuador, El Salvador, Guatemala and Mali, and it has a special treaty with Mexico. Australia and Canada will return cultural property illegally exported from its country of origin subject to proof by the country concerned that the object has been identified in the country and that it was imported after the date of the domestic legislation enacting the 1970 UNESCO convention.

UNIDROIT

As far as stolen objects are concerned, the basic legal principle in English-speaking countries was traditionally that good title could not be acquired to stolen property. In civil law countries, on the other hand, a bona fide purchaser acquires good title after a short lapse of time (ranging from immediately in Italy to five years in Switzerland). Over the years many systems have come to assume that every purchaser is bona fide and this has become very difficult to disprove. Purchasers therefore have no incentive to enquire into the previous ownership of an object: if they learned that title was dubious they might no longer be considered bona fide, whereas if they did not ask they would be protected.

This situation has clearly facilitated the passage of illegally acquired cultural objects into the licit market. The protection of the bona fide purchaser is regarded as a fundamental principle of some legal systems. But such is the current concern over the extent of the destruction of cultural sites and theft from museums, churches and private collections, that representatives of even those systems have come to recognise that it is essential to encourage buyers to exercise greater care, particularly in respect of cultural objects that are unique and easily damaged, and for which monetary compensation cannot be an adequate replacement.

This consensus has led to a new agreement: the UNIDROIT *Convention on Stolen or Illegally Exported Cultural Objects*, drawn up by the representatives of about 80 countries including all the industrialised nations. This convention provides that stolen cultural objects will be returned to the rightful owner, and that compensation will be paid to the person who is required to surrender them only if he or she can prove that he or she used diligence in checking their previous title. In deciding this, the judge will look at the price paid, the characters of the parties, the circumstances of the transaction and whether the purchaser took other reasonable steps such as consulting a register of stolen cultural property or other relevant information or documentation, or seeking the advice of appropriate agencies. These provisions refer to bodies such as the Art Loss Register in London and the International Foundation of Art Research in New York, which keep easily accessible records of important cultural objects that have been reported missing.

An important article in the UNIDROIT convention also provides that clandestinely excavated objects are considered stolen where this is consistent with the law of the country where they were excavated. In many countries, legislation provides that all undiscovered antiquities are state property. Such objects will now fall under the provisions of the UNIDROIT convention relating to theft.

The UNIDROIT convention is not yet in force, the text having been finally adopted only in June 1995. Twelve states have signed it so far (Burkina Faso, Cambodia, Côte d'Ivoire, Croatia, Finland, France, Georgia, Guinea, Hungary, Italy, Lithuania and Zambia) and it will come into force when there have been five ratifications of signature. Many other states were involved in the negotiations and are considering the convention.

The UNIDROIT convention is most effective in respect of stolen cultural objects, although it also contains provisions on other aspects of the illicit trade in antiquities. The UNIDROIT convention also applies to illicitly exported cultural objects. Although traditionally most of the big trading nations did not enforce the export prohibitions of foreign countries, this situation is changing.

Other international initiatives

In 1993 the European Union agreed on a directive and a regulation for the return of the most important cultural objects illegally removed from one EU country and found in another. That naturally includes stolen items as well as those illegally exported. The provisions of these texts are based on an early draft of the UNIDROIT convention and follow much the same philosophy. Participation in this scheme is mandatory for all EU member states.

The Commonwealth of 51 English-speaking countries has also agreed on a scheme for the return of cultural objects that have been illegally exported.

PROFESSIONAL ETHICS

Even where a collector resides in a country that is not bound by any of these treaties, the current concern over illicitly traded objects has led to a new professional ethos which it is hoped will lessen the ease with which these objects can be traded and reduce their selling price.

Museums and professional associations

The International Council of Museums (ICOM) adopted a code of ethics in 1986 which includes the following statements (Art. 3(2)):

Museums should recognise the relationship between the marketplace and the initial and often destructive taking of an object for the commercial market, and must recognise that it is highly unethical for a museum to support in any way, whether directly or indirectly, that illicit market.

A museum should not acquire any object, whether by purchase, gift, bequest or exchange, unless the governing body and responsible officer are satisfied that the museum can acquire a valid title to the specimen or object in question and that in particular it has not been acquired in, or exported from, its country of origin and/or any intermediate country in which it may have been legally owned (including the museum's own country), in violation of that country's laws.

A number of important museums have adopted acquisition policies in line with this code. This means that they will not be interested in purchasing objects where there is a suspicion that they have been illegally traded.

Furthermore, many cultural professionals will not authenticate, evaluate or conserve objects brought to them for professional services unless they are satisfied that the objects satisfy these conditions. Since these services are often required in order to get the best sale price for a cultural object, withholding them may well affect its commercial value. Some professional associations (archaeologists and conservators, for example) either have already adopted or are considering adopting such a code of conduct.

Auctioneers and dealers

The principal auctioneers and dealers in the United Kingdom have adopted a code of practice which obliges the signatories not to import, export or transfer the ownership of objects that have been acquired or exported in violation of the laws of the country where they were acquired. This applies also to objects that have been acquired dishonestly or illegally from an official excavation site or monument or which have originated from an illegal, clandestine or otherwise unofficial site. A similar code of practice has been adopted by CINOA (la Confédération Internationale des Négociants en Œuvres d'Art). A new dealers' body called the International Association of Dealers in Ancient Art has also adopted ethical principles along these lines. UNESCO has been studying the possibility of establishing an international dealers' code and has a draft in circulation.

Resale or gift

Of course, it will always be possible to buy from dealers and find restorers who are not bound by such codes. But when a collector wants to sell an item, he may find buyers wary of acquiring anything that has not passed through the hands of cultural professionals bound by codes of ethics, because this may affect any subsequent resale.

Not only sales may be affected. Many

collectors in the United States plan to obtain tax concessions by donating valuable cultural objects to museums. It should be noted, however, that the ICOM code of ethics quoted above also stipulates that museums may not acquire by gift cultural property that does not have a good provenance. This principle is reinforced by the text of the UNIDROIT convention, which provides that a museum shall not be in a more favourable position than the person from whom it acquired a cultural object by inheritance or otherwise gratuitously. This means that the recipient of such an object by gift or inheritance will likewise have to return it if it proves to be stolen – and this includes objects illicitly excavated in countries bound by the UNIDROIT convention.

CONCLUSION

In conclusion, therefore, the purchase of antiquities needs to be undertaken with prudence if subsequent challenge to title and diminution of commercial value are to be avoided. ❖

FABERGÉ: JEWELLER TO THE TSARS

Peter L. Schaffer, Vice President,
A La Vieille Russie, Inc., New York

Peter L. Schaffer, a graduate of the University of Columbia is a former member and president of the board of directors of The National Antique and Art Dealers Association of America. He has lectured and taught at the Smithsonian Institution, the Cooper-Hewitt Museum, and the Metropolitan Museum of Art, amongst other notable institutions. He is an authority on the art of the goldsmith and jeweller, specialising in Fabergé, antique jewellery and the Russian arts.

Phone: (+1-212) 752 1727
Fax: (+1-212) 223 6454

FABERGÉ: JEWELLER TO THE TSARS

Peter L. Schaffer, Vice President, A La Vieille Russie, Inc., New York

The name 'Fabergé' has become a familiar household word, but who was he and what makes Fabergé unique?

Gustav and Charlotte Fabergé settled in St Petersburg in 1842. Gustav was to become one of the city's most influential jewellers. His ancestors were Huguenots; forced to flee France upon the revocation of the Edict of Nantes in 1685, they lived and worked as jewellers in many countries, including Italy and Germany, before finally settling in Russia.

When Gustav retired in 1860 he moved to Dresden. There the education of his eldest son, Peter Carl Fabergé (b. 1846), began, and continued with studies in the city's Green Vaults (part of the royal residence of Frederick the Great, which housed his vast collection of jewelled works of art). This was Fabergé's introduction to the 'fantasies' of Dinglinger, which were to have such a profound influence on his life and work. Johan Melchior Dinglinger was court jeweller to Frederick the Great, known for his elaborate gold, enamelled and jewelled scenes.

Peter Carl Fabergé's first apprenticeship was with Friedman, a Frankfurt jeweller. He studied the treasures of Vienna, Florence, Paris and London. In 1870, he returned to St Petersburg to run the Fabergé enterprise. In Russia he was able to study the treasures of the Romanovs, both in St Petersburg and in Moscow.

OBJECTS OF FANTASY

Although the House of Fabergé continued to make jewellery and merchandise diamonds, Peter Carl Fabergé introduced what were to become known as 'objects of fantasy'. Small, beautiful and exquisitely proportioned carved animals and flowers, boxes of all kinds, frames, inkwells, letter openers, bell-pushes, cane and parasol handles, clocks, etc. were among the items on offer. All of these pieces utilised a wide range of both materials and techniques, including a revival of the 18th-century use of coloured gold, both polished and dull, and the use of multicoloured enamels over gold or silver. Fabergé also began to apply all manner of gemstones, from diamonds to agate, in the many different motifs he employed. The design was ultimately more important than the value of the stones, although these were generally of the highest quality, unless there was a specific reason for the choice of a lesser one.

Fabergé's years of apprenticeship had served him well. As his style developed, strength and stability, combined with a refined and delicate line, characterised his pieces. Fabergé excelled in the design of small objects; he fully understood their magical fascination. During his brilliant career Fabergé is not known to have drawn a single piece – he merely drew rough sketches on a blackboard – but, no matter how large the firm grew (and at times he had several hundred people working for him including goldsmiths, jewellers, enamellers, carvers and workers of all kinds), all its designs were imbued with his unique style. All major pieces were discussed at a 'round table' to decide how best to produce the individual part, and in what order the parts should be made. In fact, Fabergé could be said to have invented the quality circle.

In 1882 Fabergé's younger brother Agathon

returned from Dresden and became a partner in the firm. In the Pan-Russian Exposition in Moscow they exhibited only objects – no jewellery at all – and received the first of many gold medals, including one at the Fine Arts and Crafts Show in Nuremberg in 1885. Along with international recognition, the House of Fabergé secured the imperial warrant from Tsar Alexander III. From then on the number of objects made for the Imperial Family grew.

IMPERIAL GIFTS

In Russia the most important holy day is Easter, and the egg is the symbol of the reborn or risen Christ. It was this religious event and the assassination of Alexander II that inspired the creation of the first Imperial egg. Alexander III's wife, Maria Feodorovna, was deeply depressed at having witnessed the assassination of her father-in-law, so Fabergé designed an Easter egg which resembled a real egg – and evoked memories of her Danish childhood. The egg contained many surprises to amuse and distract the Tsarina. Fabergé insisted that the design should be secret so that it would be a surprise for the entire Imperial household on Easter Day. The egg was such a success that a tradition was born. Thereafter, the Tsar had a standing order for one Fabergé egg each Easter. Upon Alexander III's death, his son, Nicholas II, increased the number of eggs to two – one for his wife as well as his mother. Some 57 eggs were made in all, and A La Vieille Russie is pleased to have handled 26 of them since its founding in 1851 – all after 1917, of course.

Apart from the famous eggs, the Imperial Family also purchased many, many other pieces from the House of Fabergé, and the name of Fabergé became celebrated because of its association with the Russian royal family. As a result, Fabergé won orders from the courts of Norway, Sweden and Denmark, and pieces were made for the kings and queens of England, Italy, Spain, Siam (modern-day Thailand) and Greece. Also among his clients were wealthy merchants from around the world.

After the Russian Revolution in 1917 the House of Fabergé soon closed and the family moved through Riga, Berlin and Frankfurt, finally settling in Lausanne, Switzerland, where Peter Carl Fabergé died in 1920.

COLLECTING FABERGÉ

People who buy Fabergé come from all walks of life and income brackets. They range from those who want a single piece to treasure, to those starting or adding to a collection. With the price of the least expensive pieces starting at around US$3,500, owning a Fabergé piece is not an impossible dream – although prices for the most sought-after pieces can be as high as US$7m or US$8m.

What to look for

With Fabergé, as with most things, quality is a crucial factor. A quality piece in good condition will increase in value much faster than a lesser piece. With the House of Fabergé's total output estimated at between 150,000 and 200,000 pieces, items made for over-the-counter sale are the least likely to appreciate. The large pieces in silver and crystal are also best avoided if your only interest is their investment potential. Pieces in gold, enamel or stone (jade, for instance), those set with jewels and the small pieces make the best investments.

Oddly enough, the costliness of the materials used is not the decisive factor in the price of the piece. Fabergé used materials for their colour and aesthetic appeal rather than their value. But the quality of the workmanship is of prime importance.

The large eggs are the rarest pieces, and the most sought-after. Some of these were made for the Imperial Family, but so-called 'non-Imperial' eggs have sometimes sold for more than their Imperial counterparts because of their superior design and considerations of beauty. After the large eggs come human figurines, flowers (unmarked), animal figurines, frames, clocks and, lastly, objects. The smaller pieces are often the more valuable, since they are what Fabergé was renowned for.

Rising values?

A La Vieille Russie bought its first animals for less than US$2 in the 1930s. In the 1960s these items fetched perhaps US$8,000 to US$9,000, and the same pieces today would cost as much as US$100,000. The prices that the Tsars and others paid for the large eggs were between 9,000 and 11,000 roubles – a phenomenal sum given that a 50-hectare farm (with house and livestock) could be bought for as little as 150 roubles. And even more extraordinary when you consider that, with the eggs selling at around US$7m today, they have never matched their original cost in relative terms.

Beware counterfeits

It is estimated that the number of fakes or remade or remarked pieces in circulation far exceeds the number of genuine pieces made, so the collector is in very real danger of inadvertently buying a forgery. Documentation is also easy to counterfeit. The pieces documented in well-known books are the most commonly copied, but the reputable dealers and scholars in the field do know the location of most of these.

To avoid making a very expensive mistake indeed, a potential buyer should without question consult an acknowledged expert, as it takes years of study and familiarity with Fabergé's work to be sure of making a sound judgement – forgers are even able to copy the relevant marks by laser these days. A verifiable history, such as the existence of original drawings or provenance from an important collection, is of great value in supporting claims of authenticity.

Where to look

Several private collections of Fabergé as well as public ones are available for viewing. New York has the Forbes collection on display in its own museum at 12th Street and Fifth Avenue, and the Virginia Museum of Fine Arts in Richmond houses the Lillian Thomas Pratt collection. There is the Minshall collection in Cleveland; the Gray collection in New Orleans; and the Walters Gallery in Baltimore. And, of course, the collection of the British Queen which is occasionally put on display, as well as a number of collections in Russia – most notably that in the Armoury of the Kremlin in Moscow.

A La Vieille Russie offers Fabergé's work for sale in New York and, because of its association with him (he was one of its clients in Kiev), is recognised as one of the foremost experts in the market. The only other place to find truly authentic pieces is Wartski's in London. ❖

INVESTING IN DIAMONDS

Ephraim Zion, Managing Director,
Dehres International Ltd, Hong Kong

Ephraim Zion is the managing director of Dehres International Limited. Since 1971 he has been supplying the world with fine diamonds from offices in Hong Kong, Bangkok, New York and Tel Aviv.

Phone: (+852) 2521 3411
Fax: (+852) 2845 0506

INVESTING IN DIAMONDS

Ephraim Zion, Managing Director, Dehres International Ltd, Hong Kong

Diamonds have always held a peculiar fascination. Until the 15th century only kings, queens and maharajas wore diamonds – as a symbol of power, courage and invincibility. Diamonds have always been associated with strength, romance and legend: the Greeks, for example, believed that the fire in the diamond reflected the constant flame of love.

It was not until 1477, however, when Archduke Maximilian of Austria gave a diamond ring to Mary of Burgundy, that the tradition of presenting diamond engagement rings was established. Fortunately, diamonds and precious stones are no longer the exclusive domain of royalty. Over the past few decades diamonds have become a part of our culture and have acquired a unique status as the ultimate gift of love. Many people in affluent countries, and more recently in developing countries, buy diamonds as engagement and anniversary rings to symbolise their love.

Diamonds' fascination and mystique stem from their unique beauty, transparency and, above all their rarity. Diamonds are the most beautiful of all the gemstones because of their ability to refract light so that they display maximum brilliancy, lustre and dispersion, reflecting beams of light in a blazing fire of all the colours of the rainbow.

PRICING DIAMONDS

In recent years people have begun to realise that, apart from their beauty and status as a symbol of wealth and prominence, diamonds are a sound and stable long-term investment alternative. The last 25 years have proved this, as the following examples demonstrate.

In January 1972 a one-carat round diamond graded D/flawless (a criterion used to estimate a diamond's value) was worth US$2,000 per carat (p/c). Twenty-three years later, the price of this same one-carat diamond stands at approximately US$13,500 p/c, an increase of 675%. Similarly, a five-carat D/flawless round diamond was selling in January 1972 for US$6,500 p/c. Today this same piece is worth about US$50,000 p/c, an increase of 769%.

This tremendous appreciation is not confined to fine-quality diamonds. Diamonds of lesser colour and clarity, such as diamonds in the F to H colour categories rated with a clarity of VVS to VS_2, have also appreciated considerably. (Please refer to Figure 1 for an explanation of diamond gradings.)

FANCY-COLOUR DIAMONDS

Coloured diamonds have for many years attracted the particular attention and admiration of collectors, connoisseurs, investors and diamond traders. The reasons are quite clear: fancy-colour diamonds are extremely rare and beautiful, and the most precious of all the gemstones. Their scarcity means that collectors, dealers and investors snap them up as soon as they appear on the market. Some of the larger diamonds, in pink, blue, orange and green, are almost considered endangered species, and their values have skyrocketed accordingly.

The main markets for the rare fancy-colour

FIGURE 1: DIAMOND GRADING

		GIA	CIBJO/International	HK	UK	Swiss
Small mounted stones: Appearance colourless	Colourless	*D*	*Exceptional White +*	*100*	*Finest White*	*River*
		E	*Exceptional White*	*99*		*Top Wesselton*
		F	*Rare White +*	*98*		
	Near colourless	*G*	*Rare White*	*97*	*Fine White*	*Wesselton*
		H	*White*	*96*	*White*	*Top Crystal*
		I	*Slightly Tinted White*	*95*	*Commercial White*	*Crystal*
		J		*94*	*Top Silver Cape*	
	Faint yellow	*K*	*Tinted White*	*93*	*Silver Cape*	*Top Cape*
		L		*92*		
		M	*Tinted Colour*	*91*	*Light Cape*	*Cape*
	Very light yellow	*N*		*90*		
		O		*89*		
		P		*88*		
		Q		*87*		
		R		*86*	*Cape*	
	L.Y.	*S-Z*				
	F.Y.	*F.Y.*				

Flawless	*VVS_1*	*VVS_2*	*VS_1*	*VS_2*	*SI_1*	*SI_2*	*I_1*	*I_2*	*I_3*
Internally Flawless	*Very, Very Small Inclusion**		*Very Small Inclusion*		*Small Inclusion*		*Imperfect*		

** Refers to any slight imperfection in the diamond, deemed to be any item (of grit, etc.) that is located naturally inside the stone.*

diamonds have traditionally been Europe, the United States, Asia and the Middle East. In the past few years Asia, propelled by booming economies and affluence, has begun to invest quite aggressively in these gemstones, and the Asian market is increasingly important.

Colour grading

Until last year, the colour grading system for fancy-colour diamonds was somewhat limited and inadequate. Under the old system, and using pink, blue and yellow stones for illustration purposes, coloured diamonds were categorised simply as:

- Pink: Faint Pink, Very Light Pink, Light Pink, Fancy Light Pink, and Fancy Pink
- Blue: Faint Blue, Very Light Blue, Light Blue, Fancy Light Blue, and Fancy Blue
- Yellow: Fancy Light Yellow, Fancy Yellow, and Fancy Intense Yellow

The Fancy Intense grading was thus applied only to yellow diamonds. Growing interest on the part of dealers, collectors, jewellery designers and consumers prompted the Gemmological Institute of America (GIA) to embark on an extensive programme of research and investigation into the coloured stones.

The result is a new grading system for fancy-colour diamonds, which came into practice in March 1995, clarifying and strengthening the existing system. The most important change is that the GIA has introduced two more grades (Fancy Deep and Fancy Vivid) for yellow and orange diamonds and three more for the other colours (pink, blue, green, red etc.) The new system contains the following grades (using the most important colours for illustration), in order of intensity of colour:

- Pink: Faint, Very Light, Light, Fancy Light, Fancy, Fancy Intense, Fancy Deep, and Fancy Vivid
- Blue: Faint, Very Light, Light, Fancy Light, Fancy, Fancy Intense, Fancy Deep, and Fancy Vivid
- Yellow: Fancy Light, Fancy, Fancy Intense, Fancy Deep, and Fancy Vivid

Clearly, the introduction of new colour grades will contribute not only to a greater knowledge and understanding of the fascinating world of coloured diamonds, but also to higher prices for these prized gems.

Fancy prices

An examination of prices reveals the scale of the increase that has already taken place. Twenty to 25 years ago, coloured diamonds were selling to only a few individual diamond connoisseurs; at that time a five-carat Fancy Pink or Fancy Blue diamond fetched about US$20,000 to US$25,000 p/c. In sales held in New York and Geneva in October 1995 some fancy-colour diamonds fetched quite astronomical prices: a pear-shaped 1.72 carat Fancy Intense Blue/IF diamond sold for US$343,000 (US$199,709 p/c), while a slightly smaller pear-shaped Fancy Intense Blue/VS_1, of 1.15 carats, reached US$117,000 p/c. A 3-carat oval Fancy Pink/VS_1 diamond was bought for US$525,000 (US$171,617 p/c), and a 2.46-carat cushion diamond Fancy Intense Blue/VS_1 fetched US$448,000 (US$182,000 p/c).

In these same sales, some highly important, rare, intense and deep blue diamonds were sold for enormous sums to prominent dealers: for example, a Fancy Deep Blue/VS_2 13-carat heart-shaped diamond sold for US$515,500 p/c, and a smaller deeper blue oval diamond of 4 carats for US$568,740 p/c. Then came the most beautiful pink diamond ever offered at auction, which topped them all: an emerald-cut 7-carat Fancy Intense Purplish Pink, for which a dealer paid more than US$6m, nearly US$818,000 p/c.

Yellow diamonds, though less valuable, have fared no worse than the pink and blue: at the same sales in October 1995, an 8.97-carat Fancy Intense Yellow/VVS_1 round diamond sold for US$73,700 p/c, whilst a 2.53-carat Fancy Deep Yellow round diamond fetched US$38,973 p/c. At the same time a pair of emerald-cut Fancy Intense Yellow/VS_1 diamonds totalling 4.93 carats sold for US$235,400 (US$47,700 p/c), and a 12.05-carat marquise diamond Fancy Intense Yellow/SI_2 for US$50,100 p/c.

Green and red diamonds are the rarest of all. In October 1992, an extremely rare round green diamond of 1.28 carats was bought by a dealer for US$253,000 (US$197,656 p/c) after intense bidding. And at a 1987 Christie's auction a private collector paid a record price per carat for a 0.95-carat round Fancy Purplish Red at US$926,315. Nothing quite like this exquisite diamond has appeared on the market since, although in April 1995 a 2.15-carat marquise diamond in Fancy Intense Yellowish Green/VS_2 was sold for US$1.1m (US$512,791 p/c), and at a New York auction in April 1988 a 3.02-carat Fancy Yellowish Green stone brought US$1.7m (US$564,569 p/c).

From these examples, one can clearly see that the market for fancy-colour diamonds can offer marvellous opportunities for outstanding long-term capital appreciation. This segment of the diamond market will without question continue to rise, offering good investment value for those who have the foresight to keep their acquisitions for some time.

DE BEERS AND THE MARKET

The diamond organisation, De Beers, plays a significant role in controlling and maintaining an orderly world wide market in diamonds. De Beers controls about 85% of the world's rough-diamond production. The Central Selling Organisation (CSO), which is the marketing arm of De Beers, ensures that all mined diamonds pass through its main London office. From there the CSO distributes the rough stones to the cutting centres of New York, Antwerp, Tel Aviv, Bombay and elsewhere. The organisation regulates the amount of diamonds that are released onto the world market, releasing them in an orderly manner so as to avoid a glut in times of recession when demand is weak, and to prevent severe shortages when the demand is strong.

The CSO runs various advertising campaigns in its effort to maintain stability and to encourage growth in market demand. In 1994, De Beers planned to spend a staggering US$170m on advertising and promotion, spreading the notion that diamonds last forever as a symbol of enduring love. Diamonds are traded as a commodity in bourses across the world, but what other commodity can so significantly retain its value while also proving a timeless, attractive and pleasing acquisition?

THE VALUE OF DIAMONDS

Diamonds are noted for many attractive qualities, both aesthetic and financial, but one that is sometimes overlooked is their portability. Diamonds are small in size and yet they represent a most concentrated form of capital. In times of economic turmoil, political instability, currency fluctuations and monetary devaluation, diamonds have consistently proved to be not only a good hedge against inflation, but also a very stable and secure store of value. Diamonds could be considered even safer than real estate, for a few million dollars' worth of diamonds can be carried in a pocket and can be sold anywhere in the world:

provided, of course, that they are put through the right channels of distribution.

History shows us many examples of ordinary people and heads of state who have fled their countries during times of war or political upheaval, forced to abandon all their possessions except the jewels they could carry with them. Two examples are offered by the Shah of Iran's family, when they fled their country in 1979, and the ruling family of the Philippines, when they left in 1986.

Diamonds appear to be fairly immune to currency fluctuations and are considered safe currency, for they possess unique stability. They have never been the object of the kind of government measures or confiscation that tend to alter prices, nor are they the subject of devaluations or restrictions. The fact that diamonds do not bear the stamp of any government or treasury department has been crucial in making them an internationally accepted currency – and an investment far superior to gold, which is basically a speculative investment. Gold investment does not produce income and there are no guarantees of capital appreciation; in addition, gold is subject to price fluctuations and may be considered a risky investment.

Prospects for the diamond trade are excellent and the potential for diamond appreciation in the future is staggering. The increasing affluence of the wealthier classes in most developing countries means more and more people are becoming interested in buying diamonds. A pride to own and a pleasure to wear, diamonds offer the best of both worlds, and the purchaser is assured of their capital appreciation into the bargain.

For those who have the resourcefulness and foresight, it is time to make a wise and sound decision and purchase a diamond, a masterpiece of lasting value, which will afford the owner great pleasure and enjoyment for years to come. ❖

INVESTING IN WINE

John Armit, Director, John Armit Wines Limited, London

John Armit became managing director of the Queen's wine merchants, Corney & Barrow Ltd, in 1968. A year later he arranged the buyout of the business and ran it independently until 1978. In 1982 he started John Armit Wines Investments. He was adviser to the wine fund created by Dillon Read and the Group Bruxelles Lambert in 1985. One of the world's experts in wine for investment, Mr Armit is largely responsible for creating the reputation of the great Pomerols: Pétrus, Lafleur and Trotanoy. Christian Moueix, owner of Château Pétrus, considers Mr Armit one of the four greatest wine tasters in the world.

Phone: (+44-171) 727 6846
Fax: (+44-171) 727 7133

INVESTING IN WINE

John Armit, Director, John Armit Wines Limited, London

There is nothing new about the idea of investing in wine. In Britain institutions and wealthy individuals have bought large quantities of vintage port and top-quality Bordeaux since the 1920s and 1930s. They recouped their investment, reinvested in subsequent vintages and built up superb cellars. Institutions no longer invest for consumption and the days of long lunches in the City of London with copious quantities of vintage port are now long past. However, the principle remains the same. It is still possible to invest with the intention of building up a great collection, selling and rebuying the very best wines rather as great collections of art are built up. Alternatively you can invest purely for financial gain.

As with all investments timing is of the essence. Now is a particularly good time to buy because Bordeaux is the most important investment market and there is now less stock of great wines in Bordeaux than at any time in the last 30 years. The reason for this is that, in common with many others, the proprietors of the great Bordeaux châteaux invested heavily in the late 1980s in new cellars and equipment. I have a feeling that they believed they had conquered nature, that they would continue to produce great wines as regularly as they did in the 1980s and that world demand for their wines would continue to increase. Their last belief was true but nature proved itself unconquerable. Rain followed frost and there was no consistently great vintage between 1990 and 1995. Consequently the owners were obliged to sell off all their stocks of older vintages in order to finance the investments they had made in the 1980s. Since stocks are very low and there is increasing international demand for the greatest wines it seems inevitable that prices will continue to rise for the foreseeable future.

Wine can be a rewarding and enjoyable investment but you should remember:

(i) wine should only be a small part (say up to 5%) of your investment portfolio;
(ii) you should not borrow in order to invest in wine;
(iii) there are no dividends and, indeed, you will have to pay storage and insurance costs although these are modest (currently around £6 per case per year in London); and
(iv) you should allow a selling period of a month or two to ensure achieving the best possible prices, although the broking market is now quite sophisticated and you will usually receive a settlement in a month. Despite its nature, wine is not a liquid investment.

The principal reasons for including wine in an investment portfolio are that:

(i) it is an international commodity;
(ii) a finite amount of the greatest wines is produced and the quantity available is continuously decreasing as it is consumed;
(iii) there are always new markets for great wines and these markets are inevitably where economic power lies; 20 years ago the new market was the United States of America, 10 years ago it was

Japan and now Hong Kong, Taiwan, South Korea and Indonesia are the new markets; and

(iv) in the UK, wine is not subject to capital gains tax, being deemed a diminishing asset by the Inland Revenue.

THE WINES TO INVEST IN

There is no doubt that the good vintages of Bordeaux represent the best investment. There is an established secondary market led by brokers in France and Britain. The wine sales at Christie's and Sotheby's concentrate largely on the top wines from this region.

Most other wines can be quickly discarded. Vintage port is a 19th-century wine now outdated and rarely consumed. Red and white Burgundy from the greatest *domaines* do appreciate in value but are made in very small quantities and are expensive when first sold; all the other wines from this region can be ignored because of their unreliable quality. German wines are made in small quantities and the initial price is so high that there is little potential for profit. New World wines have no secondary market and no established investment record.

Which Bordeaux wines?

The best quality wines usually represent the best investment and these are the following:

(i) first growths (from the official French 1855 classification which has not been changed other than in 1973 when Château Mouton Rothschild was promoted from second to first growth): Château Margaux, Château Mouton Rothschild, Château Latour, Château Lafite and Château Haut Brion;

(ii) Top Pomerols and St Emilions:
Pomerol: Château Pétrus, Château Lafleur, Château Certan de May, Château L'Evangile and Château La Conseillante;
St Emilion: Château Cheval Blanc, Château Ausone, Château Canon and Château l'Angélus; and

(iii) second growths and their equivalents: Château Léoville Lascases, Château Pichon Lalande, Château Ducru Beaucaillou, Château Cos d'Estournel, Château Lynch Bages, Château Léoville Barton, Château La Lagune, Château Gruaud Larose and Château Montrose.

It is worth remembering that the average annual production of these châteaux is around 520,000 cases; this is very little to supply a continually increasing world demand.

HOW TO INVEST

Statistics show that the most profitable investment in wine has been to buy carefully selected châteaux of a good vintage at the best available price as soon as the market opens, i.e. in April or May following the harvest in September and October.

For example, an initial investment of £10,000 comprising a selection of wines of the 1981 to 1990 vintages would have performed as illustrated in Figure 1.

There has not been a vintage worth investing in since 1990 but at the time of writing, in October 1995, all the indications are that 1995 will be an exceptional vintage. Folklore has it that when the wind blows from the east after mass on Palm Sunday there will be an excellent vintage, so the omens were good from the beginning. Flowering in June took place in perfect conditions, the summer was the hottest and driest since 1947, but there was rain in early September. However, the rains did not

FIGURE 1: VINTAGE BORDEAUX WINES 1981-1990 EVOLUTION OF PRICES 1982-1995

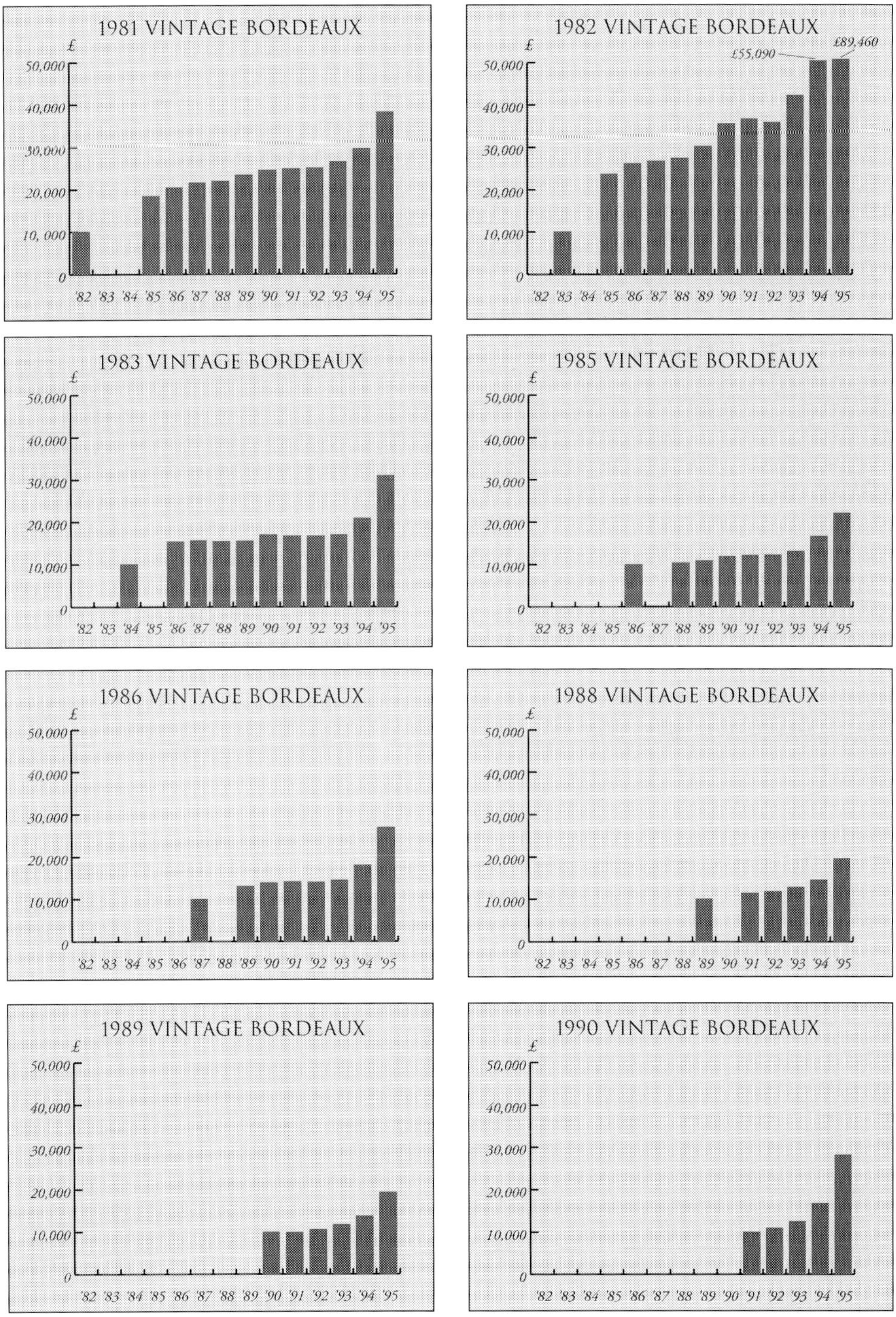

last and the early ripening Merlot grapes were picked in perfect conditions and achieved unusually high degrees of natural alcohol.

There was some concern that the later ripening Cabernet Sauvignon wines which predominate in the Médoc region might not achieve sufficient natural ripeness to make a great vintage. However, nature was once again benevolent and the important last two weeks of the harvest took place in perfect, hot weather. At present the château owners receive their guests with beaming faces and Christian Moueix, owner of the great Château Pétrus in the Pomerol region, has even drawn comparisons with the two great vintages of 1947 and 1971. It is nonetheless early days and the true quality of the wines will not be apparent until after the *assemblage* which will take place in February 1996.

The *assemblage* is the all-important selection of which wines are good enough to be bottled under the château label. I should explain that the wine from each area of the vineyard and from the different grape varieties are matured separately in vats until January or February when the owners and their advisers taste and decide which are of high enough quality to be given the accolade of the château label. Up to 50% may be discarded in the continual search for quality. You may well ask why they do not make more wine and the reason is simple: if they make wine of a poorer quality experts will notice and be less prepared to buy it, it will become less valuable and very soon the reputation of the château will suffer. It takes a long time to achieve a great reputation but it can be lost remarkably quickly.

However, I suspect that 1995 will turn out to be a very good and perhaps excellent vintage, that there will be tremendous demand for it and that those who buy it will have made an excellent investment.

The alternative investment method and one which has proved remarkably efficacious is to buy great and rare wines from great vintages in the past. The wines to buy are those that are recommended by the American wine critic, Robert Parker, whose opinions influence markets and prices. He has a unique method of judging wines, giving them a rating out of 100 – just as his American law school marked his papers. He is undoubtedly the most influential critic in the history of wine, being a very good judge of wines and totally objective in his judgements. The vintages he, and I, rate most highly are 1945, 1947, 1961, 1970, 1982, 1986 and 1990, and the best wines of these vintages will undoubtedly continue to appreciate in value.

I would offer investors one particular warning about buying older wines: you must be very careful about the provenance and be sure that you are buying from a reputable merchant. You must also check that the level of wine in the bottle is high, and that the labels and cases are in good condition. This is just natural caution because fakes are not unknown for the very rare and expensive wines, but if you are sure of your merchant you can rely on the information he gives you. His reputation, after all, depends on it.

WHERE TO BUY

As a general rule it is essential to buy the wine 'under bond' to avoid duty and taxes and to facilitate resale. In London, which is home to the international brokers, you would lose 17.5% if you bought the wine duty-paid, because this represents the VAT which would not be recoverable. If the wine is bought, stored and resold from France you risk being charged VAT on a circle of trading. In the US, auctions have only recently been legalised and the resale of wine is problematical. Therefore I would advise buying under bond from one of the specialist merchants or brokers in London. ❖

THE CHANGING CLASSIC CAR MARKET

Robert Brooks, Brooks, London

Robert Brooks was born in London in 1956. Following a distinguished 14 years at Christie's, he formed Brooks in 1989. In the ensuing years, the company has grown into the world's leading specialist auctioneers of collectors' cars, motorcycles and related 'automobilia'.

Phone: (+44-171) 228 8000
Fax: (+44-171) 585 0830

THE CHANGING CLASSIC CAR MARKET

Robert Brooks, Brooks, London

The odd thing about the international classic car market is that no one – not even the best-informed enthusiast – saw the surge in popularity coming. People had been interested in old cars for more than 30 years before prices really took off; and when they did, they took everyone by surprise.

People had been collecting old cars since the end of the 1920s, initially buying pre-1905 veterans to take part in the Brighton Run and then pre-1930 performance models to participate in events held by the Vintage Sports Car Club (VSCC). The VSCC was founded in 1934 as a kind of protest against falling standards in car design, but no one bought old cars as an investment. They were simply regarded as the cheapest way to have motoring fun and, in the early days of the vintage movement, a few pounds invested in something like a 'WO' Bentley or a 30/98 Vauxhall would buy quite astonishing performance.

EARLY INTEREST

Even in the car-starved days immediately after the Second World War, veteran and vintage cars could be bought extremely cheaply to provide entertaining, everyday transport. For example, a 1909 Napier entered in our December 1995 sale was still reliable enough in the late 1940s to be the only car of motoring journalist Ronald Barker.

But even though prices were low and a flourishing old car movement had grown in the United States, transatlantic trade in old cars was very limited and what little there was caused great controversy. When Commander W.H.C. Blake sold his 1903 Gordon Bennett Napier to his friend George Waterman of Rhode Island in 1950, for example, there was a storm of protest. The VSCC's lament that 'this unique British car is lost to our country, probably forever' was to be proved wrong in the mid-1980s when the National Motor Museum brought the Napier home to Beaulieu. However, its £300,000 price tag was many times greater than the sum that originally took it across the Atlantic!

The popularity (and prices) of veteran cars were certainly boosted by the film *Genevieve*, which opened in 1953, but most old cars remained affordable during that decade. This was aided by the fact that there were plenty of garages and breakers' yards catering for a large number of pre-war vehicles as yet unaffected by any kind of annual roadworthiness test.

In 1956, when you could still buy a 3-litre Bentley for £250 (or 'near offer'), the brief period of petrol rationing which followed the Suez Crisis saw British old car dealers starting to advertise the less economical type of antique car in American enthusiast magazines.

The first dedicated old car sale took place 35 years ago at Beaulieu, when prices were still quite modest; only two of the 115 lots, both veteran models, changed hands for more than £1,000. A drop-head Speed Six Bentley sold for a modest £330, and nobody was interested in a pair of Tipo 8A Isotta-Fraschinis or a 36-220 hp Mercedes two-seater, any one of which would have been a sure-fire show-stopper three decades later. Overseas buyers were conspicuous by their absence.

FROM A HOBBY TO A MARKET

The event that transformed the old car movement from a parochial hobby into an international collectors' market was the sale of the greater part of the Sword Collection in 1962. John Sword was a Scottish entrepreneur who began collecting old cars when they had little more than scrap value, and by the time of his death in 1960 he had more than 200 cars, commercial vehicles and motorcycles stored on his property at Balgray in Ayrshire.

Collectors from Europe and America flocked to bid at the first Sword auction and, as my late friend and mentor Michael Sedgwick reported at the time: 'Many prices soared into cloud-cuckoo land, but this should not be blamed on the American buyers, who seemed to have a clear idea of what they were willing to pay, and never allowed themselves to be browbeaten into four figures, except on outstanding cars.'

Michael, a prolific author and the former curator of the Beaulieu Museum, was as close an observer of the old car scene as any. Even so, back in 1969, he thought the market had reached its limit when reporting on a sale in which the most expensive lot was a 1920s supercharged 36-220 hp Mercedes which sold at £5,200: 'Neither an original 3-litre nor an excellent replica-bodied 4°-litre Bentley reached £3,000, and nobody wants to pay more than £500 to £550 for a good run-of-the-mill vintage light car . . . the only losers are the investors and the fast-buck merchants . . .'

His assessment is perhaps understandable, considering that only seven years before a 'very fair' 4°-litre Bentley had struggled to reach £265 at the Sword sale. So the expected market price of such a car had risen by about 10 times in that time, a rather more dramatic increase than in the price of property, which had probably just doubled. Surely such a rate of increase could not be sustained? But sustained it was; and the market, which had not long before been restricted to a small and informed fraternity that knew exactly what it wanted and was prepared to wait until it became available at the right price, suddenly found itself invaded by a new breed of collectors anxious to climb aboard the veteran and vintage car bandwagon.

This was a time of tremendous financial uncertainty, when the pound had been devalued by 14% (to the equivalent of US$2.40) by the Wilson government, inflation and prices were rising fast and the general financial attitude had become 'I want it – and I want it now.'

As Michael Sedgwick remarked at the time: 'An influx of new blood rendered the hitherto undesirable only relatively so,' and the bread-and-butter cars of the 1930s, previously scorned by the quite sophisticated market, suddenly became attractive as a cheaper (and often more useable) alternative to the older cars. Of course, prices consequently rose in this sector, and the 'hitherto undesirable' soon became the highly collectible.

THE RISE . . .

The rapidly expanding collectors' car market was accompanied by the appearance of new magazines, which reflected the changing nature of the kinds of car that were considered collectible, and the major auction houses regularly held sales of early cars, which had become elevated to the status of 'machinery of antiquarian interest'.

Price inflation throughout the 1970s was steady rather than spectacular, although Lord Montagu's 1972 record purchase price of £17,000 for a Blower Bentley with replica 'Birkin' bodywork was soon to be eclipsed by the £20,000 paid for a 1927 S-Type Mercedes two-seater the following year. As Michael Sedgwick remarked of the Mercedes, which had sold for £500 at the first Beaulieu sale: '4,000% in 13 years – what stock could give you such a return?'

Much younger cars were becoming collectible: at the same Beaulieu sale a newly imported 1957 XKSS Jaguar was bid up to £12,000, which was thought to be about the right price. A crystal ball would have been a very useful tool!

Again, in the mid-1970s it was possible to buy a Duesenberg Model J in the USA for around US$20,000, while a dozen years later vendors were asking – and getting – a million dollars for their Duesies.

Indeed, the most remarkable period for the sale of classic cars was the mid- to late 1980s, when the prices for the most desirable models suddenly took off. A buyers' guide published early in 1985 valued a Ferrari 250GT Berlinetta at £45,000; by the end of that year it had soared to £70,000 and by the end of 1987 the price of admission was £150,000. This surge in prices seemed unstoppable: the owner of a Ferrari GTO who sold it in the mid-1980s for around £300,000 saw it change hands soon after for £1m, and by the end of the decade a similar car had sold for an outrageous £6m.

Perhaps the high point of that remarkable era and the trigger for the 'final boom' of 1987 and 1989 was the sale of the Kellner-bodied Bugatti Royale which I conducted in the Albert Hall in November 1987: in just a couple of minutes, that car – which had changed hands just after the war in a deal principally involving a couple of brand-new refrigerators – soared in value to a world-record auction price of £5.5m. Other cars subsequently sold for more, but somehow the combination of the size and presence of that car and the theatrical nature of the venue made this a unique and quite unforgettable occasion.

. . . AND FALL

And then the sky fell in. The general financial downturn at the end of the 1980s was followed by a collapse in classic car values. By mid-1990, many lots – which a year earlier would have been the subject of keen bidding to increasingly astronomical levels – either failed to sell or fetched prices well below those expected.

At the bottom end of the classic car market, the effects of the recession were less keenly felt, but the elite ranks of the million-pound (or million-dollar) models were decimated. A few cars still made the magic seven figures during 1990, but the values of many others were halved. Aston Martins, which had been selling for around £140,000 in mid-1989, were lucky to reach £70,000 a year later. A Dino Ferrari which sold for £25,000 in 1987 peaked at around £100,000 in the boom years of 1988-89 then fell back to around £35,000 by 1991.

The beginning of the 1990s was uncertain and erratic, and many owners who had invested at the peak of the boom found that they had fallen into the automotive equivalent of negative equity, with costs of purchase and restoration far exceeding what the market was now prepared to pay for their cars.

SIGNS OF RECOVERY

Confidence has been slow to return but there are hopeful signs at last; our sales are now attracting a high level of entries, which are selling well. The market in general is more active than it has been since 1990, and during 1995 we have had some particularly good results. Our Goodwood and Monaco sales each included several cars that sold for substantial six-figure sums, including a 250F Maserati which sold for more than £500,000. The days of the £2m-plus sale total have returned.

However, the events of the past few years have shown that buying old cars purely as an investment is, at best, a chancy business. The golden rule must be to buy to please yourself; don't follow fashion. It is also important for you to decide how useable and practical you want your antique car to be.

There are now so many categories of collectible car that the newcomer to the field can find it quite difficult to choose which one to go for, and both the head and the heart must play a role in making that choice. You must decide what you want the car for: is it for regular use, or just for rallies or racing? Do you fancy the social aspect or is it the mechanics of the cars that fascinate you?

Veteran and Edwardian cars, built before 1919 and once the market leaders, are now by and large quite modestly priced (unless you are talking about something like a Mercedes-Simplex) and so are vintage cars (pre-1931), apart from a very few 'premium' marques like Bentley and Alfa-Romeo. However, modern sports cars produced in reasonably large numbers, such as the E-Type Jaguar or the Austin-Healey, can command high prices. It is all a matter of fashion; more modern vehicles are perceived as easier to maintain and more practical, although that is not always the case.

CONCLUSION

Restoration can be costly; if the car is a special one, preserving its original finish may be more important. You can pay for restoration, but you cannot recreate originality. Once lost, it is gone forever.

So, if you are not sure which car to buy, seek expert advice, and do not be surprised if you find that a single car cannot cover all your motoring needs. Many enthusiasts own more than one car: some even end up with their own fleet!

By now you will have some idea of what your ideal car might cost and buying it through an auction house or a specialist dealer can save a lot of fruitless searching. The more exclusive cars are sometimes sold through specialist brokers who, like the best auction houses, know just where the potential purchasers are. Or you may prefer to find your ideal car yourself, by the more laborious route of searching through advertisements in enthusiast magazines or club newsletters.

Whichever route you choose, remember to budget for restoration and maintenance costs, and provide dry, secure storage. Insurance is essential but, particularly with specialised limited-mileage schemes, can prove surprisingly cheap. These days, you cannot expect rising prices to cover the cost of expensive restoration work when you come to sell, but must regard such costs as an investment in your hobby. Remember that cars were built to be driven; they last better that way.

Buy wisely and enjoy driving your purchase. Should you ever decide to sell, you may find that you make a surprisingly good return on your money to invest, hopefully, in your next classic car. ❖

CLASSIC YACHTS

Peter Gregson, Wooden Ships, Kingsbridge, South Devon

Peter Gregson started Wooden Ships yacht brokerage in 1980 to encourage interest in the restoration and preservation of wooden boats. After taking an arts degree at St Andrews University, and spending a few years marketing in the European travel industry, he refitted and operated a 150-ton schooner and then a 50-foot Le Havre pilot cutter built in 1894. Mr Gregson has more than 25 years of experience in refitting and sailing classic vessels.

Phone: (+44-1548) 531217
Fax: (+44-1548) 531687

CLASSIC YACHTS

Peter Gregson, Wooden Ships, Kingsbridge, South Devon

Since the early days of the sport, yachting has been synonymous with wealth and status. Yacht ownership has always sent out a clear signal at all levels of society, understood in all languages and still current today. Broadly speaking, yachting has developed along two distinct lines: cruising and racing. In recent years a third dimension has been added to the sport, crossing all previous boundaries, namely the art and practice of traditional boating.

As with all forms of racing, yacht racing has pushed design methods and techniques to their limit and speed has often been cultivated even at the expense of safety and seaworthiness. In its most modern form, the search for speed has produced the familiar computer-generated designs that so often lack two vital ingredients that can only come from the human mind: grace and beauty.

For the cruising yachtsman, marketing-induced desires have all too often replaced practical common sense. He seeks the comfort at sea that allows him to make longer passages with minimum distress to his yacht, himself and his crew. In the modern yacht, considerations of comfort demand curtains, carpets and unseaworthy double beds, often in a supremely uncomfortable hull shape, tailored by the marketing profession to appeal to a perceived market.

WHAT IS A CLASSIC YACHT?

To the cognoscenti, the values of the older yachting generation are important and they appreciate the true seaworthiness, comfort, grace and beauty of the yachts from an earlier generation. Not only are these yachts beautiful to behold, they are also very fast, useable boats. Their inherent beauty of form can be enhanced by the addition of creature comforts such as carpets, hot water and the security of modern navigation and safety aids. Such innovations have produced what we now refer to as the 'classic yacht'.

What makes a yacht a 'classic' is a taxing question. The principal defining factor is probably age. A generous cutoff date would be placed somewhere at the end of the wooden yacht building period of the late 1960s whereas a more purist date would be 1939. As the years go by it is possible that future generations will include our modern designs amongst the classics. However, these modern yachts lack the features that now define a classic yacht – style, grace, varnished teak and polished brass. And of course almost all the yachts of the classic period were built in wood.

By the end of the 1930s many people believed that yacht design had reached its zenith as some of the most beautiful sailing and motor yachts were launched to race and cruise. Time, money, style, appreciation of good taste and generations of craftsmanship all resulted in some of the most beautiful things ever created by man. It is this spirit that is now embodied in the classic yacht as we define it today.

WHY BUY A CLASSIC YACHT?

Philip Jacobsen, formerly *The Times* correspondent in Paris, wrote in a recent UK Sunday newspaper: 'A private jet may display wealth but a classic yacht shows class.' This is undoubtedly one good reason for buying a classic yacht.

Buying a classic yacht appears to spring from an expression of individuality coupled with a genuine love of the boat. Add to this some really exciting sailing, the opportunity to enjoy the company of like-minded people – regardless of their means – a sense of history and a creative spirit, and we begin to see something of the classic yacht owner's character, irrespective of the size of yacht.

Design and craftsmanship

The most sought-after yachts come from the design boards of the great names – three generations of Fifes from Scotland, and their fellow countryman Alfred Mylne; several generations of Nicholsons; Fred Shepherd and the lesser-known but gifted Dallimore; and Harrison Butler and Albert Strange, designers of the most perfectly balanced little yachts. In the US, the design studios of John Alden, Sparkman and Stephens, and Philip Rhodes whose designs were as often built in Europe as at home. In Germany, Henry Rasmussen and Max Oertz were great names. Johan Anker and Bjarne Aas in Norway, and Knud Reimers in Sweden all produced great designs, but in the early period the British designers were predominant.

Only certain great yards had the craftsmanship to build to the high standards that have ensured that the yachts still survive today. These were again Fife in Scotland, Camper & Nicholson and Berthon in England, and a host of smaller European yards with equal skills: the great Abeking and Rasmussen in Germany, Anker in Norway, and Ring Anderson in Denmark. Regrettably, few of the old yards are still in existence.

RESTORED YACHTS

There is a pedigree to the classic yacht as well established as that of any thoroughbred horse. When buying a classic yacht one looks first for a thing of beauty suited in size, price and condition to the sailing one wants to do. If it is built by one of the greats, so much the better, but it will cost more and there are fine yachts from lesser-known designers and builders. Then comes the biggest decision of all: do you buy a yacht that has been restored and is in fine condition, only requiring maintenance, or do you buy a yacht in some degree of dilapidation and undertake the task of restoring it? The restored yacht will cost more but you will know what you are getting for your money.

The restoration project must be approached with caution but there is a certain satisfaction in being able to control the quality of the work done and achieving something worthwhile. To many owners, the pleasure of yachting is as much in the restoration and maintenance work as in the actual sailing. One must bear in mind that a wooden yacht is like a living organism. It may slowly deteriorate as it ages (a process accelerated by misuse or neglect) but, unlike a living organism, worn-out parts can be replaced quite simply and the yacht rendered as good as new, indeed often better. And knowledgeable, regular maintenance will keep any deterioration at bay for years.

Wood is an extraordinary material which lasts for centuries. Although we may not realise it, many people are in contact with quite ancient timber in their everyday lives in buildings and artefacts. Boats are built from many small pieces of wood held together by small bits of metal, and it is usually deterioration of the metal that causes the initial problems. However, it is a relatively simple job, though a skilled one, to take one bit out and replace it.

Such is the enthusiasm for wooden yachts amongst owners that vast amounts of time, energy and money are spent in rebuilding them and many are now infinitely better than when new. Indeed, it is not unusual to find old yachts being rebuilt in totally uneconomic exercises when it would be much cheaper and more efficient simply to build a new boat to the original design. Many rebuilt yachts have so little original material remaining it is difficult to decide if it is a new yacht or a rebuilt yacht. A naval architect recently explained his theory to me: if the rebuilt yacht occupies the same physical space as the original then it is the original yacht, but if it is a new structure occupying a new space then it is a new yacht. Do boats really have souls? Apparently, many owners believe they do.

By far the greater number of boats of the classic period are, in fact, in excellent seagoing condition. It is always a source of great surprise to me how good their condition is when one considers the pernicious environment in which they have been kept, often for many years and with little maintenance. However, every owner wants to improve and upgrade his purchase and with the current awareness of value and quality we are constantly seeing projects, some minor, some major, to restore boats to their original form rather than simply improve them.

NEW VS OLD

There has been a major movement across Europe and the US to accommodate the new market for classic yachts. New jobs are being created in boat yards, the old skills re-learned and modern materials and techniques developed and adapted to ensure strength and longevity for the yachts. The infrastructure is now in place to refit and care for these yachts, but the standards and costs of the different yards must be closely compared.

Marinas are filled with acres of white plastic, bought by unsuspecting innocents on the assurance that the new boat is maintenance-free. Experience shows that there is no such thing as a maintenance-free yacht. The dedicated classic yacht owner will tell you that a plastic boat will last you forever, provided you obey two golden rules – never put it in the water and never take it out in the sunshine. Keep your plastic boat in a darkened barn and it will last forever.

He will also tell you, correctly, that his wooden classic will last far better in the water and will benefit from overuse rather than lack of use. Both arguments are true to a point, but neither tell the full story. The spectre of maintenance rears its head in every discussion on classic boats. After more than 20 years of sailing and maintaining wooden boats, I have concluded that maintenance is more to do with an owner's personal capacity for method and organisation, coupled with a modicum of technical knowledge, than with hard work. A rotting old tub will always be just that no matter how much you varnish her; a fine classic in good condition will be a pleasure to maintain and a source of pride and satisfaction to her owner.

INVESTMENT CONSIDERATIONS

The business of buying requires detailed study and there are a handful of agents with the knowledge to assist. A buyer should read as many books as possible on the subject, starting with Dixon Kemp or the recently republished *Seglers Handbuch* in German from 1897, and taking in Heckstall-Smith's *The Complete Yachtsman*, first published in 1912; Claude Worth's definitive work on yachting, *Yacht Cruising*, first published in 1910; the later and more technical Skene's *Elements of Yacht Design* from 1948; and Arthur Beiser's *The Proper*

Yacht, which features some very attractive yachts from the postwar period.

Assuming that the classic-yacht buyer has already learnt the rudiments of yachting it would be as well to spend some time with other classic-yacht owners and get some first-hand experience. There are many classic-yacht regattas these days at which a potential buyer can see many types, sizes and conditions of yachts and most owners welcome the opportunity to talk about their boats with seriously interested people. A good yacht broker can often arrange for you to sail on a yacht of the type you are considering.

There are several useful rules of thumb for a buyer. Most important of all, always buy the best you can afford. Unless you are specifically seeking a boat to renovate, you will invariably get better value by buying the best quality in the first place. Look at the original quality of build as well as the current condition. Good varnish can cover a multitude of sins.

Always have an independent marine surveyor inspect the yacht before you buy. A knowledgeable friend can be helpful but is not enough for what is going to be a major investment of money, time and energy. Buy within your means both financially and in respect of time and energy. One of the commonest mistakes I see is people buying too big a yacht. If you cannot do what you want with a yacht then you may as well not have one. One gets more pleasure out of running a small yacht well than a large one badly. And there is no hiding the evidence!

Consider the following very carefully:

(i) how exactly you will use the yacht;
(ii) who will sail her and where;
(iii) where you will keep her both in the summer sailing season and the winter lay-up period; and
(iv) how much maintenance you will do yourself and what professional assistance is available.

The bigger the yacht, the more forward planning is required.

Remember that the biggest problem in yacht ownership is not the boat but the crew. Problems with the yacht itself can invariably be overcome with the application of a little thought and money. Crew, on the other hand, are human beings and bring their own problems. Consider carefully then if you are buying a yacht that requires paid hands.

Buying a yacht can be a frustrating business. There may be only two or three examples in existence of a design you have admired and none of them for sale. Your yacht may have to be a compromise but the acid test is whether you find her both beautiful and suitable for your lifestyle. One very viable alternative is to simply build a new yacht to the old designs. The plans, materials and skills are all available, contrary to what some people think, and a 'new' classic will embody all the features of the original with the advantages of a new yacht and, in some respects, new technology. In many cases it is possible to sail the original yacht so the buyer knows what he will get – not the case when commissioning a new design. And even in cases where there are no surviving examples quite detailed reports on history and performance are often available.

Purchasing cost

At current prices, a good quality yacht in clean, sound and original condition will cost around £1,000 per foot up to 40 foot. Above this size the number of yachts in existence decreases and the value increases. A similar quality 60-foot yacht could be around £2,000 per foot. A 75-foot 1920s Fife in fully restored condition could cost in excess of £500,000 and will probably be the only one of her type in existence. While this may sound expensive, these prices must be compared with the cost of a similar-sized, modern, top-quality yacht. Our beautiful classic then appears to be very good value, especially when one takes into account the pleasure derived from owning such a beautiful and unique boat.

The main thrust of the classic yacht movement has centred around sailing yachts but the supply of original boats to satisfy the growing market is limited and I suspect that it will shortly be the turn of the classic motor yacht. It is still possible to buy a very fine 1930s pedigree motor yacht for very sensible money and there are, of course, plenty of classic motor yachts in need of substantial restoration which can be bought relatively cheaply. We have seen restorable 60- and 70-foot pedigree motor yachts in running condition for less than £30,000. Clean, original examples in basically sound and very serviceable condition can be found for less than £100,000.

Whatever type or size of boat one buys, there is no doubt that ownership of a classic gives access to an elite club of interesting people, a secure capital investment and a lot of pleasure.

Maintenance costs

It is often said that in rough terms a yacht costs 10% of her value per annum to run. However, there are so many variables involved that this figure is almost misleading. Separate your standing costs – moorings, insurance, loss of interest, craning in and out and so on – from your other running costs: paint and varnish, regular running repairs, fuel, refitting and upgrading, and crew. Although the actual figures may seem hilarious to us today, the methods of costing referred to in some of the old books already mentioned are very illuminating. Refer the costs to the anticipated number of days you will actually use the boat and, if you are not too surprised, set against this the reason for doing it all: the pleasure factor. You should find that running a classic yacht is not significantly more expensive than running a modern yacht, provided you start with a good, sound boat, especially when you take into account the lower purchase cost.

Chartering

Many owners consider chartering their yacht as a way of offsetting costs. For the small yacht – say less than 45 foot – it is unrealistic, and probably unnecessary, unless one is setting it up as a business in its own right. Bare-boat charter of your prized yacht is a difficult business. Many of the larger yachts do charter but they have permanent crews. Stringent regulations are being introduced for the safety of yachts and their crews and must be considered before a boat is put out for charter.

With large sums being spent on corporate promotion and brand name publicity in the maxi yachts and offshore racing fleets, it is only a matter of time before the large classic yachts attract the same attention and owners derive a financial return by these means. The regattas of the Mediterranean circuit attract wide media coverage and although there has been a little resistance to such marketing, extended sponsorship will undoubtedly creep in.

WHERE TO SAIL

It has been estimated that a classic yacht owner could spend every weekend of the season sailing in a classic yacht regatta between Bergen and Monte Carlo if he had the time and could cover the distance. Regattas are great fun, enabling you to admire other boats and sail in company, comparing one design with another. Other people prefer their own company and will seek quieter seas. Scandinavia and Scotland have some of the world's most beautiful cruising waters but lack the climate of the south. Greece has been very popular for years with its islands and sunshine. Turkey is fast gaining popularity with the yachtsman because of relatively low – though climbing – prices, good scenery, numerous ports and ideal climate.

I recall a more adventurous owner telling me some years ago that he kept his boat in Turkey in the care of an old fisherman. The relative cheapness of his berth and services offset the extra travel costs. I asked him how he got there. 'With some difficulty,' he said. 'British Rail to Heathrow, a two-hour flight to Istanbul, an hour in a train, two hours on a bus sharing a seat with a goat and chickens, followed by an hour and a half walk in the sun.' 'Is it worth it?' I asked. 'It is,' was the reply, 'if the alternative is three hours driving on a crowded motorway to sail out of Brighton to Calais in the fog for the weekend.' However, do not forget that there may be a painful period of acclimatisation if you take your wooden classic from the relative humidity of the north to the searing heat of the south. Having said that, if properly maintained the yacht will suffer no ill-effects from either climate.

CONCLUSION

In the last 10 years, as classic yachting has become increasingly popular, splendid, large and often very well-maintained older yachts have changed hands for very little money and huge sums have been spent on their restoration. These yachts now change hands for vast sums but there are relatively few of them and most are well-known. We believe that all the big classics have now been discovered and restored or are under restoration.

I remember years ago inspecting the hulks laid up as house boats on the mud-flats of England's east coast – famous old yachts beached high on the tide after the war, and supposedly left to quiet and dignified ends. There are few, if any, left; they have all been identified, floated off and rebuilt, often with very little of the original yacht remaining, and at greater cost than simply building a new one to the old plan.

Classic yachts have risen significantly in capital value in the last few years. In a recent survey of yacht values we found that capital values have virtually doubled in the last 10 years. Although this does not take into account any capital investment or upgrading, but only reflects the excess of demand over supply, it is sufficient to give the buyer confidence in the enduring value of his or her purchase. However, yachting is, and always has been, a relatively expensive sport and a buyer should not be tempted into the classic yacht market solely to make a cash investment. The pleasure factor remains predominant. ❖

CHAPTER 5

MANAGING YOUR PERSONAL AFFAIRS

SUCCESSION PLANNING ISSUES IN EUROPE

John Pottage, Coopers & Lybrand, London

John Pottage is an accountant and tax adviser covering UK and international taxes. He advises on both corporate and personal tax issues arising from immigration, cross-border business investments and real estate transactions.

Phone: (+44-171) 213 8237
Fax: (+44-171) 213 8463

SUCCESSION PLANNING ISSUES IN EUROPE

John Pottage, Coopers & Lybrand, London

Many experts consider the complexity of Europe's succession laws to be something of a nightmare. As wealthy people become more mobile and more international in their outlook, it is increasingly necessary for wills to be made in more than one jurisdiction and for complex succession planning structures to be reviewed regularly. Nevertheless, a substantial minority of wealthy individuals in Europe do not have a will even in their home jurisdiction, let alone in other countries where they own properties and businesses.

The nightmare stems from a lack of harmony in tax and succession laws between European countries.

APPLICABLE SUCCESSION LAWS

What determines whether the domestic tax or legal system applies to an individual varies from country to country. In the UK or Ireland, for example, local succession laws apply to an individual who dies domiciled in that state. (Domicile has a particular meaning in Anglo-Saxon common law which may not be readily understood by all Europeans.) In contrast, in the civil law jurisdictions of Austria, Germany, Italy, Netherlands, Portugal and Spain, domestic succession laws apply where an individual dies as a national of that state. Even between civil law systems, however, there is no shared standard, as Belgium, Denmark and France apply domestic succession laws to individuals who were 'habitually resident'. To take an extreme example, the heirs of a Spanish national who dies whilst domiciled in Ireland, but who has been habitually resident in Belgium, could find three different legal systems competing to apply their own succession laws.

Moveable and immoveable property

The position is complicated further by the fact that some countries, the UK and France for instance, apply different succession rules to moveable and immoveable property. Other countries (such as Italy, Denmark and the Netherlands) apply one principle to all types of property.

Subjectivity

Even where countries agree on what criteria determine whether domestic law applies, there is a subjective element in deciding whether an individual fulfils those criteria. For example, individuals are considered to be domiciled in the UK if they intend to stay there permanently or indefinitely. While individuals are alive they can declare their intentions, but this issue can cause controversy on death. In practice, individuals' actions are taken to indicate their intentions.

Changing circumstances

Another problem is that individuals may have made their wills whilst habitually resident or domiciled in one country but by the time they die may have changed their country of residence or domicile. The Hague conference on private international law has reviewed this conflict of laws and has made a little progress towards solving some of the problems. But there is still a long way to go.

Legal and tax considerations

It is important to distinguish the legal factors determining which country's succession laws apply from those that govern the tax system: which countries have the right to levy tax when an individual dies. In France, for example, an individual can be resident for tax purposes (and therefore subject to French inheritance and gift tax law) without being habitually resident and, as such, subject to succession law. In some countries, such as Belgium, there is no such distinction.

So, what determines whether one country's tax law applies is not the same as what determines whether that country's succession laws apply. Indeed, those who divide their time between several countries should take great care where they are deemed to be tax resident. For example, residents of Monaco often spend a great deal of time in France or Italy. They should keep careful track of their visits to make sure that they do not spend too much time in either country.

Some countries, such as Germany (and until recently both the UK and Ireland), operate a test of tax residency whereby individuals who have accommodation available for their use in Germany are deemed to be tax resident if they spend any time at all in the country. The rule applies even to individuals who use the accommodation only for a week or so a year, and even if they do not own or lease the accommodation personally.

This sort of situation does not ordinarily create problems for most people but, because of the lack of a double tax treaty, residents of Monaco who have accommodation available for their use in Germany face serious inheritance tax problems.

Individuals face similar problems if they emigrate from a high-tax country to live, and eventually die, in a lower-tax jurisdiction. Individuals who emigrate from the Netherlands or Sweden are generally subject to Dutch or Swedish inheritance and gift tax for 10 years after they have left. In Sweden's case, individuals are subject to Swedish forced heirship rules even after this 10-year period, if they are Swedish nationals. In Germany, this period is five years. Individuals who have been legally domiciled in the UK are deemed to remain so for three years after they have changed their domicile. Taxpayers should also be aware that countries such as Denmark and Germany impose a capital gains tax exit charge on individuals emigrating.

Succession

Succession does not mean the same thing in every country. In common-law countries, property passes first to the deceased's personal representatives for distribution to the heirs, whereas in a civil law system the property passes directly to the heirs. In all countries the assets will normally pass in accordance with a validly constituted will, although the rules that determine what constitutes a will vary, of course, from country to country.

Intestacy

It is essential to make a will, not only to ensure that assets pass according to the wishes f the deceased, but also to avoid the delays, family conflicts and legal fees that arise from cases of intestacy.

Wills and forced heirship

In most European jurisdictions, a deceased's will may not override statutory requirements that a certain proportion of the estate should pass to either the spouse or children, if any.

In some jurisdictions matrimonial laws protect the interests of the spouse, leaving so-called forced heirship rules to protect the children. In other jurisdictions, the matrimonial law is part of the forced heirship law. This means that if there is a conflict of laws, and more than one legal system is used to devolve assets, the spouse and children might receive rather arbitrary inheritances not at all in accordance with the wishes of the deceased.

In many jurisdictions, for the purpose of forced heirship rules, the value of the deceased's estate includes, in some cases, all gifts made during his or her lifetime and, in others, all gifts made in the last few years.

Trusts

Many individuals faced with such issues plan in advance to avoid, or minimise, the impact of forced heirship rules, either by making lifetime gifts or by settling free estates on trust in a jurisdiction that will not recognise the forced heirship claims of beneficiaries from another country. Others avoid the application of these laws by converting moveable property to immoveable property. For example, it might be possible to avoid French succession problems by owning French property through an offshore company. However, one should always consider the tax consequences of such a structure.

Although trusts are unknown in civil law systems, a number of European countries have signed and ratified the Hague Convention on Trusts which means that they legally accept the existence of trusts in other jurisdictions. Moreover, many civil law countries do use foundations or other entities that resemble trusts. For example, Austrian trusts have become a popular vehicle for succession planning because of the favourable tax regime.

A considerable trust industry has evolved in response to individuals' tax and succession planning needs and, depending upon the circumstances, it may be possible to use trusts to protect assets from taxes, creditors, divorce settlements or forced heirship rules.

For example, an individual creates a family trust in Guernsey in the Channel Islands whilst resident in the UK, and many years later moves to France where he sets up home. As he is then a permanent resident of France, French forced heirship rules apply to the devolution of his assets. The French court may therefore demand that certain distributions be made to the children. The trustees, who have a fiduciary duty to act in the best interests of the children, may reasonably refuse to do this as it may constitute a breach of trust under Guernsey law. In this case, the conflict of laws will be settled by reference to the location of the assets. In other words, much will depend on whether the assets are in France or elsewhere.

Offshore holding company

The final problem is that many wealthy individuals are simply unaware of all of the assets that they own, do not know where the deeds and certificates are kept and do not know in whose name the assets are registered. In this area it is very dangerous to make assumptions. For instance, the French tax code assumes that any assets left in a safe-deposit box are the equal property of each person who has rented the box.

One possible solution is to make use of a company incorporated in a tax-haven country to own an individual's total global assets. When the individual dies, all he or she owns is some shares in a single offshore company rather than many separate assets in several different countries. This may avoid the problem of identifying the deceased's assets and obtaining probate, or its equivalent, in each jurisdiction, but it is essential that the tax consequences of using a holding company are fully investigated.

CONCLUSION

The complexities of inheritance tax and succession law for the international private client clearly present a significant challenge. In order to meet this challenge, detailed and careful planning is absolutely essential. ❖

SUCCESSION PLANNING IN FRANCE

Gail Cavat-Negbaur, Wallace R. Baker and Mathieu Lescot,
Baker & McKenzie Paris

Gail Cavat-Negbaur is a graduate of New York University Law School, and has a certificate of French legal studies from the University of Paris. She has extensive experience of international wills, trusts and succession matters. Ms Cavat-Negbaur was formerly responsible for the international legal department of Valeo Thermique Moteur, a large French company, handling licensing, joint ventures and litigation.

Wallace R. Baker is a graduate of the University of Paris Law School, the law school of the Free University of Brussels and Harvard Law School, with experience in international wills, trusts and succession law. He also has wide experience of major international litigation on product liability and industrial property, and international mergers and acquisitions including public financing. He has written extensively on international business law and, more recently, on French civil procedure.

Mathieu Lescot is a graduate of the Institut d'Etudes Politiques and the University of Paris Law School, with a special degree in taxation. He advises on French succession taxes, international tax matters, financial transactions, mergers and acquisitions, and expatriation of employees, and specialises in French-Dutch investment structures. Mr Lescot contributed to *Doing Business in the Netherlands,* a publication of the French Chamber of Commerce in Amsterdam.

Phone: (+33-1) 4417 5300
Fax: (+33-1) 4417 4575

SUCCESSION PLANNING IN FRANCE

Gail Cavat-Negbaur, Wallace R. Baker and Mathieu Lescot, Baker & McKenzie Paris

The law of inheritance and wills is contained in the Civil Code. Articles 718-892 govern inheritance and the rules relating to wills are found in Articles 895-930 and 967-1046. The statements in this article are of a general nature and may vary according to the specific situation.

WILLS

Under French law, a will is not a contract but a unilateral act; a document signed by a person who expresses an intention relating to the disposition of all or part of the assets owned by that person at the time of death. A will is always revocable prior to death. Its validity is subject to certain formal rules, such as that it must be written. Under French forced heirship rules, individuals are not free to dispose of all of their property if they leave descendants or ascendants.

Holograph will

The simplest will which can be made under French law is a hand-written will (*testament holographe*). It must be entirely hand-written by the testator, in any language. The testator may be assisted by another person. The form of the document is not important. The will must, however, be dated and signed by the testator. If a false date is used, the will may be held to be null and void.

A holograph will can, but need not, be kept by a notary, who may ensure that it is properly executed. If the will is not filed with a notary, the person who finds the will after the death of the testator should bring it to a notary.

Authentic wills

A will can also be signed before a notary, so that he can authenticate the signature, which reduces the risk of a claim that the signature is forged. All statements verified by the notary in an authentic will, which is a public document, are recognised as valid, except in very specific cases. Such a will requires the signature of two qualified witnesses or two notaries, and must contain a date and all other formal legal requirements must be valid. An executed copy (*la minute*) is retained by the notary. Neither the witnesses, nor their parents or relatives to the fourth degree, nor their spouses, nor the notary's assistant, can be beneficiaries under the will.

Such a will may be typed. It must be in the French language, even if the testator does not understand French. In that case, a translation must be read to him or her. The testator should read the will aloud in the presence of the witnesses and the notary and all should sign it.

Secret wills

A third type of will, referred to in French as '*mystique*' or 'secret' is little used. The testator submits the will to a notary, who, in the presence of two witnesses, places the document in an envelope, and stamps and seals it. This type of will may be written by the testator or by someone else, or it may be typed. The will should be signed, but need not be dated. The inscription by the notary on the envelope must be dated, however.

The purpose of this procedure is to avoid any substitution of the will by a false will. The testator should confirm in the text that it is his or her will and indicate whether it is written by the testator or by someone else and, if so, that the testator has verified the contents. These points should be mentioned in the notary's statement on the envelope.

Military or maritime wills

French law provides for military or maritime wills executed by a soldier, sailor or other person 'in isolation'. Such wills are given to a superior officer, doctor or the commander of a ship in the presence of two witnesses and a lower court judge or municipal officer. The will should be confirmed through the usual channels within six months of the return of the testator from the place where it was prepared.

International wills

Under a UNIDROIT convention signed and ratified by France on 1 December 1994, a will is valid in all signatory countries – regardless of where it is executed, the nationality, domicile or residence of the person making the will, or the location of the property which passes under the will – provided that it fulfils certain conditions. These conditions are that it is written (but not necessarily by the testator) before two witnesses of any nationality, in any language, and before someone authorised by local law to prepare a will. In France such an authorised person is a notary or, in the case of French nationals abroad, a diplomatic or consular representative.

CLAIMS BY DEPENDANTS ON AN ESTATE

The way an estate is defined in France depends on several factors, including the marital regime of the couple, the domicile of the deceased and, in some cases, the nationality of the heirs. But the nationality of the deceased is not a factor.

One peculiarity of French estates is that a designated heir is not obliged to accept an estate. For example, if an estate is subject to debt, it may be in the interest of the heir to refuse the estate or to accept it under the procedure known as 'benefit of inventory'. In this case, the heir is responsible only for debts up to the equivalent of his or her share. Currently, heirs must refuse or accept an estate within 30 years of the date of death of the deceased, although a draft law would oblige heirs to make the decision within a five-month period (extendible in special circumstances).

Matrimonial regime

There are two marital regimes under French law – community property, which has several variations, and separate property. If there is no pre-nuptial agreement, then community property rules apply. The first domicile of the couple immediately after marriage determines which country's laws will apply.

For any couple married since the definition of community property was changed in 1965, no matter which regime they are married under, inherited property belongs exclusively to the party that inherits it. Inherited property is therefore not considered part of the community estate unless specified by the donor as being bequeathed to the couple.

To the extent that a married couple is subject to a regime of community property, the surviving spouse has the right to his or her half of that property. He or she may succeed to all of the community property if the spouses are married under the universal community property regime. If not selected at the time of marriage, this matrimonial regime may be adopted any time within two years of marriage by signing a notarised contract, which must then be approved by the court of domicile of the couple. Certain courts require that the spouses furnish documents expressing the opinion of their children on the change of matrimonial regime.

Domiciliary and non-domiciliary

A French domiciliary's worldwide estate is subject to French estate law and forced heirship rules. Only real property located in France is subject to French estate law for a non-domiciled individual. Personal property is generally governed by the law of the last domicile. There are, however, specific international treaties governing this area that should be reviewed on a case by case basis.

Conflict with French heirs

If French heirs have rights which conflict with foreign heirs and part of the estate is located abroad, the French heirs may claim their share of property located in France up to the value of the property located abroad that they would, theoretically, be deprived of.

FORCED HEIRSHIP

French rules of forced heirship apply to that portion of a deceased person's property which is subject to French inheritance rules. All the deceased's ascendants (parents, grandparents, great-grandparents, etc.) and descendants, including natural and legitimate children, are entitled to a portion of the deceased's estate. For example, a single surviving child of the deceased has the right to half of the estate, two children to two-thirds, and three or more children to three-quarters.

In the absence of descendants, surviving parents are each entitled to a quarter of the estate. Brothers and sisters of the deceased, nieces, nephews, uncles, aunts and cousins do not have a mandatory right to any portion of the estate. Gifts made in excess of those allowed under the forced heirship rules are liable to be reduced by the French courts.

Spousal rights

In the Roman law tradition, the family by blood is favoured over the family by marriage. If there is neither a will nor a notarised transfer between spouses (*donation entre époux*), the surviving spouse has the right to a life interest in a portion of the estate, the amount of which depends on whether or not the deceased left children or any other protected relatives. The estate includes both what the deceased disposed of by gift during his or her life and that which he or she disposes of at death. The heirs can oblige the surviving spouse to accept an annuity rather than a life interest. It should be noted that the new draft law would, in the absence of a will, give the surviving spouse the right to a life interest in the entire estate or ownership of one-quarter of the estate should the deceased leave children (similar to the spouse's rights today if the deceased leaves a will). If there are no children, but parents or relatives with a reserve right, the life estate would cover the entire estate, and the right of ownership, half of the estate. The new law would also entitle the surviving spouse to request a contribution in order to maintain his or her present standard of living. This could include the right to a lease for the residential home (the conditions would be established by a judge) or to a specified sum.

A spouse, unlike descendants or ascendants, can be excluded from any distribution of the deceased's property by will, or by gifts to third parties. However, a deceased can maximise the rights of his or her spouse by gift or by will within certain limits.

Under a will, if there are children, a spouse may leave the surviving spouse either the 'non-reserved' part of the estate (the amount depends on the number of children and whether they are legitimate or not), a quarter of his or her property outright and a life interest in the rest, or a life interest in all of the property. The transfer between spouses, which is a simplified notarised document, can provide a spouse with the same rights as a will but has the advantage of giving the surviving spouse the right to choose between the different options (partial or total life interest, partial ownership) at the time of death of the deceased.

To the extent that a life interest is given to a surviving spouse, heirs may require that it be converted to an annuity (*rente viagère*) subject to the provision of sufficient guaranties. For example, if the deceased leaves two children, the surviving spouse can – by will or transfer – be given full property rights to one-third of the estate, or full property rights to a quarter of the estate and a life interest in the remainder, or a life interest in all the assets. If the deceased leaves only illegitimate children conceived during marriage, he or she has the right to leave the other spouse three-quarters of the estate, or half of the estate with a life interest in the rest. If there are only surviving ascendants a deceased has the right to leave by will or gift to his surviving spouse half of the property outright and a residual interest (*nue propriété*) in the remainder. If only one ascendant survives, three-quarters of the property can be given to the spouse outright, plus a residual interest in the other quarter.

INTESTACY

In the absence of a will the rules relating to the distribution of personal property, intangible property and real property are identical.

Children and children of deceased children

Where children of the deceased survive, they are entitled to the entire property. Illegitimate children, when acknowledged, generally have the same rights as legitimate children. Adopted children are treated like legitimate children depending upon the type of adoption (*adoption pleinière*). The share given to children resulting from an adulterous relationship is reduced. If a child dies before the deceased, any assets that he or she would have inherited pass automatically to his or her children.

Spouse's usufruct

In the above cases, a surviving spouse has the right to a life interest in part of the property: a quarter if there are children, and half if the deceased leaves only brothers and sisters or their descendants, a mother and father, or illegitimate children conceived during the marriage. The usufruct to the spouse is only valid if the surviving spouse has not previously received equivalent amounts during his or her lifetime.

Parents

Where no children or their descendants, nor any brothers or sisters survive, the estate goes to the surviving parents. If only one parent survives, half of the deceased's property goes to this parent and the other half to the deceased parent's parents, if surviving.

Brothers and sisters

If no descendants or ascendants survive, then brothers and sisters inherit all of the property. If both parents survive, and there are surviving brothers and sisters, then a quarter of the property goes to each of the parents, and the brothers and sisters share the remainder. If only one parent survives, he or she takes a quarter and the brothers and sisters the remaining three-quarters. If no parents or brothers and sisters survive, the closest relative in each line takes half, subject to certain conditions.

Spouses

If no descendants, ascendants or brothers and sisters survive, and the only other potential heirs are collateral heirs, the surviving spouse takes the whole estate.

Collateral relatives

If there is no surviving spouse or other close relative, the property is divided into two equal parts and distributed among collateral relatives in each line up to the sixth degree.

Escheat to the estate

If there is no surviving spouse or other relatives entitled to inherit (beyond the sixth degree), then the state takes the property of the deceased subject to certain exceptions.

THE ROLE OF THE NOTARY

There is no equivalent of probate under French law. Instead, French law requires the use of a notary for many of functions that probate provides. Notaries are appointed by the Minister of Justice and are exclusively responsible for certain 'acts' (for example, deeds of sale of real property, gifts and marriage contracts).

In order to prove one's quality as heir a notary must draw up an *acte de notoriété*, which states who the heirs are in an estate. If a hand-written or secret will exists, it must be given to the notary, who enters it in his minutes. Within amonth, the notary must send the registrar of the court a copy of both minutes and will.

Where there is real property, the notary must establish a notarised sworn statement to be published in the mortgage office.

It is generally the notary who is responsible for the division of the estate and the formal execution of the various procedures (i.e. delivering the property, providing certificates of ownership, etc.) It is not essential to use a notary to prepare the estate tax declaration. However, with large or complex estates, it is in the best interests of the heirs to consult a notary, as the formalities can be quite complicated.

Notaries are also often used to establish an inventory of personal property for tax purposes. The inventory allows the estate to avoid the forfeit amount of 5% of the value of the rest of the estate for the determination of the taxable value of the personal property in the estate.

THE ROLE OF THE EXECUTOR

In France, unlike many Anglo-Saxon countries, an estate immediately becomes the property of the heirs, and they are responsible for its administration. However, an executor (*exécuteur testamentaire*) may be named in a will. His or her function is to ensure that the legacies are carried out, to which end he or she is granted a power of seisin (*saisine*) in order to hold property and eventually sell or distribute it as necessary. The right of seisin generally applies only to personal property, and lasts for a maximum of one year and one day from the death of the deceased.

ESTATE DUTY

Taxable assets

Estate tax is levied on all transfers of property at the time of death, whether the transfer be under the rules of intestacy, by will or by gift at death. However, there is no tax due on the portion of the community property retained by the surviving spouse. If the deceased is domiciled in France at the time of death, all of his or her property, be it personal or real property, is subject to French estate (or gift) tax. Gifts made within the 10 years preceding death are also included in the estate duty calculation. However, there are a number of international treaties which treat this subject differently (in particular, those relating to double taxation).

If the deceased is domiciled outside France at the time of death, only property located in France will generally be subject to French tax. Shares in a private foreign company, if the majority of the assets consist in real property or real property rights in France, will be subject to French estate tax on the portion that covers the French real property or property rights.

The property being transferred is evaluated at its market value at the time of death.

Tax return and payment

An estate tax return must be completed in all cases except where no real or personal property is being transferred or where the transfer is being made exclusively between spouses or to direct heirs and the gross value is less than FFr10,000. The return must be filed within six months if the deceased dies in France. If he or she dies abroad, the time limit is one year.

The tax must generally be paid when the tax return is filed, but may be paid in instalments with interest over a maximum five years.

Applicable tax rate

French estates are taxed on the basis of the net value (gross value less debts) going to each heir. Rates are determined on the basis of the relationship between the deceased and the heir (see Figure 1).

Deductions

A deduction of FFr300,000 applies to each portion of the estate being transferred to a direct relative, FFr330,000 for the portion being transferred to a spouse, and FFr300,000 for the portion of the estate being transferred to handicapped heirs.

A deduction of FFr100,000 is allowed for single, widowed or divorced siblings if they are aged 50 or more, or infirm, at the time of the deceased's death, and if they have lived with the deceased during the preceding five years.

If none of the above deductions apply, then a deduction of FFr10,000 will be made.

Exonerated property

Sums payable from life assurance policies at the death of the insured are not included in the estate. For policies subscribed to after 1991, this does not include amounts paid after the age of 70, the amount of which exceeds FFr200,000. Insurance contracts with assets

FIGURE 1: TAX RATES

On the portion of the estate transferred to direct relatives (i.e. descendants or ascendants):

Amount (FFr)	*Rate*
Less than 50,000	*5%*
50,000-75,000	*10%*
75,000-100,000	*15%*
100,000-3.4m	*20%*
3.4m-5.6m	*30%*
5.6m-11.2m	*35%*
More than 11.2m	*40%*

On the portion of estates transferred to spouse:

Amount (FFr)	*Rate*
Less than 50,000	*5%*
50,000-100,000	*10%*
100,000-200,000	*15%*
200,000-3.4m	*20%*
3.4m-5.6m	*30%*
5.6m-11.2m	*35%*
More than 11.2m	*40%*

For collateral relatives, and third parties:

Amount (FFr)	*Rate*
Between brothers and sisters	
Up to 150,000	*35%*
More than 150,000	*45%*
Between relatives up to the fourth degree	*55%*
Between distant relatives more than four degrees apart, or third parties	*60%*

invested in shares may provide other tax benefits such as exemption from capital gains tax and avoidance of wealth tax if there is no cash surrender value. French insurance law provides that insurance payments due at death are not subject to the forced heirship rules provided that the cover is not excessive relative to the deceased's income and assets.

Certain other property is exempt from estate tax:

- survival annuities between spouses or direct relatives;
- three-quarters of the value of woodlands;
- a portion of the value of rural property subject to a long-term lease (under certain conditions);
- up to FFr300,000 for an investment in new buildings or those under construction between 1 June 1993 and 31 December 1994; and
- art work and other collections donated to a museum, with the approval of the state.

The following are not subject to estate taxation:

- estates of victims of war;
- estates of victims of terrorism; and
- indemnities paid to AIDS victims.

TRUSTS

France does not have a national law on trusts. Although a draft law was prepared in 1992, it was never voted on and is unlikely to be presented in the near future. As explained above, a will cannot alter French forced heirship rights. Any bequest, other than a life estate as allowed under the forced heirship rules, that would deprive an heir of the immediate right to use and dispose of assets, is not deemed valid. Thus discretionary trusts – which give a trustee the right to determine whether and when a distribution is made to a particular beneficiary – would be contrary to French public policy. However, foreign trusts are generally recognised.

Foreign trusts

Foreign trusts are generally recognised by the French courts unless they are incompatible with

French laws, such as the forced heirship rules. Under the Hague Convention of 1 July 1985, signed by France in November 1991, the signatories agreed to recognise trusts. France has not yet, however, ratified the convention.

Although trustees can generally exercise property rights relating to foreign trusts in France, when French real property is involved, the foreign trusts are valid only insofar as forced heirship rights are observed. Trusts are generally deemed transparent entities for tax purposes.

The concept of trusteeship does not exist *per se* under French law, but it has generally been assimilated and a trustee is acknowledged as having a function approximately equivalent to that of an executor or a legal representative. Difficulties can arise in determining under French law who is the actual owner of the property – the trustee or the beneficiary.

French courts tend to recognise trusts as long as they have not, in the opinion of the courts, been constituted for a fraudulent purpose, and do not contravene French forced heirship laws.

The Hague Convention will apply only to written trusts that were created voluntarily.

Under the convention, the applicable law will be that stipulated by the grantor of the trust. If this is not clear, the following factors are taken into account to determine which law is applicable: where the property of the trust is located, and where the trustee is resident. Should this result in the application of the law of a country that does not recognise trusts, the convention will not apply.

Under the terms of the convention, the parties agree to recognise trust property as separate from that of the trustee. In addition, the parties acknowledge the right of the trustee to appear as such before their courts or national administrations.

The convention specifically states that where a trust violates the public policy of a country (for example, the forced heirship rules), such country need not recognise its validity. In this case, the convention states that the national judge should try to meet the objectives of the trust through other legal means.

TAXATION OF FOREIGN TRUSTS

The tax treatment of foreign trusts is an extremely complex issue on which the French tax administration has made very few comments. Many tax implications remain uncertain and any decision or recommendation would therefore require a prior careful review of the trust agreement and the location of the assets. Consequently, the comments below are of a very general nature.

A distinction is usually made between:

(i) a revocable trust, under which the grantor is deemed to be the sole owner of the assets held in trust with all income deemed to belong to him; and

(ii) an irrevocable trust, which is characterised as such if neither the grantor nor the beneficiary has any power over the trust assets or their administration.

Inter vivos trusts

No gift tax should be due as long as the assets held in trust are not transferred to the beneficiaries.

Trusts arising on the death of the grantor

Trust property should be included in the estate of the grantor as long as the grantor is domiciled in France. ❖

SUCCESSION PLANNING IN GERMANY

Christian Brodersen, Partner, Döser Amereller Noack (Baker & McKenzie), Frankfurt

Christian Brodersen studied law at Geneva, Frankfurt and Edinburgh and was admitted to the German bar in 1975. He gained an LLM from the London School of Economics in 1976. He worked in the legal and tax department of Coopers & Lybrand for five years, before joining Döser Amereller Noack (a member firm of Baker & McKenzie) in 1981, where he has been a partner since 1987. In 1991 he was admitted as a Notary Public.

Phone: (+49-69) 299 080
Fax: (+49-69) 2990 8108

SUCCESSION PLANNING IN GERMANY

Christian Brodersen, Partner, Döser Amereller Noack (Baker & McKenzie), Frankfurt

INTESTACY

German law provides for statutory succession rules in cases of intestacy. These rules differentiate between relatives according to their relationship to the deceased. Children of the deceased and their descendants are classified as the first group; the parents of the deceased and their descendants form the second group; and the third group is composed of the grandparents of the deceased and their descendants, and so on. All members within one group of heirs inherit equally and any heirs in one group exclude all those in more distant groups.

The surviving spouse is in a group of his or her own and has a statutory succession right to 25% of the estate if there are heirs in the first group or 50% if there are heirs in the second group only. The entire estate will fall to the surviving spouse if there are no heirs in either the first or second group, or any grandparents.

If there are no heirs whatsoever, the entire estate falls to the state.

WILLS

The will is an instrument by which estate planning can deviate from the above statutory rules, albeit subject to certain limits.

Estate planning can be conducted by means of a will or an inheritance contract. Whereas the will is a unilateral declaration which can be revoked at any time, the inheritance contract is an agreement with third parties which binds the deceased like any other contract. The inheritance contract is only valid if concluded in notarial form.

The most usual form of estate planning is by way of a will. A will can be established in public or private form. A public will is declared in the presence of a notary public, who will establish a deed on the will and place it in a public depository.

The most common form of will is the private will. There is only one formal statutory requirement that must be met by a private will: it has to be written by the deceased in his or her own hand from beginning to end and must close with his or her signature. No witnesses are required to attest to a private will and there is no need to include the place and date of signing. However, it is totally invalid if it is not hand-written or if it is unsigned.

It follows that such a will can meet the formal requirements of those countries which demand statutory attestation by witnesses and the requirements applicable in Germany, provided that it is hand-written and signed and that the signatures of witnesses are attached. Such signatures do not detract from the validity of the will under German law.

The will can provide for many different kinds of succession. It is, for example, possible to appoint persons as heirs '*pro temp*' with the effect that such persons are only heirs for as long as they live or for any duration as specified in the will. Once the period stipulated in the will is over, the estate will fall to another heir as nominated in the will ('heir subsequent'). The heir *pro temp* is under a number of statutory obligations (which can be expanded or reduced by the will) with respect to the handling of the estate and is accountable to the heir subsequent. Any disposition of real estate encroaching upon the rights of the subsequent heir is invalid without the consent of the heir subsequent.

The system of heir *pro temp* and heir subsequent is supplemented by the 'replacement heir'. In the event that an heir should not be alive at the time when the estate falls due, the will can specify a replacement heir. The same procedure is advisable in the event of an heir declining the inheritance.

The will can specify that a person, or group of persons, shall inherit the entire estate. It is also possible to rule on a partial or specific endowment and to place certain requirements on heirs, such that they have to pay a part of the inheritance to named recipients.

CLAIMS BY DEPENDANTS ON AN ESTATE

The ability to allocate the estate is subject to certain statutory limits. These limits are imposed by the forced share in the estate available to both dependants and the spouse. Statute decrees that the forced share shall be equal to 50% of the statutory inheritance right. The forced share is only available to children (by birth or adoption) and the spouse.

The statutory inheritance right of the spouse amounts to 25% of the estate if there are children, who will share the residual 75% in equal proportion. However, the spouse can choose a share of 50% of the estate, the additional 25% constituting a lump sum settlement of all rights arising in the event of the termination of the marriage. Children will again share the residual amount in equal proportion. This additional 25% lump sum settlement is only available to the spouse if the marriage is subject to the standard statutory matrimonial regime. If the spouses have elected a different matrimonial regime, the additional 25% may not be available.

Once the statutory inheritance rights available to dependants have been determined, the estate planning can take the forced share entitlement into account and provide accordingly.

Another area for which the will should make detailed provision concerns succession planning for businesses. Any such planning must be supplemented by appropriate rules in any partnership agreements providing for the right by certain partners to appoint a successor in the partnership through a will.

The forced share entitlement can be revoked in certain exceptional circumstances. This would be the case if the entitled party should:

(i) conduct his or her life in a manner that is 'without honour' against the will of the deceased, unless the heir should permanently renounce by intent and action his or her previous ways at the time of the inheritance; or
(ii) intentionally have violated his or her statutory obligation of financial support towards the deceased; or
(iii) have been guilty of a crime against the deceased or the spouse of the deceased; or
(iv) have intentionally committed bodily harm to the deceased or the spouse of the deceased; or
(v) have made an attempt on the life of the deceased, the spouse or any child of the deceased.

In summary, courts will only accept a particularly serious malfeasance on the part of an heir as justification for withholding the forced share. The forced share does not entitle the party to a share in the estate *in rem*, but only the right to demand cash payment. In view of this statutory requirement, meticulous estate planning is necessary in order to avoid placing an undue burden on the estate as a result of cash outflows. The same concern, of course, applies to estate duty (see below).

PROBATE

The probate court is charged with investigating all rights to the estate. The probate court will question all known relatives of the deceased in order to determine entitlements to the estate and investigate the validity of any will submitted.

On the basis of its investigations, the probate court will issue an inheritance certificate. This certificate carries the statutory presumption that the persons shown are in fact the legitimate heirs. All public bodies will accept the inheritance certificate as sufficient proof of entitlement.

The inheritance certificate can be contested on the basis of additional or subsequent evidence and any inheritance certificate shown to be incorrect can be cancelled by the probate court. The period of investigation by the probate court is typically three months or less in straightforward cases. More complicated situations may, however, require a longer period of investigation by the probate court.

EXECUTORS

The deceased is free to appoint an executor to handle all aspects of the estate and its administration. The maximum period of appointment is 30 years. It is generally advisable to appoint not only an executor, but also a number of replacements in the event that the original executor should no longer be available.

The appointment of an executor is imperative if the estate is complicated to handle or if it includes business interests which require advice from an experienced party. This would be the case, in particular, if the heirs were young, inexperienced or unreliable. The appointment of executors is a routine matter in such circumstances.

ESTATE DUTY

The estate duty statute applies to both inheritance and gifts. The duty is payable if the deceased, at the time of death, or the heir, at the time of inheritance, is resident in Germany. In addition, estate duty will be levied if the estate is located in Germany (for instance, real estate, business assets, shares of at least 25% in resident corporate entities and certain loans).

The groups of heirs discussed above also feature in the determination of estate duty, as follows:

(i) group I includes the spouse and children;
(ii) group II includes descendants of group I;
(iii) group III includes parents, grandparents, siblings and their children, in-laws and any divorced spouse; and
(iv) group IV includes all others.

Each group benefits from certain exempt amounts, as follows:

(i) DM500,000 for the spouse and DM90,000 for each child;
(ii) DM50,000 for members of group II;
(iii) DM10,000 for members of group III; and
(iv) DM3,000 for members of group IV.

In the case of gifts, these exempt amounts are available once every 10 years.

The estate duty rates are listed in Figure 1 (sample only).

There are specific rules applicable to the valuation of assets forming part of the estate.

FIGURE 1: ESTATE DUTY TAX RATES

Groups	I	II	III	IV
Value in DM	%	%	%	%
500,000	*7.5*	*15.0*	*24.5*	*38.0*
1,000,000	*10.0*	*20.0*	*32.0*	*48.0*
6,000,000	*14.0*	*28.0*	*40.0*	*56.0*
25,000,000	*21.0*	*36.0*	*50.0*	*62.0*
100,000,000+	*35.0*	*50.0*	*65.0*	*70.0*

These provide, in particular, for real estate to be valued at approximately 25% or less of the fair market value. Similarly, shares not publicly traded and partnership interests are subject to complex valuation rules, whereas publicly traded shares are valued on the basis of their fair market value. For this reason, it used to be highly beneficial to convert the estate to real estate, for instance, prior to making a gift or in preparation for the passage of the estate. However, the constitutional court, in a widely noted judgement, recently rejected these valuation rules as unconstitutional on the basis that they violate the equal treatment rule. As yet, however, the legislator has not formulated new valuation rules. Therefore, from 1 January 1996, the fair market valuation will be applicable to all elements of the estate.

The heir is subject to a statutory duty to notify the tax office of any acquisition liable for estate or gift duty. Banks and insurance companies are subject to a separate obligation to notify tax authorities of any inheritance of which they have knowledge.

DISCRETIONARY TRUSTS

German law does not incorporate any specific statutory regulations on trusts, although the function of a trustee is provided for. The only structure remotely similar to the Anglo-Saxon trust is provided by the option to set up an incorporated trust (*Stiftung*). This requires state consent and is generally only used for the administration of substantial estates, primarily for public purposes.

The treatment of a trust established pursuant to a foreign law in Germany depends on the arrangement in the particular trust deed and on the law of the country of establishment. It is difficult to give general rules on such treatment. Normally, however, a foreign trust will be treated on a see-through basis, thus allocating trust assets and income to the parties eligible under the terms of the trust, particularly in the case of an endowment trust. The procedure may vary, however, in the case of discretionary trusts. ❖

SUCCESSION PLANNING IN THE UK

Paul Stibbard, Partner, Baker & McKenzie, London

Paul Stibbard graduated from Cambridge with an MA in natural science and law in 1969, and from INSEAD Business School in France with an MBA in 1981. He joined Baker & McKenzie's London office in 1988 and became a partner in 1991. He is a member of the firm's private client department.

Phone: (+44-171) 919 1000
Fax: (+44-171) 919 1999

SUCCESSION PLANNING IN THE UK

Paul Stibbard, Partner, Baker & McKenzie, London

WILLS

A will deals with the disposition of property upon a person's death. Until then it may be revoked or changed at any time and, during the testator's lifetime, he or she may dispose of property as he or she pleases. A beneficiary under a will who survives the testator will be entitled to a share in the testator's estate.

Formal and essential validity of wills

The formal validity of a will depends upon whether the formalities governing its execution have been observed. Under English law a will should be signed by the testator in the presence of two independent, adult witnesses, both of whom should be present at the same time, and who should sign their names as witnesses.

The essential validity of a will depends on the outcome of certain issues arising from conflicts of law. In particular the bequests made in the will (other than those for English immoveable property) must comply with the laws of the testator's country of domicile that govern the rights of family members. For example, if a French-domiciled man with three children makes an English will leaving all his English moveable property to a charity, his children may apply to the English court to amend the will so that it complies with the principles of the French 'forced heirship' rules. If French law provided that the children should receive 75% of the estate between them, the will would be amended to comply with this requirement.

Property held in joint names

In general, testators may dispose of any property they own at the time of their death. The main exception to this general rule is joint property. The law governing joint property is technical in nature.

People who jointly own immoveable property may hold it either as 'joint tenants' or as 'tenants in common'. Property held as a joint tenancy passes independently of the will (or intestacy) to the surviving joint tenants. The share of property held as tenants in common devolves as part of the estate in accordance with the terms of the will or intestacy.

Where a person owns moveable property in joint names, this may, subject to certain exceptions, pass by survivorship to the surviving joint tenant(s).

Foreign immoveable property

Where a testator domiciled in England owns immoveable property in France, the devolution of that property must comply with French forced heirship rules, and the will should be executed accordingly. In the above example of a man leaving three children, French law would disregard any provision of the will that did not comply with French forced heirship rules.

Deeds of variation

It is open to any beneficiary who has reached his majority, who receives a benefit from a deceased person's estate, to redirect the benefit for tax purposes. This may be useful for tax planning after the date of death.

INTESTACY

The property of people who die intestate (without making a will), will be distributed to their surviving relations according to statute, under the intestacy rules.

If the spouse and children are still alive, the spouse takes any personal chattels, such as cars, jewellery and furniture (but not goods owned for business purposes) together with the sum of £125,000, which may be disposed of by the surviving spouse as he or she wishes. The remainder of the estate is then divided in half. One half will be held on trust for the spouse for life, with the capital passing to the children on his or her death (provided that they have by then reached the age of 18). The other half will be held upon trust for the children and pass to them when they reach the age of 18.

Where the deceased's spouse survives but there are no surviving children, the spouse takes the sum of £200,000 together with any personal chattels. The remainder of the estate is divided in half. The spouse takes one half of the estate absolutely (i.e. both capital and income) and the other half passes to any surviving parents of the deceased or, if the deceased's parents are no longer alive, is shared equally between any surviving siblings.

Partial intestacy

A partial intestacy arises where part of a person's estate remains undisposed of under the terms of his will. In these circumstances, the intestacy rules are amended to take into account any property that is effectively disposed of. For example, if the deceased's spouse is surviving and he or she acquires, under the will, property other than the deceased's personal chattels, then the lump sum to which he or she would be entitled on total intestacy is reduced by the value of that property.

CLAIMS BY DEPENDANTS ON THE ESTATE

Where the devolution of a deceased person's estate does not adequately provide for a surviving spouse or dependants, they may have a right under the provisions of the Inheritance (Provision for Family and Dependants) Act 1975, to apply to the court for an order for financial provision. This act only applies where the deceased was domiciled in England and Wales.

Spouses, former spouses, children, cohabitees (in certain circumstances) and any person who was maintained by the deceased at the time of death may, within six months of the date of the grant of probate or letters of administration, apply to the court for financial provision. In making financial provision for a surviving spouse, the court may have regard to what provision would have been awarded in divorce proceedings, but this is not the sole criterion. The overriding consideration is what is reasonable in the circumstances.

It is at the discretion of the court whether to make any financial provision for a spouse or dependant, and the court will take into account circumstances such as the applicant's financial resources and needs, and any obligations that the deceased had towards the applicant. When determining what sum, if any, should be awarded, the court will also consider the financial resources and needs of the beneficiaries and of the other applicants. This is to ensure that other actual or potential beneficiaries are not deprived of their entitlements from the estate. When considering the assets of the deceased available for distribution, the court will also take account of joint property.

INHERITANCE TAX

Inheritance tax (IHT) is charged on the total value of a person's estate at the time of death, together with the cumulative total of lifetime transfers made by him or her. In determining the value of an estate for IHT purposes, the net value of all property that the deceased owned, all trust property in which he or she had a life interest, all property gifted up to seven years before death and any property gifted in which he or she reserves a benefit, are aggregated and the rate of IHT applicable to the various items is determined. Under current law, the first £154,000 of the value of an estate is exempt from IHT. The balance is subject to IHT at the rate of 40%.

Exemptions

There are a number of exemptions from IHT. For example, transfers between spouses (except in the case of a transfer from a domiciled spouse to a non-domiciled spouse) or to charities will generally be exempt from tax.

There are also a number of exemptions that can be used during the testator's lifetime, but cannot be taken advantage of at the time of death. For example, gifts of up to £3,000 per person per year will be exempt from IHT. If transfers in one year fall short of £3,000, the amount by which they fall short may be brought forward to the next year. Any shortfall unused at the end of the second year is lost.

Similarly, gifts of up to £250 made to any one person in any one year are exempt. The exemption cannot be used to exempt the first £250 of a larger transfer. Gifts made out of normal expenditure out of income may also be exempt from IHT.

Agricultural property relief

Agricultural property relief is available, both on lifetime transfers and on death, in respect of agricultural property, such as land and buildings occupied in association with it. The relief operates by reducing the value of the property transferred and thus the tax payable.

The reduction in the value of the property is 100% where the transferor had the right to vacant possession immediately before the transfer, or the right to obtain vacant possession within 12 months of the transfer. Where the transferor does not have such rights, relief will be applied at 50%.

The right to agricultural property relief applies in only two cases:

(i) if the transferor occupied the land for agricultural purposes for the two years before the date of the transfer; or

(ii) if the transferor owned the property for seven years prior to the transfer, and it was occupied for agricultural purposes by the transferor or another person throughout that period.

Business property relief

Relief is also available for 'relevant business property'. As with agricultural property relief, it reduces the value transferred. 100% relief is available for the transfer of certain types of business property, including businesses, controlling shareholdings in unquoted companies and minority shareholdings of more than 25%.

Other forms of business property, such as controlling shareholdings in quoted companies, attract relief at a rate of 50%.

In order to benefit from relief, the transferor must have owned the relevant business for two years prior to the transfer. Where the transfer is made before death, and the transferor dies within seven years, the relief is only available if the property continued to be business property in the hands of the transferee.

Potentially exempt transfers

The Inheritance Tax Act of 1984 introduced the concept of the potentially exempt transfer (PET). Any gifts made up to seven years before the deceased's death, after making allowances for any relevant exclusions or exemptions, will be subject to IHT. On the other hand, gifts made more than seven years prior to death are potentially exempt.

Gifts with a reservation of benefit

Under IHT legislation, if a person makes a gift but is not entirely excluded from the enjoyment of the property, then the gift is deemed to be ineffective for IHT purposes and the property is treated as forming part of the donor's estate. However, if the reservation ceases before the donor's death, it is deemed to become a PET as of that date, and will be subject to the rules relating to PETs. An example of such a gift is where a woman gives her house to her son but continues to live there, rent free, for life.

Foreign property

The total global assets of any person domiciled in the UK will be liable to IHT. For people not domiciled in the UK, only property located in the UK will fall within the IHT net.

The rules relating to foreign property can result in a person being liable to IHT in two countries. For example, a person who is domiciled in the UK but owns a house in France, is potentially liable for IHT in the UK. He or she may also be liable for tax on the house under French law. So tax becomes payable in both countries, on the basis of the location of the property in one country, and the testator's domicile in the other. UK law may, subject to conditions, grant relief from such double taxation either unilaterally or, where applicable, under relevant double tax treaties.

DISCRETIONARY TRUSTS

A discretionary trust gives the trustee a discretion to distribute income amongst a class of beneficiaries named in the trust. Under the terms of such a trust the beneficiaries do not normally have any vested entitlement to the income or capital of the trust fund; the trustees have the power to make distributions in accordance with the terms of the trust. Where the settlor of property retains no interest in the trust, the assets of the trust will not be aggregated with his assets, but will instead be taxed under a separate IHT regime.

IHT on gifts to discretionary trusts

Any transfers to a discretionary trust over and above the exempt threshold may be subject to an immediate charge to IHT at the lifetime rate of 20%.

10-yearly periodic charge

Property held by discretionary trusts is revalued upon each 10th anniversary of the trust.

The top rate at which IHT is charged on the trust property is 30% of the lifetime rate of 20%. The maximum rate of IHT payable on the 10-yearly periodic charge is therefore 6%, i.e. 30% x 20%.

Exit charge

IHT will also be payable when property ceases to be subject to the terms of the trust: for example, when it is transferred to a beneficiary. IHT will be charged on the value transferred on a time-apportioned basis from the date of the creation of the settlement or the 10th anniversary of the trust, whichever is the most recent. For example, if five years (out of 10) have elapsed since that time, the maximum IHT payable, based on current rates, will be 3% of the value transferred out of the trust.

PROBATE AND ADMINISTRATION

To obtain a grant of probate or, in the case of intestacy, letters of administration of a deceased person's estate, the personal representative must either prove the will of which he is appointed as an executor or, if no executor is appointed, show that he is the person legally entitled to administer the estate.

A grant of probate or letters of administration is an order by a judge of the Family Division of the High Court to the deceased's personal representatives, who are placed under a statutory duty to administer the deceased's estate. This will involve paying the debts, expenses and other liabilities of the deceased and collecting and realising his or her assets. Once this has been done, the estate must be distributed to the persons entitled under the will or intestacy rules. The personal representatives will also be liable to account for IHT chargeable on the estate.

In the case of a person dying domiciled in England and Wales leaving an English will, the procedure is as follows:

(i) The executor must swear an oath for executors with the will.

(ii) The executor must disclose, in an Inland Revenue account, details of the deceased's assets and liabilities, together with trusts in which he or she had an interest, and gifts made up to seven years before death.

(iii) IHT on the estate must be paid before a grant of probate can be obtained.

Where a person dies intestate outside the UK, the procedure may be more complicated:

(i) Besides the oath to be sworn by the administrator, an affidavit of law is often required; or a court-certified copy of the probate documentation may be required where title to the estate assets has already been obtained in another jurisdiction. If this is in a language other than English, the document will need to be translated by a competent expert, who must also provide an affidavit in support.

(ii) Not only will the Inland Revenue require an account disclosing details of the deceased's assets in the UK, but the administrator will have to prove the deceased's domicile to the Capital Taxes Office before the probate papers can be submitted to the probate registry. ❖

DIVORCE AND PROTECTING YOUR WEALTH

David Davidson, Partner, Charles Russell, London, with the assistance of Karen Bruce-Lockhart, Brodies WS, Edinburgh, and Tim Amos, Barrister

David Davidson, admitted as a solicitor in 1972, is a partner in the family law department of Charles Russell. A founder member of the International Academy of Matrimonial Lawyers and a member of the Procedure Committee of the Solicitors Family Law Association, Mr Davidson has written and broadcast on taxation and divorce. Mr Davidson has more than 20 years' experience as a specialist in family law.

Phone: (+44 171) 203 5000
Fax: (+44 171) 203 0205

This article has been prepared with the assistance of Karen Bruce-Lockhart and Tim Amos. Ms Bruce-Lockhart is a writer to the signet at Messrs Brodies WS, a family law firm practising mostly in the Court of Session (Scotland's highest court) for almost 25 years.

Phone: (+44 131) 228 3777
Fax: (+44 131) 228 3878

Mr Amos is a member of the English bar specialising in family law, and is also a member of the British-German Jurists Association.

Phone: (+44 171) 797 7837

DIVORCE AND PROTECTING YOUR WEALTH

David Davidson, Partner, Charles Russell, London, with the assistance of Karen Bruce-Lockhart, Brodies WS, Edinburgh and Tim Amos, Barrister

This article is a discussion of the way the courts in six major jurisdictions deal with financial applications following divorce.

CROSS-JURISDICTION CASES

People of substance who are employed by international organisations, and who face marital difficulties, are likely to be faced with a choice of jurisdictions in which to file for divorce. Following is a list (which is not exhaustive) of factors that a couple's respective advisers will have to consider at the outset.

- Which country has jurisdiction to entertain divorce proceedings?
- What are the permitted grounds for divorce and do they have any bearing on financial provision?
- Which country's law will apply to the financial consequences of divorce?
- What is the effect of any pre-nuptial agreement on property matters?
- Will a decree in one country be recognised in the other countries?
- What is the nature and extent of financial provision allowed under each country's law?
- Is it possible to enforce financial orders made in one country in the other relevant jurisdictions?
- What powers does the court have to order disclosure of financial assets and to make orders restraining the parties from disposing of or dealing with their assets?
- Does the court have the power to make orders for costs in respect of:
  (i) the divorce, and
  (ii) the financial proceedings?
- Does the court have power to make orders for interim provision?
- Would the court give preference to proceedings in a foreign jurisdiction if there were concurrent proceedings in two different jurisdictions?
- What taxes will there be on transfers of property?

ENGLAND AND WALES

England and Wales is one of the most sophisticated jurisdictions in its resolution of money matters on divorce. The Family Division judges have a wide-ranging knowledge and understanding of the affairs of high-net-worth individuals, company accounts and complex tax-planning schemes. There is a similarly experienced core of individuals at the family law bar and in the family law departments of leading solicitors' firms.

The court's powers

The court has powers to make orders for payments of periodic maintenance and of lump sums of capital, and for adjustment of property rights, and it may require a fund to be secured to ensure payment of maintenance.

The jurisdiction is discretionary in that no fixed formulae are applied to the distribution of capital and income on divorce. The court has wide-ranging powers to set aside dispositions made by a spouse with the intention of defeating the other spouse's financial claims. Procedural rules strictly enforce the requirement to make full and frank disclosure of all the family's income and assets; for example, production of accounts, bank statements and tax returns going back several years. A court order can be set aside if one spouse has materially misrepresented his or her means.

Trusts and inherited wealth

If one spouse is a beneficiary of a trust, the court will look at the history of distributions to determine the underlying reality of the conduct of the trust's affairs and will assume, unless there is compelling evidence to the contrary, that the trustees will continue a similar distribution policy in future years. The court looks at the reality of the situation.

In cases involving large estates that have been passed down through a family for generations, the court is alive to the intention that it should not be broken up to the detriment of the next generation. Overall resources usually permit a solution to be found that provides fully for the spouse who has not inherited whilst keeping the estate intact.

Illiquidity

Genuine illiquidity may be a husband's best ally in minimising a wife's financial claims.

Duration of maintenance orders for wives

In many cases there is a requirement on wives to take proper steps to enhance their earning capacity. This is particularly relevant to the duration of maintenance orders, which are subject to review upon changes in the spouses' financial circumstances. The court is reluctant to dismiss wives' claims for periodic payments whilst children are still under the age of 18.

The 'clean break'

In most high-net-worth cases the court will impose a 'clean break' between the parties if there is sufficient money to capitalise the wife's future maintenance entitlement. Broadly, the approach is, in the context of all the couple's circumstances, to ascertain the wife's reasonable housing requirements (which may in some cases be for two or more houses), to provide her with other necessaries such as cars and, finally, to calculate her annual requirement and capitalise it on an actuarial basis. Of course, the wife's own assets and income are offset against the overall capital needed in fixing the provision the husband must make.

The effect of this approach is that the greater the wealth the smaller the percentage that will be awarded to the wife. In practice, no wife has been awarded more than 32.6% of the family assets in a contested financial-provision case involving large amounts of capital. Of course, higher-percentage awards have been achieved in negotiated settlements.

Changes in the law

Although changes to the divorce procedure are likely to be put through Parliament in the near future, no changes are contemplated in respect of financial provision.

SCOTLAND

Scotland is also a sophisticated jurisdiction when it comes to resolving financial issues on divorce but, with a smaller population, fewer cases relating to high-net-worth individuals reach the Scottish courts than the courts south of the border.

The court has the power to make orders for payment of capital, transfer of property and maintenance of the other spouse. It can also freeze property pending the outcome of the divorce, and can strike down transactions or disposals that have the effect of defeating the other party's claim, if they were made within five years of the application for provision.

The court has a very wide discretion, but will look first at achieving a fair division (one that is reasonable with regard to the parties' resources) of net matrimonial property acquired during the marriage and existing at the date of separation (not divorce). It can take into account special circumstances: inter alia, that such assets were 'sourced' from outside the marriage, such as by inheritance or as a gift from a third party, or simply owned prior to the marriage.

Other special circumstances include the terms of any agreement between the parties as to the division, destruction or dissipation of the property, the nature of the property, and the extent to which it can be divided or used as security. The court can, however, look at the overall assets of the parties at the time of divorce in certain clearly defined circumstances: unfair financial advantage or disadvantage during the marriage, the financial burden of the children (in addition to maintenance), a period of adjustment, and severe financial hardship caused by the divorce. A capital settlement is always to be preferred to an income settlement.

The laws of Scotland differ from those of England in that financial provision for the spouse must be made at or before divorce. If divorce is granted without financial provision being made, the right to a capital settlement is lost and maintenance can only be claimed on limited grounds on a material change of circumstances.

Any agreement (whether made before or after marriage) in respect of the financial consequences of divorce, that was not fair and reasonable at the time it was entered into, can be struck down in whole or in part on divorce.

Inherited wealth is better protected from attack than in England.

FRANCE

Property rights during marriage

In French law, spouses are subject to either separation of property or community of property. Whenever a foreign element arises, it is advisable that the applicable regime be set out in a notarial contract prior to marriage. Otherwise the spouses will, in most cases, be deemed to have adopted the normal regime of the country where they established their first stable matrimonial home, and this may not be what they intended.

Grounds for divorce

Divorce can be by consent, or on the basis of fault or six years' separation.

The effect of divorce

Except in six-year separation cases, compensatory orders called *'prestation compensatoire'*, based on equitable distribution, can be made to compensate for the difference in standards of living after divorce, and to be provision over and above the 'community' distribution (when applicable). They are final orders and cannot be varied, save where failure to vary would cause exceptional hardship.

Payment may be in the form of a lump sum, an indexed annuity or a life interest in property.

The court takes into account the (actual or probable) division of matrimonial assets according to the applicable matrimonial property regime. The maintenance award will take into account any wealth, including community of property rights.

There is no fixed formula for calculation of the annuity and it is rare for capital sums to be provided in lieu of an annuity. The use of, or right to occupy, the matrimonial home for life is a frequent provision in consent orders.

When deciding on quantum, judges do not try to treat each spouse equally, but try to provide for the financially weaker party so as to allow him or her to have a similar standard of living to that enjoyed during the marriage.

In fault divorces, the sole 'guilty' spouse is not entitled to equitable distribution, but there is a hardship clause allowing monetary compensation after divorce in special circumstances, if to refuse it would be contrary to equity.

The matrimonial home

The matrimonial home cannot be sold, let or mortgaged unless the other spouse agrees. Where it has been purchased jointly by both spouses during marriage it can either, by agreement, be sold and the proceeds divided, or a judge can order that it be attributed to one of the spouses either immediately or after a period during which the judge has authorised the other spouse, usually with children, to live in it but to pay a sum in lieu of rent, e.g. half of the normal rent. At the end of this period, the judge can order that the home be transferred to one spouse in return for a sum equal to the value of the transferor's share.

Where the home is owned by one spouse, the judge can order the owner to let it to the other spouse, and can determine the rent if the divorce is based on separation and the non-owner is not cohabiting, or if the non-owner has custody of children living with him or her.

Clean breaks

French family law encourages clean-break settlements. When *prestation compensatoire* is in the form of an indexed annuity, it can be for life or be limited to a fixed period depending on all the circumstances, e.g. the age of the wife, her earning capacity, the existence of children. In practice it often runs for between one and 15 years (more often for five to 10 years) when the wife is less than 50 years old. An older wife is usually entitled for life.

In general, in a six-year separation divorce, the right to maintenance as opposed to an annuity continues for life. It can be varied because of changed circumstances and usually ceases on remarriage.

Tax consequences

Important note should be taken of the impact of French tax laws on the implementation of a court order. Transfers between husband and wife may lead to tax on the value of the asset (not the value of the consideration) and it is very important to take local advice. Transfers between strangers (and this would apply to former spouses after a final divorce) are taxed at a higher rate.

Disclosure and injunctions

Although there is a duty on both parties to disclose all relevant financial circumstances and the court has a general power to order the disclosure of specific information by third parties, there is no effective process for enforcing disclosure in the face of a refusal to give it.

Where an order has been made by consent, it cannot be set aside for non-disclosure.

The French courts have the discretionary power to grant injunctions for non-disposal of assets in exceptionally urgent cases, but it is rarely exercised.

GERMANY

Property rights during marriage and the effect of divorce

Unless otherwise provided for by agreement in the form of a notarial deed, spouses live under a scheme called 'community of accrued gain'. This means there is no automatic community of property; every spouse remains 'owner' of his or her property and upon the dissolution of the marriage a compensation claim is made. This will take into account the amount of the gain made by either spouse (subtracting inheritances, gifts and so on, but adding back sums intentionally dissipated within the last 10 years). The gain of each spouse is then compared with the other and the spouse with the larger gain is required to pay 50% of the difference to the other. It is, of course, perfectly possible for spouses to reach their own separate contractual agreements on divorce.

Maintenance and support law is full of provisions that give the court a wide discretion, but the starting point is spousal self-support. Child support is based on income bands set out in the 'Düsseldorf Table'. Spousal maintenance, where appropriate, tends to be a quota of three-sevenths of total income for the payee. In high income cases, the quota approach is replaced by a needs-based assessment.

There is a complex system of splitting rights and expectancies derived from public and private pensions and insurance. It works in a manner similar to the system of community of accrued gain, and is thus useful to housewives but less so to career-women.

Monthly maintenance payments are subject to variation, if required. Orders for property transfer do not exist except in relation to household effects. Preservation of the matrimonial home where necessary is usually regulated by transfer (or creation) of tenancy or occupation rights. Lump-sum orders exist only in the sense of being the means of achieving the equalisation of accrued gain referred to above.

Disclosure and injunctions

There is no general duty to disclose, but there is a 'right to information about relevant data' in relation to income and to those assets relevant to determine the calculation of accrued gain. If the spouse required to give information fails to do so properly, there are financial penalties and a possibility of intervention by the court bailiff. If the contents of the information given are disputed and there is sufficient reason to doubt their accuracy, a sworn affidavit may be requested. If that is found to be incorrect, the only sanction is criminal proceedings.

A well-organised person will generally find it fairly easy to hide assets or to 'park' funds without the knowledge of his or her spouse.

Injunctions preventing the disposal of assets or related financial dealings are available but rarely used prior to the final financial order. A statutory ground for such an injunction exists whenever the judgement would otherwise have to be enforced abroad.

ITALY

Property rights during marriage

Spouses enjoy common ownership of property, which ends on separation – when each spouse becomes the owner of 50% of the family assets. It is possible to avoid common ownership by agreement. In general, real and movable goods fall under common ownership whether purchased jointly or individually by the spouses during the marriage, but there are exceptions under the Italian Civil Code. Individual ownership may be achieved when the purchasing spouse uses money that was already his or her exclusive property.

There is no proper 'marriage contract' in Italy, but spouses can make marital-property agreements, which regulate the economic status of the family and which must be sealed by a notary or public official.

Financial orders

In determining the amount of alimony the court aims to maintain approximately the lifestyle that the economically weaker spouse enjoyed before the separation or divorce. Each spouse's income and assets, including property deriving from separation of ownership, are taken into account. If a spouse is able to work and support him- or herself but is unwilling or fails to take measures to do so, alimony may be reduced or even denied.

The grounds for divorce, the respective personal and economic contributions made to the family and to commonly and individually owned assets, and the duration of the marriage are all taken into account in determining alimony.

It is possible to have a one-off lump sum payment to the spouse entitled to support only if the parties are in agreement and the judge approves the amount.

On divorce or separation, community of property comes to an end and the community property must be shared equally.

Discovery and injunctions

There is an obligation to provide income statements and other documentation pertaining to revenue or personal and commonly held property at the first hearing in any cases of separation or divorce. In cases of dispute, the judge may order an investigation of income, assets and lifestyle, and may turn the matter over to the excise and revenue police. This police division may carry out its investigations even through financial institutions and companies. Proving the existence of a substantial hidden income is a difficult task.

One can obtain an injunction restraining a spouse from disposing of family assets before the final order of divorce is issued.

Interim orders

During the first hearing, the president of the divorce tribunal can issue temporary decisions which have immediate effect with regard to urgent problems, such as child custody, who is to have the family home, maintenance and child support. These decisions are confirmed or amended when the final sentence of divorce is issued.

SWITZERLAND

Divorce and family law is federal law, but the procedural law governing claims based on it will vary from one canton to another.

Most couples will be governed by the current law on financial provision, which came into effect on 1 January 1988. This article does not deal with the 'old law', which applies to a couple married before that date, and who decided before 1 January 1989 to remain governed under the old law, or who had a marriage contract in accordance with the old law and did not wish, before 1 January 1989, to accept the new law.

Property rights during marriage

If one spouse is not a Swiss citizen or the marriage has been celebrated abroad, the applicable law is that chosen by the spouses. If they have not made a choice of law, their marital property is governed by the law of the state in which both are domiciled simultaneously or, if this is not the case, the law of the state in which both spouses were last simultaneously domiciled.

Where Swiss matrimonial property law applies there are three systems:

(i) The ordinary system called 'participation in acquisitions'. This consists of each spouse's own property and property acquired as income under the matrimonial property system. In the event of a divorce each spouse is entitled to half of the other spouse's matrimonial profit, being the difference between the total value of the acquisitions after deduction of debts by which they are encumbered.

(ii) Community of property can be put into force only by domestic contract, publicly authenticated. Under community there are three separate slices of property: the common property and each spouse's own property. The common property belongs to both spouses jointly and neither spouse may dispose of his or her share in it. In the event of a divorce each spouse is entitled to his or her own personal property and to half of the common property. It can be difficult and complicated for the court to separate the two, if there is insufficient evidence regarding the provenance of assets.

(iii) The system of separation of property provides that each spouse will administer, use and dispose of his or her property as he or she wishes and can only be established by domestic contract.

The court's powers

Where a divorce prejudices the innocent spouse in his or her pecuniary rights or expectations, the court can make orders for a lump sum payment, transfer of property, maintenance and damages. The court has a wide discretion in fixing the amount of the indemnity for damages. Besides considering the spouses' property, the court must consider every circumstance which could influence their financial situation, for example, their age, their health, their professions and how many children they have.

Disclosure, injunctions and interdicts

Each spouse can request information as regards income, fortune and debts from the other spouse; at a spouse's request the judge can order the other spouse or a third party to give the information required and to produce the necessary documents.

If it is necessary to protect industrial or commercial secrets, the court can obtain and consider a document without showing it to the other party.

The court does not have coercive powers in respect of the production of information but, if a spouse does not produce the documents required, the court will draw its own conclusions as to his or her failure to do so.

The court has the power to grant injunctions or interdicts, during or before a proceeding, to restrain a spouse from transferring assets if they are necessary to protect the rights of the other spouse, i.e. to avoid defeating a wife's claim.

GENERAL

This is only a brief résumé of the issues that can arise in dealing with the financial consequences of divorce in major European jurisdictions. It is absolutely essential to seek advice from a lawyer resident and practising in the jurisdiction in which you may be contemplating bringing proceedings. ❖

MANAGING YOUR PUBLIC PROFILE

Jeffrey Schultz, Burson-Marsteller, Hong Kong and Rob Koreneef, Burson-Marsteller, The Hague

Burson-Marsteller is a leading international communications consultancy. Jeff Schultz is managing director, corporate and financial, at Burson-Marsteller in Hong Kong.

Phone: (+852) 2880 0229
Fax: (+852) 2856 1101

Rob Koreneef is a consultant with Burson-Marsteller's office in The Hague.

Phone: (+31-70) 362 4861
Fax: (+31-70) 360 5800

MANAGING YOUR PUBLIC PROFILE

Jeffrey Schultz, Burson-Marsteller, Hong Kong and Rob Koreneef, Burson-Marsteller, The Hague

Those who play an important financial role in society today have many options open to them, but obscurity is not one of them. The media's appetite for stories, and the public's curiosity about those who affect their lives even peripherally, are insatiable. Eventually you, too, will be thrust into the limelight in one guise or another.

Even if you cannot be approached directly, your background can be researched. You might refuse to answer requests for interviews, but your friends might not, and your enemies most assuredly will not. Nor do you have control over how information about you is used. You do, however, have a choice about how you present yourself to the world. While you cannot avoid the public glare completely, you can most certainly endeavour to manage your public image.

First, you must decide upon the kind of public profile you wish to adopt. Few people seem to choose to have an extremely high profile, and the reasons for this are obvious. In Europe, the drawbacks of fame can range from mere inconvenience to threats to personal safety. In some countries the price of notoriety is paid in heavy taxes and exposure to the risk of kidnapping.

These risks will deter most people from seeking a permanent spot in the limelight. However, even if you are intent on keeping a decidedly low profile, it pays to set the rules of the game yourself. Proven techniques exist which will enable you to manage your profile. In marketing, one of the most powerful of these is called positioning, that is the deliberate creation of competitive advantage.

THE PRINCIPLES OF PERSONAL PROFILE MANAGEMENT

From Julius Caesar to François Mitterrand, history shows us that there are many ways to handle your public persona. It is wise, however, to bear in mind three important principles:

(i) conservatism;
(ii) a proactive approach; and
(iii) positioning.

Conservatism

There is much to be said for the old adage that less is more. You probably do not need to be quite as famous as you think in order to achieve your aims. And in any case, the appearance of effortlessness will often actually help your cause. Thus, it is wise not to seek a higher profile than you need. This helps you avoid the 'tall poppy' syndrome, by which some might resent your new prominence, and possibly even seek to detract from it. It is a lamentable truth that the media, having thoroughly covered a remarkable rise, find nothing so interesting as a precipitous downfall. Such a situation becomes more than just an irritant when your creditors join the clamour. For instance, few would doubt that Donald Trump's high profile contributed to his subsequent troubles.

However, if you are determined to achieve a very high profile, the slowness of the conservative approach might be discouraging. If that is so, go 'over the top' by all means. There is nothing wrong in doing so, provided you are prepared to take the risks, which may include, but are not limited to, public ridicule.

One advocate of the more hard-hitting approach is Silvio Berlusconi, one of Italy's leading business and political figures, who was in a hurry to establish a powerful personal profile. Berlusconi enhanced and used his status as a successful businessman to reach the apex of Italian political power in a matter of months. Regrettably, his rapid rise to power only served as a prelude to his fall from public grace. Having promised to strengthen the Italian economy by running his government like his businesses, he soon learned a harsh lesson about the difference between private enterprise and government, resulting in his resignation as prime minister. He has been suffering in the media and the polls ever since.

A proactive approach

Some people do indeed have greatness thrust upon them, but Shakespeare's aphorism makes no mention of timing. One would not expect financial gain simply to materialise as if by magic, and it is no more rational to expect a public image, much less a desirable one, to form spontaneously and on schedule. You must take control of the situation and think strategically, whether you are seeking a high or a low profile. You must make the best possible use of the environment and the tools at hand. In short, you should decide what to do with the media before they decide what to do with you.

Positioning

In using the term 'positioning', no one is suggesting that you should be 'packaged' and marketed as something you are not. I would, however, suggest that many of the principles of product and corporate brand management apply equally to individuals. The reason is that all three share the same challenge: there is limited space in the public mind. One cannot be all things to all people; one can, however, be some things to all people. Positioning is the methodical assessment and management of how one is perceived in the public eye.

Begin with some research on your current positioning (and yes, you probably have one, even if you have never issued a press release in your life). It is doubtful whether you can, or indeed should, employ a market research firm to do this. In all probability, a very honest and perceptive self-assessment will suffice.

The first question to ask is: how do the media, your community, your colleagues and your public see you? Continue by asking yourself: what do I really want from my personal profile – appointment to the board? Romantic conquest? A knighthood? Or perhaps just a little respect (and if so, from whom?) Next question: who do I need to influence to get there – the other members of the board? My best friend? The general public as a whole?

Finally, what influences are likely to affect the behaviour of these target groups? Newspaper reports? TV? The family? The annual report? Trade gossip? There is no need to confine yourself to what is rational here; often emotional factors, such as fear, chauvinism or family loyalty, can drive people more powerfully than the most persuasive rationale.

Only after you have plumbed your own objectives honestly, identified the individuals or institutions whose behaviour you will need to influence, and found the best means to get them on your side, can you be said to have begun to position yourself. Notice, however, that none of this has to do with pulling the wool over anyone's eyes. The objective is not to change or misrepresent yourself, but to focus the attention of the world on those aspects of yourself that you most want them to see. If it helps, look on public opinion as a newspaper: do you want to be on the front page, the sports page or in the lifestyle section?

One of the most effective ways in which to develop personal positioning is through personal 'branding'. Most people are aware of individuals whose 'branded' personal style is inseparable from their public image: Winston Churchill and his cigar, or Richard Branson in his jeans. If this tactic works for you, and it comes naturally, by all means use it. However, such a personal trade mark cannot be faked or forced, and it is far better to make use of your unique assets. If you have an illustrious family, for instance, the society pages may be your route to prominence. Equally, if you race cars or yachts, you have a ready-made profile builder.

DEALING WITH THE MEDIA

The media today have an image problem of their own. All too often, the press and television are seen as enemies to be held at bay. Sometimes they are considered, with equal inaccuracy, easily duped accomplices in every conceivable publicity stunt. It would be wiser to view the media collectively as a group of friends, intelligent and independent-minded friends to be sure, but by no means hostile or disagreeable.

To deal effectively with the media, it is important to be armed with a few essential facts. The 'European media' do not exist. Europe consists of many different cultures, and this inevitably has an impact on the way the media work in different countries. For example, in the UK and in southern Europe it is permissible to invite journalists on an all-expenses-paid trip to tour a factory. In Germany and the rest of northern Europe, much depends on the policies of the individual media.

In most European countries media relationships are crucial, and you will need to know how to establish and maintain them. The press luncheon is a time-honoured way of building relationships with newspaper and television people in a non-threatening environment. The contacts you make will prove useful when your name crops up in the news later; it helps enormously to know with whom you are dealing.

Once you have honed your sense of positioning, you will find yourself varying your message slightly from one reporter to another, and from one situation to another. A reporter who is a seasoned specialist in your field will demand different treatment from an inexperienced general features writer.

In this regard it is possible, and advisable, to prepare for your public life through media training and role plays. Most people are surprised at the number of traits and mannerisms that can be caught on video, criticised and improved upon. With proper training, you can become a veteran before your first interview; you may still lack Henry Kissinger's mastery in dealing with the press, but you will certainly be more at ease and in relative control.

One advantage of dealing with the media is that it is possible for you to speak to a great number of people through a very small number of contacts. However, do not be misled by the small number of people in this circle of journalists: to gain useful exposure to international publications, one must know precisely what they are after. Here an experienced public relations (PR) agency is your most useful tool. A PR firm will know what it is that the US and Asian papers are looking for, and they can help you gain publicity there without leaving Europe.

Another consideration is that the expression 'as old as yesterday's news' is no longer applicable today. Yesterday's news is alive in electronic databases, which journalists access regularly for research. These days, it is much more difficult to escape past mistakes! Reports of minor indiscretions, and even innocent but misleading facts, may well pop up in an article reviewed years later. If you spent a year as a used-car salesman before beginning 20 years of industrial entrepreneurship, rest assured that it will be uncovered and published, and possibly weighted with derogatory implications.

TACTICS

Once you have chosen an appropriate profile to present to the public and have made a start at positioning and understanding the role of the media as a delivery mechanism, you can turn your attention to tactics.

Issue management

One effective tactic for projecting your profile is the deliberate management of public issues, or rather your participation in such issues. Espousing a worthwhile cause can single you out from the pack in the same way as a strong corporate culture can create an outstanding corporation, or unique operating features can promote a product.

Your chosen issue should be something you passionately believe in and which you are knowledgeable about. It should be flexible enough to be adapted to a number of public platforms. Choose your cause carefully, and pursue it tenaciously.

What issue should you select? Depending on the country you operate in, politics can be risky, but extremely effective. In the right hands, however, and with a high degree of commitment, such political issues can play an important part in establishing a personal reputation. Environmental causes, and those who espouse them, also attract a great deal of attention from the media in some countries. Once you have chosen a field in which to be active, pursue it vigorously. Make speeches, write articles and host forums with others active in associated organisations. If there is a governmental or civic panel dedicated to the issue, you will want to be on it. You will benefit from personal promotion, while the cause will also naturally be publicised.

Philanthropy

Perhaps less popular than the championing of public issues, but just as effective as a means of profile development, is philanthropy. Personal philanthropy is prevalent in America: in Europe, corporate sponsorship is more usual.

Like corporate sponsorship, private donations can yield substantial benefits – to individuals, their families and the corporations associated with them, if done strategically and over a prolonged period. To this day, the names of the Rockefeller family and the Chase Manhattan Bank are closely associated with philanthropy in the arts and education. Although the family's managerial influence and ownership interest in the bank dwindled to insignificance many years ago, it serves the interests of both to maintain such associations in the public mind.

Investor relations

If you have a public company, do not overlook the importance of investor relations. From the point of view of news editors, public companies are easily twice as newsworthy as unlisted companies. Until the start of Operation Centurion in 1990, the Philips company seriously neglected investor relations, even though shareholder communications events present invaluable opportunities for profiling the controlling shareholder. Under the reign of Jan Timmer, however, Philips' investor relations improved dramatically, amplifying the positive effects of his restructuring efforts.

Government relations

Government relations constitute an essential part of your multi-faceted positioning strategy. In fact, government is probably one of your most important target audiences. Although many business people prefer an indirect approach to government, this is becoming ever more difficult as governments become increasingly supranational and complex. You should, as in all positioning exercises, begin by familiarising yourself with your target audience; you must understand the predispositions and biases of the officials and departments you will be dealing with, and be patient with red-tape and officialdom if they delay your access to the higher realms of government.

Networking

Perhaps the most widely employed positioning tactic of all is old-fashioned networking – the pursuit and cultivation of personal contacts. This is not always as easy as it sounds, however, even if you are a gregarious person. You will need a strong stomach for the endless procession of lunches, dinners, cocktail parties and teas which are essential to the process of cultivating contacts! It may take years of thankless wining and dining to develop personal contacts in government or the media. Nevertheless, such contacts are integral to every other tactic mentioned above.

THE PUBLIC RELATIONS FIRM

At some point in your career, you are likely to contemplate seeking professional assistance in establishing your personal profile. It is worth considering the advantages and disadvantages of employing a public relations firm.

Positioning yourself can be a daunting task, as the preceding summary indicates. A public relations firm will offer experience, a tried and tested approach and, perhaps most important of all, ruthless objectivity. The firm will also offer a host of ready-made contacts. Professionals know which publications and outlets are the most effective for your particular purposes, and they will approach them in a systematic and disciplined fashion. With their experience, they know who is receptive and who is not, and where the greatest effect will be achieved. Importantly, the reputations of PR professionals also depend on their discretion.

You must be careful to choose only the best PR firm. There are many amateurs who call themselves public relations professionals, and while they may cost less than top firms, they may offer poor value for money. Remember, your reputation is your most valuable asset. How many of your other assets would you entrust to amateurs?

The public relations firm you choose should be generalist enough to grasp the overall task at hand, but have the kind of specialist capabilities – financial and government relations, for example – that will make it adept at dealing with the specific audiences you need to reach. And note that, even with a large agency, you must be careful to deal with senior members of staff at all times in order to ensure that your particular interests are properly represented.

CONCLUSION

In summary, managing your personal profile requires a combination of common sense and highly professional positioning techniques. If you are Margaret Thatcher or George Soros, there is little that anyone will be able to teach you on this subject. On the other hand, most people will find the methodology of positioning – so valuable to marketers of toothpaste and sugary water over the years – a very useful starting point.

The media can be a valuable delivery mechanism, yet media contacts are no substitute for substance, strategy and incisive analysis of those whose behaviour you are trying to influence and the methods of reaching them. Public opinion is like a newspaper: you must decide whether you want your face to be found on the front page, in the financial pages or in the society pages. ❖

BRITISH INDEPENDENT SCHOOLS

David Woodhead, National Director, ISIS

David Woodhead has been national director of the Independent Schools Information Service since 1985. He was a journalist (the *Cambridge Evening News* and the *Sunday Telegraph*) from 1965-75 and chief press officer, Inner London Education Authority, 1975-84. He was educated at Queen Elizabeth Grammar School in Wakefield and at Leicester University.

Phone: (+44-171) 630 8796
Fax: (+44-171) 630 5013

BRITISH INDEPENDENT SCHOOLS

David Woodhead, National Director, ISIS

British boarding schools have a long tradition of educating foreign and expatriate pupils. Anyone who has represented such schools at education exhibitions throughout the world is soon made aware of how high a reputation they enjoy.

Parents elsewhere in Europe or many thousands of miles away in East Asia are equally aware of the quality which attracts about 7,000 overseas pupils every year to the British independent schools that are members of the Independent Schools Information Service (ISIS). These include all the best-known British schools whose names are synonymous with educational achievement. Many others come to Britain to attend independent further education colleges or independent schools outside ISIS.

In recent years the number of overseas recruits to British boarding schools has increased substantially: by 10% in both 1993 and 1994. The biggest boost has been from mainland Europe (particularly Germany): an unprecedented single-year rise of 17%. While German, Spanish and French pupils lead the European field in numerical strength, there is rapidly increasing recruitment from Eastern Europe, particularly Russia. Outside Europe, the largest numbers come from Hong Kong (which still provides about 1,700 new pupils each year to ISIS-member schools), the rest of East Asia (in which Taiwan has become an increasingly important market at both school and higher-education levels) and North, Central and South America.

Why do they come? The chief attractions are undoubtedly high academic standards and intellectual rigour, expertise in English language teaching and unparalleled extracurricular opportunities, all underpinned by a caring environment.

So the British boarding school, born of the monastic tradition, transformed in the public school boom of the 19th century and updated dramatically in the 1980s, provides a unique blend of traditional educational values in a contemporary social setting. With teaching and facilities of the highest standards, these schools equip pupils for the 21st century in whatever part of the world they make their careers.

'Public school?' We talk about independent rather than public schools, not least because calling schools 'public' when they are in the private sector is rather confusing. The term 'public school' has no legal definition. It probably derives from the fact either that schools admitted fee-paying pupils as well as those to whom the school gave a free education, or that these schools were open to boys (and it was only boys then) from all over the country rather than just those living in the neighbourhood.

A LONG TRADITION

The history of independent schools in Britain is largely the history of school education for all but about 120 years out of the past 14 centuries. That is because until the last quarter of the 19th century there were no schools provided by the state. And that is what we mean by independent schools: those which are neither controlled nor financed by the state or by local education authorities. They are run by their head teachers and staff, controlled by their boards of governors or proprietors, and financed by the fees paid by parents.

The King's School, Canterbury, can claim to be the oldest independent school in the UK. Its roots go back to 600 AD, when St Augustine founded his monastery. King's School in Rochester and St Peter's, York are almost as old. But it was at Winchester in 1382 that a school was first founded on the boarding pattern that was to be emulated by many others in the centuries to come. Eton followed in 1440.

In the 16th and 17th centuries there occurred a great flowering of what were often called grammar schools, because they concentrated on teaching Latin and English grammar. The next great expansion of schools took place in the 19th century in response to the needs of the prosperous middle class created by the industrial revolution. It was not until some 150 years ago that independent schools for girls began to appear, although a small number do date back to the 17th or 18th century.

A WIDE RANGE OF OPTIONS

Today there are 2,400 independent schools throughout the UK. They educate 600,000 children between the ages of two and 19, representing 7% of the total school population. Of these, about 100,000 are boarders.

The choice of schools is very wide: girls' schools and boys' schools for parents who want their children educated in a single-sex environment; girls' schools that take boys at the lower end of the age range and boys' schools that admit girls from the age of 16; and coeducational schools for those who want boys and girls to be educated together.

There are day schools, boarding schools and many with a mix of day and boarding pupils. There are schools catering for the entire age range, junior and preparatory schools (the latter so called because they prepare boys and girls for senior schools) and senior schools taking children from the age of 11 or 13. And there are specialist schools for musically gifted pupils, choir and ballet schools and schools providing for a wide range of special educational needs.

HIGH ACADEMIC STANDARDS

Not only do independent schools have smaller teaching groups than their counterparts in the state sector, but they are often smaller in overall size. The average ISIS-member school has about 350 pupils. Both are attractive features to many parents and pupils. Among parents' reasons for choosing independent schools are their high academic standards, proven success in encouraging children of all abilities to achieve beyond their expectations, good discipline and broad curriculums.

Independent schools are preeminent in examination performance. Pupils at all ability levels achieve better results than those at other schools. More than 80% of independent school pupils gain five or more passes in the General Certificate of Secondary Education (GCSE) at the top grades – nearly twice as many as the national average.

A similar proportion achieve three or more passes at GCE Advanced Level (A level), compared with a national average of 58% (and the national average is inflated by independent school results). About 90% of school leavers with A levels go on to higher education degree courses. And whilst only 7% of children attend independent schools, more than a quarter of British university students are independent school educated.

Comparisons of academic success are now measurable not only between senior schools: preparatory and independent junior school pupils have been shown to be up to six times more likely to achieve top grades in government classroom tests for 11-year-olds than children in other schools. No other schools in the country have more experience than preparatory schools of the benefits of regularly testing pupils' academic progress.

Preparatory schools' success also depends on 'setting' pupils by ability, so that the brightest are not held back and the slower learners are given more time; on specialist subject teaching; manageable class sizes; giving pupils a thorough grounding in basic skills; and, perhaps above all, having high expectations.

CAREFUL MONITORING

Parents need to be confident that independent schools are regularly monitored. Every new school must be registered with the Department for Education and Employment and is visited by one of Her Majesty's Inspectors (HMI) for this purpose. Independent schools are also regularly inspected by HMI and the Office for Standards in Education (OFSTED).

Schools belonging to ISIS and its parent body, the Independent Schools Joint Council (ISJC), provide parents with an important additional assurance of quality: they and their constituent associations of governors and heads must be accredited by the ISJC's Accreditation, Review and Consultancy Service. This involves a thorough initial inspection and review visits are made thereafter to ensure that standards are being maintained.

OVERSEAS PUPILS

Most parents sending their children to British independent schools from overseas want them to experience a British education. That means mixing predominantly with British pupils rather than being in an international-style school which they could probably find in their own country. Because many parents from overseas do not want their children to be educated among too many others of their own nationality (which could adversely affect their English-language skills, for example), schools that take foreign children normally limit their numbers to between 10% and 15% of all boarders, and encourage a mixture of nationalities.

Foreign children are admitted to British boarding schools at various ages. The best times to enter schools are: before nine years of age for a junior school, at 11 or 13 for boys' senior schools, and at 11 or 12 for girls' senior schools. It is not recommended that pupils enter a new school between the ages of 15 and 17, but arrangements can be made even in these cases. The years between 11 and 13 are particularly beneficial in a preparatory school and enable a good choice of senior school to be made. At 14 years of age, new arrivals normally embark on the two-year GCSE course and at 16 years entrants usually begin the two-year A-level course. Many stay on in British education after the age of 18, entering British universities.

Junior and preparatory schools are able to help younger children improve their English if it is weak. Children who cannot read English fluently in textbooks for their age group need schools that make special provision for the teaching of English as a foreign language.

The school year starts at the beginning of September and finishes in early July, with month-long holidays starting in December and March, and short mid-term breaks. A few schools accept new pupils from overseas in January.

Parents who live overseas are normally required to appoint a guardian in Britain to be responsible for the child during mid-term breaks and longer holidays if the child cannot go home. Guardians are often friends or relations living in the UK; otherwise, schools recommend suitable people who live locally or ISIS can suggest guardianship agencies. There are agencies that provide responsible adult escorts to accompany younger children while travelling from home to school.

FINDING THE RIGHT SCHOOL

There are more than 800 independent schools with boarders in the UK, so overseas parents need expert help to find the right ones for their children. ISIS International in London has many years of experience in placing pupils from overseas in British schools. It offers placement and consultancy services for which parents pay fees (ISIS takes no commission from schools) and a clearing service for which a nominal fee is charged. In each case, the most important considerations are the needs of the child and the requirements of the parents.

Boarding fees vary: at preparatory or junior schools they range from £6,000 to £9,000 a year; at senior schools from £7,500 to £13,000. These are basic yearly fees. Extras can add considerably to the bill (perhaps another 10%), depending largely on which activities an individual child chooses to take part in (such as instrumental tuition and school trips). Almost all schools have a uniform. Schools will explain what items are additional to the basic fee. Most schools increase fees once a year, usually in September; at present annual increases are in the region of 5%.

ISIS has recently carried out two major surveys of what pupils and parents think of boarding schools. The verdicts of both groups were overwhelmingly enthusiastic. The pupils' views were encapsulated in the title of their survey – *One Big Family* – and the parents' views in the title *The Best of Both Worlds.*

'One big family in which each individual counts' is a good way to sum up the British boarding school ethos. ❖

CHAPTER 6

STRUCTURING YOUR WEALTH

EUROPEAN TAX PLANNING ISSUES

Coopers & Lybrand

This article was prepared by the following:

FRANCE

Marc Crетté is a partner in the French member firm of Coopers & Lybrand International. He specialises in advising on foreign investment in French real estate, and structuring the ownership of private clients' assets to minimise their liability to income, wealth or inheritance tax.

Phone: (+33-1) 4572 8353   Fax: (+33-1) 4572 8701

GERMANY

Bjorn Ahrenkiel is a partner in the German member firm of Coopers & Lybrand International. He is a lawyer specialising in the tax affairs of owner-managed businesses and high net worth individuals.

Phone: (+49-211) 952 1337   Fax: (+49-211) 498 0494

ITALY

Marcello Romano is an accountant in the Italian member firm of Coopers & Lybrand International. He specialises in advising private clients and businesses.

Phone: (+39-51) 345 762   Fax: (+39-51) 341 883

SWITZERLAND

Patrick de Balthasar is a partner in the Swiss member firm of Coopers & Lybrand International and heads its legal and tax department. He specialises in advising private clients from all over the world.

Phone: (+41-21) 728 3750   Fax: (+41-21) 728 7526

UNITED KINGDOM AND EUROPEAN TAX RATES

John Pottage, the coordinator of Coopers & Lybrand's international private client network, is an accountant and tax adviser covering UK and international taxes. He advises on both corporate and personal tax issues arising from immigration, cross-border business investments and real estate transactions.

Phone: (+44-171) 213 8237   Fax: (+44-171) 213 8463

EUROPEAN TAX PLANNING ISSUES

Coopers & Lybrand

To invest savings in one's home country alone is unduly restrictive today when overseas markets can offer excellent, if sometimes temporary, opportunities for capital growth. Investment in any region, however, requires consideration not only of factors such as political and currency risks but also of the tax laws, and perhaps more importantly still, the methods available for keeping taxes to a minimum. The advantage of depositing funds in a country paying high interest can be completely eliminated by a swingeing withholding tax, for example; if the remedy is to seek relief through the use of a tax treaty with the interest-paying country, the question then arises of which treaty provides the best protection.

Such double taxation agreements often provide relief for a resident of a country covered by the agreement, but a private individual would often have many non-tax reasons for choosing to live in a country that is not ideal for tax purposes. A decision then needs to be made on a suitable vehicle into which savings could be transferred. If it is to be a company incorporated in a country covered by the required tax treaty, what are the rules of residence in that country? What effect will incorporation have on tax liability there? Is liability to tax based on source alone or on both source and residence? What anti-avoidance measures exist in its tax laws and to what extent are its tax laws overruled by the tax treaty under consideration?

A company may not be the best vehicle in which to hold a long-term investment in real estate if avoidance of estate duty on the death of the beneficial owner is the ultimate aim. Ascertaining whether the potential real estate investment is situated in a jurisdiction that recognises trusts could therefore be an important first step in any tax planning exercise for a real estate investment.

But these are complex issues, and space does not permit a full exploration here. When considering an investment abroad, you should always consult a qualified tax adviser who will take into account all the specific circumstances of your position.

The following are brief descriptions of important tax jurisdictions in Europe. They provide the basic tax facts and answers to questions that an individual considering investment in those countries may ask.[1]

FRANCE

Who is liable for income tax?

Individuals who are resident in France for tax purposes are liable for income tax on their income sourced in France unless otherwise provided by a double taxation treaty. Individuals are deemed to be resident in France for tax purposes when:

- they have their home in France or France is the place of their principal abode; or
- France is the place where they perform their main professional activities; or
- France is the centre of their economic interests.

[1] *This introduction, by Antony L. Brown of Price Waterhouse Hong Kong, is reprinted from the Asian edition of* The Credit Suisse Guide to Managing Your Personal Wealth.

An individual need meet only one of these criteria to qualify as a taxpaying resident under French domestic law.

This principle can be overridden only by international double tax treaties, which take precedence over domestic law.

Income tax rate

The income for each category is combined and the total taxed at progressive rates ranging from 0% to 56.8%. Nonresidents are taxed at the same rates, but their total income tax bill may never be less than 25% of their French taxable income.

The total taxable income is divided by a coefficient allocated to taxpayers according to their family status, before the income tax is calculated.

Married taxpayers are taxable on a common basis (*foyer fiscal*). The joint tax return reports the combined incomes of husband, wife and dependants. Spouses may, however, be taxed separately under certain circumstances, such as divorce.

Tax on interest

Interest received by French residents is treated as ordinary income and is thus normally subject to personal income tax. However, individuals resident in France may elect to be subject to a prepayment tax which constitutes complete satisfaction of their tax liability. The prepayment is levied at source by the French bank paying the interest at rates ranging from 15% to 35%, depending on the nature of the interest. On top of the prepayment tax come additional contributions totalling 4.4%, made up of a 1% social contribution, a 1% social levy, and a 2.4% general social contribution.

Interest sourced in France paid to nonresident individuals is subject to a mandatory withholding tax at rates ranging from 15% to 35%. The additional contributions are not levied. However, most double taxation treaties signed by France reduce the rate of the withholding tax to 10% and some even reduce it to 0%.

Tax on rental income

Rental income is taxed as ordinary income after the deduction of the following:

(i) a fixed 13% allowance for depreciation and insurance expenses. On the acquisition of a new dwelling which is to be rented out as a principal residence in France for a minimum period of five years, this allowance rises to 25%;
(ii) mortgage interest;
(iii) management expenses; and
(iv) property taxes.

Rental losses deriving from expenses, other than financial costs, are creditable against all other income up to a limit of FFr50,000. Rental losses exceeding this limit can be carried forward for five to 10 years, depending on the type of property let.

Property tax and inhabitant tax

Owners of real estate are liable for property tax at rates fixed annually by each town council. The basis for calculating this tax is equal to the rental value of the property as determined each year by the Land Registry.

An inhabitant tax is levied on any individual who occupies a residence on 1 January in any year, whether he or she is the owner, a lessee or even living there rent-free. The tax is levied on the rental value of the property, at rates fixed annually by the local authorities.

Capital gains tax on real property

Generally, capital gains realised by French residents are treated as ordinary income and are subject to tax at progressive rates after the deduction of various allowances, which depend on the duration of ownership, inflation, repairs, improvements and mortgage interest. Some transactions are expressly tax-exempt. This is the case notably for gains deriving from the first sale of a secondary residence (subject to certain conditions), and gains from the sale of any real property held for more than 22 years.

Under certain conditions, nonresidents may also benefit from the exemption for very long-term gains and for gains on the sale of their residence in France. These exemptions apply only if the individual has been a tax resident of France for at least one year and has had full use of the residence since its acquisition, or for at least three years.

The net taxable gain is calculated in the same way as for French residents, but is subject to a 33.33% withholding tax in complete satisfaction of their French tax liability on that income. Even if the transaction is exempt or no taxable gain is realised, they are still required to file a special return and to appoint a tax representative who is responsible for paying taxes and any tax penalties.

Capital gains tax on transfers of securities

As a general rule, capital gains realised upon the sale of stock or securities by French resident individuals are taxed at a flat rate of 19.4% when the annual amount of gross sales exceeds a certain threshold which amounts to either:

(i) FFr50,000 in 1995 for sales of certain securities; or

(ii) FFr336,000 in 1994 for other sales.

Moreover, capital gains on the sale of shares in companies liable for corporation tax are also subject to the 19.4% tax provided the seller has held, either directly or indirectly, on his or her own or with his or her spouse and their ascendants or descendants or both, more than a 25% interest in the company at any time in the five years prior to the sale.

Nonresidents are liable to pay a 16% final tax on capital gains realised upon the sale of substantial interest (25%) in French companies, unless an applicable tax treaty provides otherwise.

Any other capital gains on securities (other than gains on shares in French real estate companies which are taxed according to the same rules as property gains) are exempt from French tax.

Taxation of dividends

Dividends received by French residents are subject to French personal income tax as ordinary income.

When distributed by a French company, the taxable amount of the dividends is grossed up by a tax credit equal to 50% of their amount, the so-called *'avoir fiscal'*, which is credited against income tax. The *avoir fiscal* tax credit is normally granted only to taxpayers who are resident in France for tax purposes. However, many tax treaties provide for the extension of the *avoir fiscal* to nonresident individuals through a direct refund from the French treasury.

France does not recognise tax credits for foreign withholding tax unless there is a double taxation treaty. As a consequence, dividends received by a French resident taxpayer from non-treaty countries are taxed on their net amount. Dividends received from a treaty country are grossed up by the tax credit corresponding to the foreign tax withheld at source at the treaty rate. This tax credit is offset against the corresponding French income tax. Dividends distributed by a French company to nonresident taxpayers are subject to a 25% final withholding tax, unless a tax treaty provides for a reduced rate (usually 15%).

Both the dividend itself and the refund of the *avoir fiscal* are subject to withholding tax.

Wealth tax

Individuals domiciled in France on 1 January of a given year are liable to pay French wealth tax on their worldwide assets, unless provided otherwise by a tax treaty. Nonresident individuals are liable for tax only on their assets located in France (e.g. properties located in France, claims on French debtors, shares and securities issued by French resident companies or French real estate companies).

The wealth tax is levied on net wealth exceeding a certain amount (FFr4.53m in 1995) according to a progressive schedule ranging from 0.65% up to 1.65% for net wealth in excess of FFr43.94m (1995 figures). Various exemptions are provided, notably for:

(i) antiques;

(ii) woodlands and interests in woodland companies. These are exempt for up to three quarters of their value, subject to certain conditions;

(iii) professional assets, and notably interests in certain companies when the interest is at least 25% and the taxpayer carries out a managerial activity in the company; and

(iv) nonresidents' financial assets.

Gift and inheritance tax

Gift and inheritance tax is a unified tax.

If the donor or the deceased is domiciled in France, estate duties are due on all assets, whether moveable or immoveable, and whether located in France or abroad. If the donor or deceased is a nonresident of France for tax purposes, only property located in France is subject to French inheritance or gift duties. Claims on French debtors, shares and securities issued by French resident companies and shares in foreign companies whose assets mainly consist of French real property, are deemed to be located in France for the purposes of such taxes.

Gift and inheritance tax is levied according to tax schedules which vary depending on the family relationship between the beneficiaries and the donor or deceased. Between spouses and for transfers in direct family line, the rates range from 5% to 40% for the part of the assets exceeding FFr11.2m. Between non-related parties, the maximum rate is 60%.

An allowance of FFr330,000 applies to the surviving spouse, while each of the ascendants or children benefits from a FFr300,000 allowance.

A favourable tax regime is granted for gifts given by parents to their children as a distribution of property from their potential estate. Such gifts – the so-called '*donations partages*' – benefit from a 25% reduction of the gift tax when the donor is less than 65 years old and 15% when the donor is between 65 and 75 years old.

GERMANY

Who is liable for income tax?

Individuals whose residence or place of abode is in Germany are liable for German income tax on their worldwide income. A residence is any house or apartment which is available for personal use and is actually used at any time during the year. The accommodation does not have to be owned or rented by the individual; even a single room in a family home which is reserved for the individual may constitute available accommodation. It does not matter how long the person stays there. For example, a hunting lodge in the Black Forest, which is used for only a fortnight every year or every other year, may constitute available accommodation so that the individual is liable for income tax on worldwide income.

Individuals are deemed to have a place of abode in Germany if they stay in the country for more than 183 days during a fiscal year.

Individuals without any residence or place of abode in Germany are subject to German income tax only on German-sourced income, such as:

(i) farming and forestry in Germany;
(ii) income from a business operating in Germany;
(iii) capital gains from the disposal of shares in German corporations where the individual owns more than 25% of the shares;
(iv) freelance work (e.g. lawyers and doctors) if the work is performed, or benefited from, in Germany;
(v) employment income if the duties are performed in Germany;
(vi) interest and dividends if paid by a German creditor;
(vii) rental and leasing income deriving from Germany;
(viii) speculative dealings, i.e. capital gains from the disposal of property if the period between acquisition and sale does not exceed two years; and
(ix) licence fees derived in Germany.

This rule only applies fully if there is no relevant double taxation treaty. If, under such a treaty, income is taxable in another country, Germany generally exempts such income from taxation. In most cases, however, such exempted income is included when the individual's income tax rate is calculated. If there is no double taxation treaty, the taxpayer is entitled to credit the tax paid abroad against the German tax, provided that the foreign tax is calculated on a comparable basis to German income tax. If it is not, the tax may be deducted as an expense.

According to most of the double taxation treaties between Germany and other countries, individuals only become subject to unlimited taxation in Germany if their place of abode and main residence is located in Germany. Therefore a residence, such as an apartment used for holidays, does not constitute an unlimited tax liability for individuals covered by such treaties.

Income tax rate

The income tax rate in Germany is between 18% and 53% (or between 18% and 47% for business income), depending on the amount of income. For example, an unmarried person with a taxable income of DM200,000 has to pay about 42% income tax. As already indicated, foreign income that is exempt from German taxation may be included in the calculation of the income tax rate. If, for example, the individual in the above example earned DM150,000 of German income and DM50,000 of foreign income which was exempt in Germany, the DM150,000 would be taxed at the rate normally applicable to DM200,000.

At present, and probably until the end of the century, there is an additional solidarity surtax payable because of the special cost of the reunification of Germany. This surtax is 7.5% of the actual income tax to be paid.

Members of certain churches (such as the Catholic Church) are subject to church tax, which is levied by the federal government and amounts to 9% of the tax due. This tax may be reduced for higher-income individuals in some areas of Germany by particular churches.

Land tax and rental income tax

German real estate is subject to land tax levied by the local government. This does not normally exceed 0.5% of the value of the land.

Rental income in Germany is calculated as the excess of the rents collected during the tax year (which is the calendar year) over allowable expenses. Losses can be offset against other German income or carried back or forward. If a long-term (more than 20 or 30 years) forecast shows that profits are unlikely, there is no tax and the activity is treated as a hobby.

Property transfer tax

Any transfer of German land is subject to a tax of 2%, but there are exemptions for transfers to close relatives. Property transfer tax is normally calculated on the basis of the fair market value of the land. According to German law, as a building is legally attached to the land, the fair market value of a building has to be included in the calculation. If land is sold to an unrelated party, the gross transfer price (before the deduction of liabilities assumed by the purchaser) is the basis for the tax.

Capital gains tax

The German capital gains tax regime is quite liberal. Capital gains from the disposal of privately owned assets are exempt from tax, regardless of the size of the gain, provided they have been owned for at least 24 months in the case of real estate, or six months for any other assets. With regard to shares in a German corporation, this exemption applies only to holdings of less than 25% of share capital.

If individuals frequently buy and sell real estate, they may be taxed as though they are trading. In this case, the 24-month period does not apply and all profits are taxable.

Taxation of dividends

There is no preferred tax rate for dividends in Germany but there is an allowance per individual of DM6,100 per year for income from interest and dividends. Dividends paid by a German corporation are linked with a tax credit, which increases the taxable income and can be credited against the income tax liability. In some cases, it will be reimbursed. The tax credit is available only to recipients of dividends who are subject to unlimited taxation in Germany, or who own the shares through a business activity carried out in Germany.

Foreign passive income and extended tax liability

Individuals subject to unlimited taxation in Germany may be liable for tax on their share of profits from a company located in a tax haven, even if the company does not distribute dividends. In addition, individuals who have been subject to German unlimited taxation for more than 10 years, and who are German citizens, may remain taxable in Germany on certain income sourced in Germany if they emigrate from Germany and, within a period of 10 years, establish their main residence in a tax haven country.

Stamp duty

There is no stamp duty, capital duty or comparable tax or duty in Germany.

Wealth tax

Individuals subject to unlimited taxation pay a wealth tax of 1% on worldwide assets. The rate is reduced to 0.5% for certain assets, such as business assets. Certain deductions can be made for liabilities. Individuals not subject to unlimited taxation in Germany are taxed on assets producing income sourced in Germany.

Gift and inheritance tax

A federal tax is imposed on gifts and estates. As a rule, both kinds of transaction are treated alike. With few exceptions, the tax rates for gifts and inheritance are the same. They vary between 0% and 70% depending on the value of the transferred assets and the relationship between transferee and donor. For example, if a father transfers DM5m to his son, the rate will be about 35%; if to an unrelated individual, 7%. All chargeable transfers within 10 years are cumulated to calculate the applicable rate.

Gift or inheritance tax is due only if donor or recipient is resident in Germany. However, the transfer of certain assets (such as real estate and businesses operating in Germany) is subject to inheritance or gift tax even if neither the donor nor the recipient is resident in Germany.

Unless otherwise provided by a double taxation treaty, anyone liable for German gift or inheritance tax is entitled to credit such tax paid abroad against their German tax liability.

ITALY

Who is liable for income tax?

Individuals are considered to be tax-resident in Italy if they are physically present in the country for more than 183 days in a given fiscal year (the same as the calendar year). For the purposes of this test, a day is counted if the individual is present at midnight, days of arrival are counted, but days of departure are not.

Personal income tax is levied on Italian residents on their worldwide income and on nonresidents on their Italian income only.

The Italian tax system operates on the basis of six categories of income or profit:

(i) income from land and buildings;
(ii) income from employment;
(iii) income from self-employment;
(iv) investment income;
(v) business income; and
(vi) other income.

Whether or not a tax applies depends on which category the income or profit belongs to. Similarly, the type of annual return or certification that an individual needs to make depends on his sources of income and profits.

Generally, income earned by individuals is subject to both personal and local income tax.

Personal income tax is imposed at progressive rates on all income and on certain short-term capital gains. The rates for 1995 range from 10% for income up to 7.2m lire, to 53.3% for income over 300m lire. Tax on certain types of income (such as interest on specific types of bank account) is limited to tax withheld at source.

Local income tax is applied at the flat rate of 16.2%. Local income tax generally applies only to income earned from businesses and investments. In practice, expatriates working in Italy will not be subject to local income tax because their income (a salary as an employee, self-employment income and investment income such as dividends, interest and capital gains from securities, for example) would fall into categories that are exempt from local income tax. The position is obviously quite different for a wealthy individual with significant unearned income that is subject to local income tax.

Italian legislation provides for the following tax deductions:

(i) family allowance. This allowance is not available for nonresidents.

(ii) fixed tax allowances where the income is derived from either the individual's salary as an employee or from freelance work, as follows:

- A reduction of 22% in the amount of certain costs incurred by the taxpayer, such as interest paid on loans guaranteed by mortgages on the principal residence, medical expenses and premiums for voluntary life and accident insurance, for a total amount not exceeding 2.5m lire.
- Compulsory social welfare and benefit contributions paid in compliance with the law.

Property tax, land tax and taxation of rental income

Land and buildings in Italy are taxed on a lump-sum basis according to the type of construction. For example, the tax on farmland is fixed on a deemed agricultural income, whereas buildings are taxed based on their cadastral value. Tax on commercial buildings is based on the cadastral value or 85% of the annual rental, whichever is the higher.

Income from land and buildings is no longer subject to local income tax. Instead, a new tax applies called ICI, which is calculated on the value of land and buildings. This tax is a local property tax calculated as a percentage (ranging from 0.4% to 0.7%) of the value of the building. This is determined by multiplying the cadastral income by a certain coefficient, which is 100 in the case of residential property. The ICI tax is also payable on land according to different criteria. It is not deductible for the purposes of personal income tax.

Gross income from land and buildings located abroad but owned by an Italian resident must be declared in Italy unless otherwise provided in a double taxation treaty. The taxpayer is entitled to deduct foreign taxes from tax due in Italy. This deduction is limited to the proportion of Italian tax corresponding to the ratio between all foreign and total income.

Taxation of dividends

Dividends distributed by companies resident in Italy form part of taxable income. A tax credit will be due to the recipient amounting to 56.25% of the profits that form the total shareholders' income. For example, if the company makes a profit of 100, there will be a net dividend of 67.6 (after deducting corporate income tax of 32.4) which is treated as 100 of income for the shareholder. Shareholders may deduct the 32.4 from their tax liability of say 51, leaving 18.6 of additional tax to pay.

For nonresidents, the dividend withholding tax is 32.4% unless otherwise provided for in a double taxation agreement. Italy also gives a tax credit on foreign taxes when the taxpayer also pays income tax abroad. In this case, the credit is due in the same proportion as the ratio between foreign income and total income.

Stamp duty

Stamp duty is payable on deeds, documents and registrations as indicated by the legislation.

Capital gains tax

Capital gains earned on the sale of shares and capital investments are subject to a tax of 25%.

Capital gains are the difference between the sales price and the purchase price of an asset, revalued according to the inflation rate.

Capital gains arising from the sale of an individual's main private residence are exempt. Other real estate sales are exempt provided they have been held for at least five years, otherwise, the 25% tax rate applies.

For non-qualified investments, i.e. sales of less than 2% of a quoted company, sales of less than 5% of an unquoted company, or share capital of less than 10% of investments not in the form of shares, it is possible to choose a different calculation method, in which case the tax is reduced to 1.05% of the sale price.

Wealth tax

No wealth tax exists in Italy.

Gift and inheritance tax

Chargeable transfers on death or by *inter vivos* gift are subject to tax. The tax is applied at scale rates on taxable values and the rates vary from 3% to 33%. If the deceased makes several donations during his or her life, these must be added up when calculating the tax rate to be applied.

For the purpose of inheritance tax, the taxable base is the difference between the assets and liabilities existing at the date of the inheritance (excluding money, jewels and furniture which are presumed to be 10% of the taxable base). In the case of transfer of assets worth less than 250m lire for spouse and children, or 100m lire for brothers and sisters, no tax is due.

Italian government bonds provide a useful tax-planning opportunity as they can be transferred as a gift or upon death without being liable for gift or inheritance tax.

SWITZERLAND

In Switzerland, taxes are imposed at federal, cantonal and local level.

Federal taxes

At the federal level, there are both direct taxes (income tax and withholding tax on income) and indirect taxes (e.g. VAT and stamp duty).

Liability for taxation

Liability can be limited or unlimited. Unlimited tax liability applies to individuals who are resident in Switzerland, or who stay in the country for at least 30 days a year (90 days in some cases). In such cases, the individual's worldwide income and assets are subject to tax.

Limited tax liability applies to nonresidents who own a business or engage in commercial activity in Switzerland, or who own or lease real estate there. Such individuals are subject to tax at source and withholding tax.

Income tax

Withholding tax is deducted at source from income and moveable assets. At the cantonal level, there are direct taxes on income, wealth and immoveable property. Some cantons also levy death duties and gift tax. Local authorities usually levy the same income taxes as the canton.

Expenditure-based (lump-sum) taxation

Under certain conditions, foreigners who are resident in Switzerland, but not engaged in business activities there, are taxed on the basis of their expenditure rather than on their income and wealth. In this case, annual living costs, not income, are taken as the basis of taxation. Living costs are calculated according to actual expenditure, but they must be at least five times the annual rent of the taxpayer's accommodation or its equivalent. In addition, the income tax computed must be at least equivalent to the tax payable on gross income received from sources in Switzerland.

Income from foreign sources may also be included if the taxpayer has asked the authorities to exempt these amounts from foreign taxes under a relevant double taxation treaty.

Taxation of individuals without a Swiss residence permit (C permit)

Foreign nationals without C permits are taxed at source; for example, at their place of employment in Switzerland.

Direct income tax

Income tax is not generally deducted at source in Switzerland, except for employees who are not resident for tax purposes. Income from employment includes salaries, bonuses, commission, benefits and stock options.

Necessary expenses are deductible, generally on a lump-sum basis. These allowances include travel costs (from home to place of work), and the cost of professional literature, training and clothing.

The following types of income from moveable property are taxed:

(i) interest on assets;
(ii) dividends and all distributions of profits;
(iii) redemption or sale of bonds; and
(iv) income from investment funds, patents and licences.

No tax is levied on private capital gains from the sale of moveable property, except in the canton of Grisons.

Income from immoveable property

Revenue from real estate includes mainly the rental income and the rental value of the apartment or house owned and occupied by the taxpayer. Deductions are available on the following:

(i) maintenance costs;
(ii) insurance premiums for third-party liability and damage cover;
(iii) operating costs for rented property;
(iv) property taxes; and
(v) mortgage interest.

A lump-sum deduction can be applied instead of actual maintenance, running and management costs.

Pensions

Switzerland has three key parts to its retirement system:

(i) Membership of the federal old-age and survivor's insurance fund (AHV) and of the federal disability insurance scheme (IV) is compulsory. The contribution is tax-deductible, but the pension paid is taxable.
(ii) Membership of an occupational pension scheme is compulsory for employed persons but voluntary for the self-employed. The contribution is tax-deductible and the pension is taxable.
(iii) There also exist various types of private pension schemes which are divided into

two categories: a restricted-access scheme in which contributions are usually deductible under certain conditions, and a free-access scheme in which the contributions are partly deductible on a lump-sum basis. In each case, the income arising from the scheme is taxable.

For recurring income, such as annuities, where such benefits were wholly financed by the recipient or by a related person, 60% of such income is subject to tax.

Capital gains

Capital gains are the difference between the proceeds of a sale of an asset and its cost price or book value. Such gains are treated differently depending on whether the asset disposed of is a business or a private asset. A distinction is also made between moveable and immoveable property.

Outside the canton of Grisons, tax is not levied on private capital gains from moveable property. However, where the capital gain is derived from a business transaction, it will be considered as a regular income and taxed accordingly.

Private capital gains on immoveable property

No direct federal tax is levied on private capital gains from immoveable property. However, such gains are subject to a special tax in all cantons, albeit at varying rates.

Wealth tax

Wealth tax is not levied at federal level, but the cantons do charge wealth tax on all moveable and immoveable property. Moveable property includes securities, which are valued at their stock-market quotation.

Some shares and income from companies based in the canton are partially or totally exempt from tax, but again this varies from canton to canton. Patents, authors' rights, insurance policies (on their surrender value), claims and precious metals are also taken into consideration when calculating net wealth.

Maximum rates for net wealth tax are around 1% or less, depending on the canton.

Income from self-employment

All such income is taxable. This includes income from independent professions, sole proprietorships and partnerships.

Withholding tax

Federal withholding tax is deducted at source mainly on income from moveable assets. Its primary aim is to the payment of taxes where income might not be declared.

Swiss residents can ask for their withholding tax to be refunded, but only if the income on which the tax was paid has been duly declared. 35% of the gross amount due to the recipient must be deducted at source. Interest on private loans is not subject to this tax, but bonds and debt certificates issued by a Swiss resident are. Unlike many other countries, interest on accounts (in any currency) with domestic banks in Switzerland is also subject to withholding tax.

Dividends, whether declared or hidden, including any distribution of profits, bonus shares and free increases in the nominal value of shares, are all regarded as income for withholding-tax purposes when paid by a Swiss company. Furthermore, distributions paid by

Swiss investment funds are subject to a 35% withholding tax, except when the fund distributes capital gains by way of special dividends.

Withholding tax is levied at source at the rate of 15% on insurance benefits and on annuities of more than SFr500 a year and at 8% on capital payments of more than SFr5,000 a year from a life-insurance policy.

Federal stamp duty

Stamp duty is payable on:

(i) **Issue of securities.** As of 1 January 1996, this tax is due only when more than SFr250,000 worth of shares are sold. In addition, the rate has been reduced from 3% to 2%. This tax is applicable to shares in a Swiss joint-stock company or limited liability company as well as to certificates, participation certificates and dividend-right certificates in a Swiss cooperative company. In the case of a merger a reduced rate is payable.

(ii) **Securities dealing.** This duty is levied on professionals who trade in securities. For documents issued by a Swiss entity, the rate is 0.15%, compared with 0.3% for those issued by a foreign entity.

(iii) **Insurance premiums.** This duty ranges from 1.25% to 5% depending on the type of insurance and includes life, medical and travel insurance policies.

Death duties and gift taxes

There are no federal death duties or gift taxes. Every canton, however, levies such taxes in a different way and on a different basis. Only one canton (Schwyz) levies no death duties or gift taxes at all.

Liability for taxation is determined by the individual's place of residence. If the individual is resident in Switzerland, the estate, inheritance or gift is fully taxable at the last place of residence, except for immoveable property, which is taxed at the place where it is located.

Liability to taxation rests with the heirs in the case of inheritance tax, with the estate in the case of estate tax, and with the recipient in the case of gift tax.

The amount of tax charged varies significantly from one canton to another. Both the relationship between the donor and the recipient, and the size of the gift or inheritance are relevant in calculating the liability. The local authorities usually levy the same death duties as the cantons.

FIGURE 1: RATES OF INHERITANCE TAX ON A BEQUEST OF SFR1M

CANTON	CHILDREN (%)	NON-RELATIVE (%)
Zurich	*5.3*	*36.0*
Nidwalden	*0.5*	*20.0*
St Gallen	*5.1*	*42.0*
Vaud	*3.4*	*25.0*
Berne	*1.8*	*37.1*
Basel-City	*4.2*	*37.7*
Ticino/Lugano	*7.4*	*51.3*
Geneva	*5.2*	*54.1*

UNITED KINGDOM

Who is liable for income tax?

In the UK liability for income tax depends on an individual's place of residence, ordinary residence and domicile. Individuals are regarded as resident in the UK if they spend 183 days a year or more in the country, or if they make habitual and substantial visits averaging more than 90 days a year over a four-year period. Until recently, individuals were treated as residents for tax purposes if they had accommodation available for use and set foot in the country at any time. This rule was seen as a disincentive to foreigners investing in and coming to the UK and was therefore abolished. Domicile is a technical legal concept referring to where an individual intends to permanently or indefinitely reside. An individual may come to the UK for many years without acquiring a UK domicile.

Individuals who are taxpaying UK residents are liable for income tax (at a maximum rate of 40%) on worldwide income, subject to double-taxation treaties and one very important exception: individuals who are resident, but not domiciled, in the UK are liable to pay tax on UK income; however, foreign income is taxable only if it is remitted to the UK. It is therefore possible for individuals who are not domiciled in the UK to become UK tax residents, but not to pay any income tax. This is because the assets are maintained offshore and the individuals live on income or capital sourced in the UK, but do not remit overseas income to the UK. It should be stressed that what constitutes a remittance is not always obvious. Detailed advice should be sought when considering how to structure assets outside the UK to ensure that there is no inadvertent remittance of income or gains to the UK which would result in a tax charge.

Nonresidents are only liable to pay tax on earnings sourced in the UK. However, on a concessionary basis, the Inland Revenue does not generally tax interest arising on UK bank accounts of nonresidents if this is their only source of income in the UK. This practice has now been formalised and such income is only subject to withholding tax, if any. Similarly, dividends are subject to a type of withholding tax (advance corporation tax).

For UK tax residents, there exist a number of tax-free investment options as well as investments in particular types of company, or geographical areas which themselves attract tax relief.

Property tax

The occupier, or in some cases the owner, of residential property must pay an annual council tax, which has replaced the community charge (popularly known as the poll tax) and the former rates system. Commercial property is subject to the uniform business rate.

Rental income from UK property is taxable for both UK residents and nonresidents. It is, however, normally possible to deduct both loan interest and other expenses relating to the property in order to reduce the tax liability. To obtain a tax deduction for interest payments, it was formerly necessary for the property to be located in the UK, and the borrowing to come from a UK bank or the UK branch of a foreign bank. Both of these limitations have been abolished. In certain circumstances, a tenant must deduct UK tax when paying rent to an overseas landlord.

Taxation of dividends

When a UK resident company pays a dividend, it must pay 20/80ths of the net dividend to the tax authorities as advance corporation tax. This tax can be deducted from the final corporation tax liability of the company. For example, if a company makes a profit of 100, it will have a corporation tax liability of 33, leaving 67 available for distribution. On the net dividend of 67, advance corporation tax of

17 would be due. This 17 would be deducted from the mainstream corporation tax liability of 33, leaving just 16 payable. Individual shareholders are deemed to have received a gross dividend of 84 (net dividend of 67 and tax credit of 17), which is subject to 40% tax. The net receipt for individual shareholders is therefore a little more than 50% of the profits of the company. It should be noted that the corporation tax rate in the UK ranges from 25% to 33%, depending on the level of the profits.

Tax credits under tax treaties

The UK has one of the largest networks of double-taxation treaties in the world and these can be very useful in reducing foreign taxes for UK residents.

In general, the UK double-taxation treaties give relief from double taxation by using the credit system. Thus, interest which has been received net of foreign taxes would be grossed up and subject to UK tax, with a credit being given for the foreign tax suffered against the UK tax liability which arises. It is not generally possible to have excess foreign taxes repaid or to carry them forward.

Where no double-taxation treaty exists or applies, UK legislation will normally give a unilateral tax credit.

Stamp duty

Stamp duty in the UK is levied on stampable documents rather than specifically on transactions. However, stamp duty normally arises on the sale of company shares, on the purchase of a business or on the purchase of real estate. Stamp duty is charged at 1% on the acquisition of real estate and 0.5% on the acquisition of shares.

When large amounts of stamp duty are payable, it may be possible to avoid or defer the liability by complex planning involving an offshore declaration of trust.

Capital gains tax

In the UK capital gains tax is levied on persons who are resident, or ordinarily resident, in the UK. The rate will depend on an individual's level of income, but will normally be 40%. Persons who are not resident or ordinarily resident are not subject to capital gains tax, unless they are trading in the UK.

An individual who is resident in the UK is liable to pay capital gains tax on worldwide gains, subject to an important exception. An individual who is resident, but not domiciled, in the UK is only liable to pay tax on gains arising from assets located in the UK and from other gains remitted to the UK. It is therefore possible for an individual who is resident, but not domiciled, in the UK to make substantial capital gains on offshore properties and overseas companies, but avoid paying capital gains tax by not remitting the gain to the UK.

For UK residents, there is a system of relief and exemption designed to reduce the impact of capital gains tax on private businesses. For example, when an individual sells his or her business on retirement, provided that certain conditions are met, part or all of the capital gain will be exempt from tax.

Transactions between spouses are deemed to have taken place with proceeds that resulted in no gains or losses.

Subject to various detailed rules, residents of the UK are not liable to pay tax on the disposal of their principal private residence. It is therefore unusual for tax to arise on the sale of the family home. However, second homes are subject to tax. In certain circumstances, gains arising from gifts of business assets can be deferred.

In some cases, complex tax planning can be carried out using offshore structures. It is often possible to reduce a potential capital gains tax liability from a rate of 40% to 25%, or even 0%, and the earlier advice is sought, the easier this is to achieve.

Wealth tax

There is no wealth tax in the UK.

Inheritance and gift tax

Apart from a modest exemption (currently £154,000), inheritance tax is charged on death at the single flat rate of 40%.

An individual who is domiciled in the UK is liable to pay inheritance tax on worldwide assets. At present, it is possible to take out insurance policies to cover the risk of not surviving the seven-year period.

An individual who is not domiciled in the UK is only liable for UK inheritance tax on directly-owned UK real estate. Thus, a foreign national who owns a property worth £1m, for example, will owe approximately £340,000 in inheritance tax on his or her death. It would be possible to save this by using an appropriate offshore entity to structure the purchase. In some circumstances, it may be more cost-effective to take out a life-assurance policy to insure against one's tax liability.

Lifetime gifts are also subject to inheritance tax, although gifts between UK-domiciled spouses, gifts of business property, and most cases in which the donor survives for seven years after giving the gift, are exempt. In the rare situations in which a tax liability arises immediately, it is provisionally charged at half the tax rate arising on death.

There are complex rules governing how various transfers to and from trusts are subject to inheritance tax. In general, an individual who is entitled to receive income from a trust is deemed to own the underlying capital which gives rise to the income. A discretionary trust, on the other hand, is a taxable entity for the purposes of inheritance tax, although this is commonly used either to ensure that the nil rate band of £154,000 is used, or by individuals not domiciled in the UK.

EUROPEAN TAX RATES

Income tax

The highest marginal income tax rates applicable to individuals in Europe range from 30% in Russia to nearly 70% for high earners in Finland. The average rate is around 50% for the highest earners. The level of income at which the highest rate applies varies enormously (see Figure 2).

Corporate tax

The highest marginal rates of tax applicable to corporations range from 23% in Hungary to 45% in Germany. The average is around 35% (see Figure 2).

Inheritance and gift tax

In many cases, the marginal inheritance and gift tax rates are substantially lower for gifts or inheritances which pass to close relatives, and many countries offer a generous system of relief and exemption. The rates vary from 21% in Hungary to 90% in Denmark, and average nearly 50% (see Figure 2).

FIGURE 2: EUROPEAN TAX RATES

(%)	Income tax	Corporate tax	Inheritance and gift tax	Capital gains tax	Net assets/ wealth tax
Austria	*50*	*34*	*60*	*25*	—
Belgium	*56.65*	*39*	*80*	*17*	—
Czechoslovakia	*43*	*41*	*40*	*43*	—
Denmark	*63.5*	*34*	*90*	*40*	*1*
Finland	*69.85*	*25*	*63.5*	*69.85*	*0.9*
France	*56.8*	*36.6*	*60*	*19.4*	*1.65*
Germany	*53*	*45*	*70*	*26.5*	*1*
Hungary	*44*	*23*	*21*	*10*	—
Ireland	*48*	*38*	*40*	*40*	*2*
Italy	*53.3*	*37*	*33*	*25*	—
Luxembourg	*51.25*	*40.29*	*48*	*34.64*	*0.5*
Netherlands	*60*	*35*	*68*	*20*	*0.8*
Norway	*41.7*	*28*	*30*	*28*	*1.5*
Poland	*45*	*40*	*45*	*45*	—
Portugal	*40*	*39.6*	*50*	*10*	—
Russia	*30*	*43*	*40*	*30*	—
Spain	*56*	*35*	*34*	*56*	*2.5*
Sweden	*56*	*28*	*30*	*43*	*1.5*
Switzerland (Zurich)	*44*	*28.5*	*36*	*0*	*0.75*
UK	*40*	*33*	*40*	*40*	—

Capital gains tax

The table shows the highest marginal rate of tax which can apply to the sale of all the shares in a family company. Rates are often higher for other disposals (e.g. land) and lower where small shareholdings are sold. Many countries operate exemptions for other assets held outside a defined speculation period. The rates vary from 0% in Zurich (Swiss taxes depend on the canton) to nearly 70% in Finland, and average 32%. Gains are generally taxed at lower rates than income.

Wealth tax

As shown in the table, just over half of the countries in the survey levy a wealth tax (up to 2.5% of net assets per annum in the case of Spain). Different rates sometimes apply to different classes of asset. ❖

THE ESTABLISHMENT AND OPERATION OF OFFSHORE TRUSTS

Douglas H. Moore, Credit Suisse Private Banking, Hong Kong

Douglas H. Moore is the head of business development of Credit Suisse Private Banking in Hong Kong, and is responsible for all marketing activities. The department is active in Southeast Asia and has responsibility for Thailand and the Philippines. Mr Moore is a member of Credit Suisse's Asian Regional Investment Committee and works closely with the CS Fides Trust Group in Southeast Asia. Following his graduation from Osgoode Hall Law School, Mr Moore commenced the practice of law with the tax group of a large full-service law firm. Mr Moore is admitted as a solicitor in Hong Kong, England & Wales and Canada (Ontario).

THE ESTABLISHMENT AND OPERATION OF OFFSHORE TRUSTS

Douglas H. Moore, Credit Suisse Private Banking, Hong Kong

Offshore trusts have become widely accepted vehicles for structuring and holding the wealth of individuals of high net worth. The offshore trust industry has matured both in the range and quality of its services and in the sophistication of the jurisdictions available. In today's ever-changing world, the management of one's personal wealth must take account of the opportunities presented by offshore trusts. An offshore trust is not an investment *per se* but can be an excellent vehicle from which to make investments in a secure and tax-effective manner.

The word 'offshore' has a number of connotations. It is clear that there must be some kind of multi-jurisdictional element to the trust. For example, the beneficiaries might be in a different country from that of the trustee. It might also be possible that the trust fund is in a different country from that of either the trustee or the beneficiaries. The word 'offshore', however, implies more than a cross-border dimension. It also implies that the tax burden will be reduced and that the offshore party, usually the trustee, will be in a more favourable legal or regulatory environment than the 'onshore' parties (usually the beneficiaries). In this sense, the word 'offshore' is often equated with 'tax haven', and the word 'onshore' with 'high tax jurisdiction'.

TYPES OF TRUSTS AND STRUCTURES

The diverse needs of high net worth individuals and the fertile imaginations of battalions of professional advisers have created an almost unlimited range of combinations and permutations of trusts, Continental European solutions and offshore holding companies. Anglo-Saxon trusts are well understood and widely accepted in the common law jurisdictions.

The use of trusts in England dates back to the 11th century. Anglo-Saxon trusts are available in most offshore jurisdictions. Taxing authorities and other regulatory bodies generally understand Anglo-Saxon trusts and have developed rules applicable specifically to them. Discretionary, reversionary, testamentary and charitable trusts are of the greatest relevance to high net worth individuals. In addition to Anglo-Saxon trusts, offshore companies and Continental European solutions are also well worth looking at.

Discretionary trusts

The position of a beneficiary under a discretionary trust can be contrasted with that of a beneficiary under a 'fixed trust' or non-discretionary trust under which the share of the beneficiary is specified and certain. Where tax-planning or asset protection objectives are to be achieved it is invariably the case that a discretionary trust will be more advantageous than a fixed trust.

Under the terms of a discretionary trust, the beneficiaries have no entitlement to anything from the trust fund unless the trustee exercises its discretion to make a distribution to a beneficiary. The beneficiary in this circumstance cannot be said to have any vested interest in the trust. At best, the beneficiary has an expectation that he or she might at some point in the future receive a distribution from the trust.

Some discretionary trusts also have the power to add beneficiaries at a later stage. These 'prospective beneficiaries' are subject to a double layer of discretion (i.e. that they may be added as beneficiaries and that, if added, they may receive distributions). Sometimes the definition of beneficiaries also includes a list of people who are excluded from ever benefiting from the trust. The nomination of 'excluded beneficiaries' is invariably motivated by the tax implications of that particular beneficiary (usually the settlor) tainting the trust.

Reversionary trusts

Under the terms of the trust agreement the settlor:

(i) may take back the trust fund from the trustee, in which case the trust will be a reversionary trust; or

(ii) may not be entitled to take back the trust fund, in which case the trust will be a non-reversionary trust.

Reversionary trusts provide the settlor with a measure of control as an unhappy settlor can simply take back the trust fund. Most taxing or estate duty statutes, however, look through a reversionary trust and regard it as not having a separate existence from the settlor.

If a settlor cannot create a reversionary trust without adverse tax or estate duty implications but would still prefer not to lose total right to the trust fund, there are two possibilities. The first is that the settlor could become a beneficiary of the trust. (One must consider the implications of this alternative in the light of the particular taxing statute.) The second is that the settlor could fund the trust by way of a loan or sale of the assets forming the trust fund and take back from the trust a promissory note. If the note-holder were to demand payment of the note this would effectively return to the note-holder the initial amount of the trust fund.

Testamentary trusts

A testamentary trust is created by the testator upon his or her death. In some jurisdictions, more favourable tax treatment is extended to testamentary trusts than to *inter vivos* trusts. The rationale would appear to be that it is quite drastic tax planning to die! Other jurisdictions take the opposite approach and apply fairly harsh rules to testamentary trusts on the basis that they do not want the survivors to benefit from the death of the deceased.

Charitable trusts

The purpose of this type of structure, which sometimes takes the form of a 'purpose trust' or a foundation, is to benefit a charity or a particular cause. Usually individuals are excluded from benefiting from such a trust in their personal capacity. A purely charitable trust can have quite favourable estate duty and tax implications, as the settlor is considered to have 'given away' the assets at the establishment of the trust. The family will often wish to remain involved with the allocation of funds to various charities.

Offshore companies

Offshore companies often provide a convenient vehicle for transacting business or holding certain types of assets. In many offshore jurisdictions, the offshore company will be permitted to issue bearer shares. This does not achieve quite the same objective as a trust, however. An offshore company held by a trust often provides a most desirable combination: the offshore company can transact business while its ownership remains confidential and secure.

Continental European solutions

For those who are more comfortable with civil law, a variety of alternatives are available.

Liechtenstein is an independent principality nestled between Switzerland and Austria. Liechtenstein entered into a Customs Treaty with Switzerland in 1924 and enjoys the advantages of the Swiss infrastructure and currency while retaining political independence. The Liechtenstein foundation is a legal entity with its own independent existence. Like the settlor of a trust, the founder of a foundation can be either the person who is behind the foundation or a nominee founder who can be utilised to provide maximum confidentiality.

A Liechtenstein-registered trust differs from an Anglo-Saxon trust in that the trust fund of a registered trust is actually held in the name of the trust as opposed to the name of the trustee. The registered trust has an existence as an entity separate and distinct from the trustee. A registered trust can also have settlor's rights that provide a high measure of control to the holder of those rights. Sometimes the settlor's rights are constituted to expire on the death of the settlor.

The Continental European solutions provide the establisher of the structure with the option of a high degree of control over the assets of the structure. In addition, the Continental European solutions are not subject to any particular life expectancy and can therefore have unlimited duration.

ESTABLISHMENT OF AN OFFSHORE TRUST

The establishment of an offshore trust involves a number of steps and there are numerous decisions to be made. Generally speaking, one is motivated to consider establishing an offshore trust by reason of a particular concern, life event or business development. The initial perceived purpose in establishing the trust is thus usually reasonably clear. However, some thought may still be required to determine one's real objectives in establishing the trust.

Once the decision is made to establish a trust and the reasons for doing so are well understood, it will be necessary to select a trustee and an offshore jurisdiction. It will also be necessary to consider carefully the terms of the trust agreement to ensure that the trust will achieve the purpose for which it is being established.

FIGURE 1: TRUST STRUCTURE

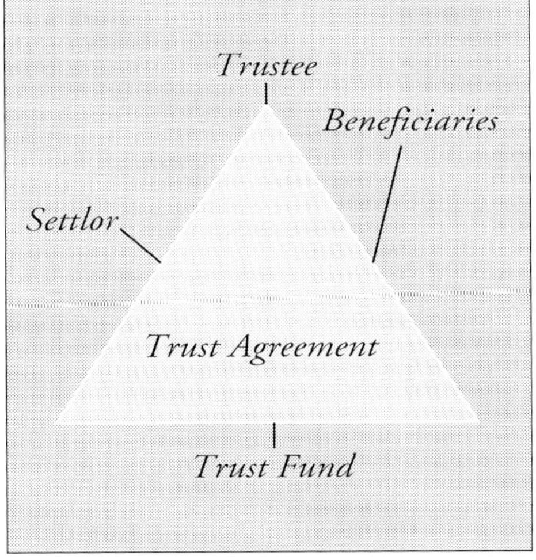

Purpose of the trust

By far the most important element in establishing an offshore trust is to identify carefully the reason for establishing the trust. 'Locker room' planning must be subjected to careful scrutiny; establishing a trust because one's friends are doing so is not likely to lead to a satisfactory conclusion of the exercise. Establishing a trust for a somewhat nebulous reason such as 'asset protection' is also likely to lead to failure. It is necessary to identify exactly what the perceived threat is in order to ensure that the trust will offer a measure of protection against it.

A fairly typical reason for establishing a trust is the recognition of one's own mortality. This can often be brought on by the death of a loved one or contemporary. If one takes stock of one's assets and what would happen to them in the event of death, it often becomes evident that a trust would greatly improve the situation. In a number of jurisdictions, probating an estate can be time-consuming and expensive, lead to unwanted complications in the disposition of assets and involve high rates of estate duty. Quite often these difficulties can be eliminated or greatly alleviated by the establishment of a trust. In addition, 'putting one's affairs in order' can often bring peace of mind. Thinking about a dispositive scheme is an important exercise for which many people seem hard pressed to find time.

This raises the question of the relationship between a trust and a will: the two are mutually exclusive. The trust deals with assets that are in the trust fund and does not extend to assets outside the trust fund. The will deals with assets that are owned by the deceased at the time of death and by definition cannot include assets forming part of the trust fund, as such assets are legally owned by the trustee. The establishment of a trust is often an opportune time to reconsider one's will, as taking inventory of assets and formulating a dispositive scheme are necessary for both. The will and the trust should work together to ensure that the overall entitlements are desirable and all objectives are met.

Another event that might trigger a review of the way personal assets are held is immigration, which can involve the liquidation of assets in the old country and assessment of opportunities in the new. Pre-immigration tax planning often involves the establishment of a trust that is not resident in the immigration destination. This type of trust is often tax-sensitive and must be structured carefully.

Business developments, too, can often lead to consideration of a trust. For example, if a private company is going to be listed on a stock exchange, the controlling shareholder might consider the manner in which his or her shares are to be held. Once a company becomes publicly listed, regulations on stock exchange filings, insider trading, connected transactions and a host of other rules, become applicable. If an investment is to be contemplated in another country, it is necessary to consider the manner in which it will be held. A trust can often provide favourable tax treatment.

Role of the professional adviser

Once the purpose of the trust has been identified, it is essential to ensure that appropriate specialist advice is obtained to tailor the trust to the desired objective. In addition, the professional adviser might well be able to help with a realistic definition of the purpose of the trust. If the adviser and the client enjoy a close and continuing relationship, the adviser will be in a position to assist the client with the design and the establishment of the trust.

Even when one does not have an adviser, specialist advice can be retained on an 'as-needed basis'. Offshore trusts are often motivated by tax considerations and it can be necessary to obtain specialist advice in one or more jurisdictions. This can be also be true when matrimonial or asset protection issues are to be considered.

Selecting a jurisdiction

For most routine offshore trusts the selection of the trustee is more important than the selection of the jurisdiction. A good trustee in an average jurisdiction will still result in a happy client, while a substandard trustee in a good jurisdiction will result in an unhappy client. If you have selected a trustee, you will probably find that the trustee has operations in more than one jurisdiction. Generally, levels of service will be best in the trustee's principal centre of operation.

However, jurisdiction becomes relevant if the trust has a very specialised purpose. For example, if protection from creditors is a primary motivating factor, some jurisdictions make it more difficult than others for creditors to attack the trust. In this situation, a careful review of the attributes of the different jurisdictions will show that some are more suitable than others. Another consideration is whether or not the trust will have a 'public profile' as a result of the nature of the trust fund (for example a controlling interest in a publicly listed company or a real estate or business interest in a high-tax country). The jurisdiction must be solid and respectable in the case of the public company. A tax-treaty country may also be a wise choice in the case of investment, for it may provide for lower rates of withholding tax on payments from the source country of the investment, where the recipient is in a treaty jurisdiction.

Assessing various choices can result in a lesson in geography. Major centres are located in the Channel Islands (Guernsey and Jersey), the Isle of Man, the Caribbean (Bermuda, Bahamas, Cayman Islands, British Virgin Islands), Europe (Liechtenstein, Gibraltar) and Asia (Cook Islands, Labuan). Relevant considerations can often be as simple as the time zone in which a jurisdiction is located, or a preferred area of travel.

Terms of the trust

Most trustees provide standard trust agreements. Generally speaking, these agreements are not drafted with any particular purpose in mind other than to create a valid trust. Accordingly, they may or may not achieve certain tax objectives in certain jurisdictions. This would be in the domain of a specialist adviser. For many clients, a standard trust agreement is perfectly adequate and can be utilised with satisfactory results. It is important to ensure that the document is adequately explained and that its terms are fully understood.

For other clients, highly specialised trust agreements are required and are prepared by specialist advisers. These trust agreements often contain many provisions that are solely motivated by tax or other legal considerations. The effect of the provisions must be carefully explained to the client by the specialist adviser. Tailor made trust agreements tend to focus the settlor's attention on the terms of the trust agreement and are in this respect preferable to standard trust agreements, which are often not subject to the necessary scrutiny by the client.

Forced heirship

Historical and religious factors have resulted in different jurisdictions adopting different approaches to the issue of the transfer of wealth from a deceased individual to succeeding generations. Some jurisdictions allow a high degree of testamentary freedom and a person can do with his or her estate largely as he or she chooses. Other jurisdictions do not allow such latitude. Forced heirship allows certain individuals the absolute right to a certain share of a deceased's estate. Forced heirship rules vary considerably from jurisdiction to jurisdiction. The variables include the definition of the forced heirs, whether entitlement is automatic or must be claimed, the class of heir that is favoured by the rules and the extent and operation of 'claw-back' rules.

The offshore trust is the primary device available to someone faced with forced heirship laws that create an undesirable dispositive scheme. Establishment of an offshore trust by the settlor may well attract the operation of claw-back rules, which can be used by a disgruntled heir to launch a claw-back claim either in the jurisdiction of forced heirship (compensating adjustment), in the jurisdiction of assets (if forced heirship), or against the trustee in a forced heirship jurisdiction in which the trustee has operations. Account must be taken of applicable claw-back rules, as they differ from one jurisdiction to another.

Although it has not been established that a forced heirship claim would prevail in an offshore jurisdiction that does not provide for forced heirship, a number of offshore jurisdictions have put in place 'anti-forced-heirship legislation' to provide further protection to trusts established in the jurisdiction.

Flee clause

The purpose of the so-called 'flee clause' in a trust agreement is to provide a measure of protection in the event of an adverse development in the jurisdiction of the trustee. A jurisdiction is chosen because it is favourable in that there are no taxes, no exchange controls, an acceptable level of political stability and no restrictive laws. If a change in any of these essential elements is likely, one would hope that this would occur with enough advance warning to allow the trust to move out before such change became effective. The fear is that such change might occur overnight with no advance warning and there would be no chance for the trust to escape. It is to this situation that the flee clause is addressed.

The flee clause provides that in the event of an adverse development in the jurisdiction the trust is immediately relocated to a new jurisdiction with a new trustee. It is triggered automatically when one of the specified adverse events occurs. Whether or not a flee clause is likely to be legally effective is the subject of some debate, but the consensus view is that it is better to have one than not if there is any concern regarding the jurisdiction.

Funding the trust

A question usually on the mind of someone establishing a trust is: 'What assets do I transfer into the trust and what assets do I leave out?' The answer to this question depends on the nature of the assets and the purpose of the trust.

Practical concerns often limit the suitability of certain assets for injection into a trust. If an asset has an accrued gain and its transfer by the settlor or contributor would trigger an inherent tax liability, this 'transaction' cost will often result in the asset not changing hands to avoid crystallising the tax liability. Other transaction costs include transfer taxes, stamp duties and legal fees, and some types of asset are subject to restrictions on foreign ownership. It can sometimes be difficult to transfer private company shares as the 'driving force' behind the business needs to be seen as a shareholder by lending banks, customers or employees.

The purpose of the trust can have a tremendous impact on what type of assets are injected into it. For example, a pre-immigration trust will not include all assets as the immigrant will want to take some assets into the destination country and pay a reasonable amount of tax. Fixed assets in the destination country, such as real property, are often more efficiently held directly by the immigrant.

The type of assets injected into trust structures includes just about everything that one can imagine. Standard types of assets include cash, stocks, bonds, real estate and shares in private companies, as well as others such as yachts, works of art and jewellery. Where liability is a concern, such as in the case of an operating business, the assets are usually held by a corporation that can be owned by the trust.

OPERATION OF A TRUST

The operation of a trust subsequent to its establishment should take into account those areas that can either enhance the objective of the trust or undermine it. Once an offshore trust is established there is a tendency to forget about it. Everything tends to run smoothly and there never seems to be any need to address one's mind to the ongoing operation of the trust. This is often a mistake, for the trust needs to evolve with the beneficiaries.

Residency of the trust

The residency of a trust is generally relevant for tax purposes. It is usually desirable for residency to be established in an offshore jurisdiction that is not subject to tax. The beneficiaries or the settlor (or both) may be resident in a high-tax country. If the tax authorities in that country have occasion to challenge the trust, one of the grounds might be that the trust is resident there. What are the grounds of attack and how does one ensure that the trust remains offshore?

The residency of a trust is rarely subject to any specific test. It is usually a question of fact to be determined with regard to all of the surrounding circumstances. No one fact is likely to be determinative and it is largely an exercise of weighing up the favourable and adverse facts, and judging which prevail. Factors to be considered are the governing law of the trust, the residency of the trustee, the location in which the trust assets are administered and located, any terms of the trust agreement which suggest that someone other than the trustee has influence (such as a protector for example), and the conduct of the beneficiaries.

If the trustee is a major professional trustee with a substantial operation in the offshore jurisdiction, challenges regarding the substance of the trustee will be eliminated. Or, if the trustee is a shell with no real capability to discharge its responsibilities, there may be an argument that no weight should be attached to the fact that the trustee is located offshore, as nothing can happen offshore in any event. If the trustee does not undertake any independent decision-making, the challenge might be made that the trustee is merely an agent of the settlor or beneficiaries and that the trust is resident where the principal resides.

In short, if offshore status is necessary to the efficacy of the trust, care must be undertaken to ensure that the offshore status is protected and connections 'onshore' must be minimised.

Attacks on the trust

A trust may be subject to various attacks. For example, disgruntled beneficiaries might attack the trust with a view to capturing a greater share of the trust fund. Creditors might attack the trust with a view to forcing payments out of the trust or setting aside transfers to the trust. Estranged spouses might attack the trust in order to include the value of the trust fund in the pot to be divided upon the breakdown of a marriage. Tax authorities, too, might try to attack the confidentiality of the trust.

Will the trust withstand these various attacks (should they materialise), or will a 'trust buster' succeed? The answer depends very much on the circumstances surrounding the establishment of the trust, and its subsequent operation. Certain jurisdictions provide a greater measure of protection than others. It is also important to bear in mind that 'onshore pressure' on the beneficiaries or settlor can cause disclosure or settlement even where the trust would have withstood an 'offshore attack'.

CONCLUSION

Establishment of an offshore trust is an exercise that requires clear thinking. The greater the care invested in establishing the structure, the more likely the trust is to achieve its objective and provide peace of mind for both the settlor and beneficiaries.

The operation of an offshore trust should be handled with enough care to make sure that the objective of the trust is preserved and that this objective is constantly assessed against changing circumstances during the lives of the beneficiaries. An offshore trust that is well-structured and carefully administered can achieve many benefits during the lifetime of the settlor and for generations of successive beneficiaries. ❖

ASSET PROTECTION TRUSTS

Philip Marcovici, Partner, Baker & McKenzie, Zurich

Philip Marcovici is a partner with the international law firm Baker & McKenzie, and is resident in the firm's Zurich office. Admitted to practise in New York, British Columbia, England & Wales and Hong Kong, Mr Marcovici recently moved from Hong Kong where he spent 12 years as a member of the firm's international tax group. He has written and lectured widely on international tax and private-client matters.

Phone: (+41-1) 384 1414
Fax: (+41-1) 384 1284

ASSET PROTECTION TRUSTS

Philip Marcovici, Partner, Baker & McKenzie, Zurich

The term 'asset protection trust' has taken on a particular meaning among professionals involved in the structuring of trusts, and now commonly refers to trusts established to take advantage of the asset protection legislation of various offshore jurisdictions. In general, this legislation makes it difficult for an individual's creditors to gain access to assets held in trust, notwithstanding any personal obligations the individual may have to his or her creditors.

The ways in which trusts can help protect one's assets go well beyond this restricted meaning of the term, however, and this article will therefore consider not only how trusts can be structured to minimise the risk of assets being exposed to the claims of creditors, but also the use of trusts in asset protection generally.

Wealth is created in a limitless variety of ways, and there are any number of creative approaches by which it can be preserved. The starting point for any well-thought-out asset protection structure is the identification of the dangers to an individual's ability to preserve (and increase) accumulated assets. There is no 'off-the-shelf' product or approach that can address every individual's situation. The establishment of an asset protection trust for an individual with connections to one or more jurisdictions in Europe will have to take into account the risks peculiar to those countries.

People's personal circumstances may also affect their ability to preserve their wealth. Connections, such as citizenship or residence, to high-tax jurisdictions in North America, Europe or elsewhere may be a concern, whether it is the individual who is so connected or family members who may benefit from the assets held in trust.

Many different people may be interested in finding ways to gain access to an individual's assets – a former spouse, perhaps, on the break-up of a marriage, or an heir under applicable forced heirship legislation. Illegitimate children, extramarital companions and others may all be able to formulate claims which might threaten the individual's wealth.

Since the use of trusts for asset protection can adequately be addressed only with regard to individual circumstances, this article will discuss asset protection approaches based on a particular fictitious situation. It is hoped that the reader will gain some understanding of the important role that trusts can play in asset protection, and of why customer and adviser must cooperate closely if the goal of maximum asset protection is to be attained.

CASE-STUDY: ALBERT

Albert is 48 years old and lives in Spain with his wife and two children. Born in Germany, he moved to Spain with his parents in 1949. The family ceramics business, established by Albert's grandfather in Slovakia, recently shifted to the Czech Republic with the family's move. Albert joined the family business after obtaining an engineering degree from a US university. During the period of his studies Albert obtained a US 'green card' giving him the right to reside permanently in the USA.

The family business expanded over the years and now comprises operations in a number of countries in Europe and elsewhere. Part of the family business is held by a Bermuda holding company, for which the family plans to seek a listing. Albert personally holds 15% of the shares of the holding company. Other assets owned by the family include property, securities, cash and an interest in a fast-growing private company involved in the development of high-performance ceramic engine components.

Albert has developed an intimate relationship with a woman he met through business colleagues and is concerned about how he will deal with the situation.

One of Albert's personal investments is in a large tract of raw land in California that he and his partners hope to develop. Albert identified the land during a stay with his family at a home he owns nearby. Unfortunately, he recently learned that part of the land had been used for chemicals storage, and that seepage into a nearby fresh-water supply may have occurred over a period of years.

While Albert is aggressively building his wealth, he has mentally earmarked a sum representing the portion of his assets that he no longer wishes to maintain at risk and which he views as essential to preserve for the long term.

CHOOSING ADVISERS

Although Albert realises that he needs help to devise an asset protection plan to suit his circumstances, he is not sure whom to approach. He has a number of close contacts practising law and accountancy and has developed relationships with several private banks, some of whom have suggested to him the use of a trust.

Despite the apparent complexity of his circumstances, Albert's position is far from unique. The issues that he must deal with will involve the tax laws of several jurisdictions: Germany, Spain, the Czech Republic, the United States and possibly others. The domestic relations legislation of Spain and possibly elsewhere may also come into play, given Albert's extramarital relationship.

It would be difficult, if not impossible, for Albert to find any one adviser, whether a lawyer, banker or accountant, who was able to address competently all of the issues pertinent to his situation. The key for Albert is to work with an adviser who can help him to identify the issues that any structure established for him should deal with, and to seek advice from specialists in those issues. The adviser to avoid is the one who plays down the importance of those issues he is unable to deal with himself, relying only on secrecy, perhaps, to avoid exposure to creditors or tax collectors. Sadly, this kind of adviser is all too common, his aim being to get the job done, collect his fees and move on to the next client.

WHAT IS A TRUST?

Before considering how he might go about choosing a suitable trustee and structuring the transfer of assets to be held in trust, Albert should gain an understanding of what a trust actually is and, in particular, to absorb the fact that a trust is an extremely flexible vehicle for holding assets. Their flexible nature means that there are any number of ways that trusts can be set up. If he is to give effective input on the final structure, the individual whose assets are to be put into a trust must understand the different ways in which the trust can be structured.

A trust is a relationship, recognised in law, involving the transfer of assets from their original owner to one or more individuals or entities (the trustees) who are to hold the assets for the benefit of other persons – who may or may not include the original owner. The trust is created by a 'grantor' or 'settlor' who transfers assets to the trustee at the outset. It is important to note that the named settlor does not have to be the individual whose assets are to be the main subject of the trust.

For example, depending on Albert's objectives, it may be to his benefit to have the trust that is to hold his assets settled by an individual or entity other than himself, with perhaps a nominal sum settled into the trust to establish it. The named settlor can be a nominee paid for the purpose, or a friend or relative who cooperates in the establishment of the trust.

How Albert's assets are subsequently transferred to the trust will depend on the objectives of the trust and its settlement and may involve, among other possibilities, a transfer of assets by gift or sale. The trustee will be charged with the responsibility of holding the assets in accordance with the wording of the deed governing the trust, and in accordance, also, with the trust law of the jurisdiction where the trust is domiciled.

Choosing a trustee

The choice of trustee will be an important one for Albert. The trustee will be holding assets for the benefit of the beneficiaries of the trust and, depending on the circumstances, may be involved in Albert's affairs over a long period.

In choosing a trustee Albert should be aware of the considerable number of options available to him, including the possibility of having friends or family members act as trustees. While this approach could save money, difficulties can often arise. Individuals, as opposed to trust companies, can pass away or migrate to jurisdictions whose tax laws or other legislation could possibly harm the trust. There may also be difficulty in enforcing the terms of the trust, particularly after Albert's death. For these and other reasons, Albert would probably be well advised to consider the use of a professional trustee.

In choosing a professional trustee, Albert may wish to consider the size of the institution involved and, in particular, whether there would be any risk of the trustee itself not surviving in the long term. Trustees are by definition fiduciaries, meaning that the trust assets would not be subject to the claims of the trustee's creditors. However, Albert would gain considerable extra peace of mind by choosing a reputable trustee with a record of effective trust administration.

The choice of trustee will also involve choosing the law governing the trust, as this is generally dictated by the trustee's place of residence. As we will see, where the protection of assets from the claims of creditors is an objective, choosing the legal system that is to govern the trust will be of the utmost importance.

Trusts are creatures of the common law and are, therefore, generally established under the laws of common-law jurisdictions. The most popular jurisdictions for the purpose are those that do not tax trust assets or the earnings generated thereby. Whether or not a vehicle established under the laws of a civil-law jurisdiction would be appropriate for Albert depends upon a number of circumstances that would need close review. Before using a foundation or similar civil-law entity, for example, Albert should consider carefully whether such an entity would achieve the specific objectives of his asset protection plan.

Is the trust a trust?

In any asset protection plan involving the use of a trust it is extremely important to ensure that the trust is actually recognised as a trust.

Where the trustee is no more than a nominee following the specific instructions of the original owner of the assets, there is a strong risk that the trust could be challenged. For example, if Albert uses as a trustee a relative or a nominee company provided by a professional firm, there may be a tendency for the named trustee not to provide any independent input in the administration of the trust, but to follow Albert's instructions strictly, even when contrary to the language of the applicable trust deed.

The key to achieving the tax and other asset protection benefits of a trust is to respect the division of legal and beneficial ownership of assets that trusts achieve. The trustee becomes the legal owner of the assets held in trust: a fundamental point if the original owner of the relevant assets is subsequently faced with a claim on the assets. Where there is no effective independent administration of the trust, the door is open to a challenge of the effectiveness of the trust.

Fixed and discretionary trusts

A fixed trust is generally one in which one or more beneficiaries have a fixed entitlement to an ascertainable part of its income or principal. For example, if Albert's trust provides that his wife will receive 20% of its annual net income, it is considered to be a fixed trust.

More common in the context of asset protection trusts, however, is the discretionary trust, whose beneficiaries, such as Albert's wife and children, have no absolute current right to the assets of the trust, any distributions to them being at the discretion of the trustee.

When a trustee is given discretion in this way, the trust deed is generally accompanied by a letter of wishes, a non-binding document expressing the wishes of the creator of the trust. For example, through a letter of wishes Albert would be able to inform the trustee of his desire that the trustee exercise his discretion to distribute assets of the trust to particular individuals at specified times. While such an expression of wishes is not binding, a professional trustee is generally guided by it.

TRUSTS AND ASSET PROTECTION

As the use of a trust allows legal title to assets to pass to someone other than their original owner, such holder of legal title not being the ultimate beneficiary, trusts are very attractive tools for asset protection planning. In very simple terms, Albert can transfer assets he seeks to protect to a trust, setting out to whom such assets should be transferred in future.

Assuming that the trust has been properly structured, Albert should then be able to show that the assets held in trust are not his and should not, therefore, be subject to the claims of creditors or others who may have an interest in gaining access to them.

Once Albert has decided on the trustee and chosen his advisers he can move towards structuring the trust and arranging for the transfer of assets to the trustee. Albert should be sure that his asset protection trust addresses the following issues, among others:

- Tax minimisation
- Family break-ups
- Political risk
- Protection against claims of creditors

Tax minimisation

A highly important aspect of asset protection involves ensuring that the trust is structured to minimise (or, if possible, eliminate) exposure to tax, whether on the transfer of assets into the trust or at any other time.

In Albert's case, the tax laws of Spain and

the United States are of particular relevance, as are those of other jurisdictions in which he may directly or indirectly hold assets. (Albert's options with regard to forced-heirship issues, tax planning and overseas property ownership are detailed elsewhere in the book, and for this reason these issues are not covered here. Suffice to say that in structuring his trust and organising the transfer of assets into the trust, Albert should take into account what opportunities the trust may offer to help him minimise his overall tax exposures and to achieve his estate-planning and other objectives.)

The fact that Albert holds a US green card is an important issue and, on careful reflection, Albert may decide to relinquish his right to reside in the United States, given the US tax cost of such a right (subject to the effect of any tax treaties Albert may be able to take advantage of).

Albert does not require multiple trusts in order to achieve his various tax minimisation objectives. Indeed, the overall trust he establishes will be all the more effective if it incorporates all of these objectives; not only because he will have solved problems on several fronts, but also because anti-avoidance legislation in a number of jurisdictions focuses on there being a primary motive of tax avoidance under the laws of that particular jurisdiction.

A single well-planned trust, taking into account Albert's tax minimisation objectives in several jurisdictions, can therefore be a more effective planning tool than separate trusts each focusing on one jurisdiction.

Asset protection trusts and family break-ups

Albert is a little concerned about his potential exposure to claims against his assets by his wife, on a possible break-up of his marriage, or by his mistress.

It is worth noting that, as in Albert's situation, when deciding on the structure of a trust that might provide asset protection in the context of spousal claims, the first step is to determine the jurisdiction whose laws will govern any such claim. If Albert's wife discovered his extramarital relationship and decided to make a claim, would she do so under Spanish law? What of the home Albert maintains in the United States? Could this residence, under applicable state law, create the connection needed for Albert's wife to make a claim in the United States? It could well be that Albert's trust will have to contemplate the possibility of attack under the family relations laws of a number of jurisdictions. Albert may want to rely on the secrecy an asset protection trust can offer, but a well-structured trust will also be designed to withstand review by the courts in the event of a claim.

Political risk

One of the more interesting issues to be addressed in the context of an asset protection plan is the question of political risk. In analysing this, it is important to look far beyond the jurisdiction of the trustee, which unfortunately is all too often the only jurisdiction examined.

It is certainly appropriate for an asset protection trust to provide for an immediate and automatic change of trustee should dramatic political changes take place in the jurisdiction of the original trustee. In Albert's case, however, he may be more concerned about political risk in the place of his residence than of the risk of Bermuda or some other tax haven running into difficulty. He would more than likely also be concerned about the possibility of unwanted political changes taking place in any of the countries in which he has investments.

As a general principle of international law, a state has the right to expropriate property belonging to its nationals if the property is within its jurisdiction. Whether compensation was payable would be a matter of domestic legislation.

Various countries have asset protection agreements, which generally provide, among other things, protection against the expropriation of assets belonging to residents of the contracting states. For Albert, protection would only extend to those assets located in countries with which Spain has an asset protection agreement (and not to his assets in Spain itself).

A well-structured asset protection trust set up for Albert might include the ownership of his assets through companies incorporated in countries that have asset protection agreements with the countries where he has investments, allowing the companies he indirectly owns to receive compensation in the event of an expropriation.

Areas of risk

In developing an asset protection plan Albert should, among other things, address three specific areas of political risk:

(i) With respect to assets owned by Albert outside his country of residence, the trust mechanism he employs might involve ownership of the relevant assets directly or indirectly by the trust. In the event of any dramatic political change, any beneficial interest retained by Albert would automatically shift away from him, thus affecting his interest. A change in law prohibiting the private ownership of property should not affect assets properly owned by the trust.

(ii) The trust should contemplate the possibility of disputes developing between Albert's country of residence and third countries whose own nationals may suffer expropriations or similar occurrences. For example, under US law, there are circumstances where the United States, in retaliation for an expropriation of US assets, can freeze or vest assets located in the United States belonging to nationals of the expropriating state. This would mean that assets personally owned by Albert that are located in the United States could be at risk unless owned through a carefully structured trust.

(iii) Albert's asset protection plan should seek to minimise the risks associated with assets physically located in high-risk countries.

Further protection can be achieved through the use of an intermediate ownership entity incorporated in a jurisdiction which has an asset protection agreement with the country where the assets are located. Albert may be tempted to seek to leverage assets located in high-risk locations free of personal guarantees, effectively passing some of his political risk to his lenders. This is often difficult to achieve, however, given the increasingly conservative approach of lenders, who often insist on guarantees in such circumstances.

Albert might also explore the possibility of obtaining political risk insurance, which is usually available, albeit at a cost.

In developing an asset protection trust for Albert, considerable care will need to be taken to ensure that changes in the trustee, and shifts in beneficial interests, do not attract stamp duty, transfer taxes or other exposures. Similarly, asset protection plans involving companies, particularly publicly listed companies, need special care to avoid the risk that they might affect the rights of the companies' minority shareholders or creditors. If they do, there may be a basis for a challenge to the efficacy of the asset protection plan.

Protection against claims of creditors

As mentioned at the outset, the term 'asset protection trust' has come to refer to the structuring of trusts in a manner that might protect assets against the claims of creditors. Because of the risks involved, however, some trustees shy away from acting as trustees in such circumstances. Nonetheless, considerable business has been developed in the field – mostly from the United States, where professionals and others have found there is a substantial risk of litigation associated with their activities. For example, a surgeon concerned that a professional-malpractice action may result in a loss of personal assets may seek to settle those assets into a well-structured trust to keep them out of the reach of potential creditors.

Whether or not a trust will be effective will depend on: (i) how it is structured; (ii) when it is settled in relation to the timing of potential claims; and (iii) the operation of law in each of the jurisdictions in which any claims are brought and the jurisdiction whose laws govern the trust.

STRUCTURING

A well-planned trust will take a number of issues into account and will generally involve the original owner of the relevant assets giving up a direct interest in exchange for an interest in the trust, perhaps as a discretionary beneficiary. Depending on the circumstances, the trust may actually exclude the original owner of the assets from any potential benefit, this being restricted to members of his family.

TIMING

The timing of the settlement of a trust is of enormous importance to its effectiveness in protecting assets from any claim by creditors. Depending on the law applying to the transfer, this will often depend on its having taken place before the circumstances arise upon which any claim might be based.

LOCATION

Choosing the location of the trust is also a crucial component in the development of an effective asset protection plan. A number of tax havens have adopted legislation specifically designed to attract asset protection trust business. A creditor seeking to make a claim on assets settled into an asset protection trust will often attempt to challenge the efficacy of the transfer of assets to the trust, relying on the laws of the jurisdiction in which the creator of the trust lives. For example, a transfer would generally be considered invalid under the laws of the jurisdiction of the creator, if it was made at a time when the transferor would have been considered technically insolvent on the basis of a review of his assets and liabilities.

The use of an asset protection trust, under the laws of a suitable offshore jurisdiction, can introduce another legal system for a creditor to deal with: a jurisdiction, hopefully, whose laws will stretch further to protect settled assets. Offshore jurisdictions with laws that have been amended to attract asset protection trust business often have laws that do not permit a transfer to the trust to be set aside unless it was made with an intent to defraud creditors. Obviously, this is a higher threshold than the simple requirement that the transferor was technically insolvent at the time the original transfer was made.

Other attractive features of offshore legislation include the burden of proof being on creditors seeking to challenge a transfer to a trust, short limitation periods for claims, strong protections for the interests of good-faith beneficiaries and cost protections to trustees acting in good faith.

For Albert, the use of a well-structured trust, designed to take into account the potential claims of creditors, would be a worthwhile endeavour, particularly in view of his potential long-term exposure to claims in the United States relating to his property interest there. Careful selection of the jurisdiction whose laws will govern the trust will also help Albert to isolate all, or part of, his assets from claims by his wife or others on the basis of applicable family relations legislation.

CONCLUSION

The structuring of asset protection trusts can be an interesting and fruitful undertaking for an individual living in Europe. It is far from uncommon for families to have connections to several different countries, some with almost confiscatory tax regimes, making it necessary for the creator of a trust to take a close look at a number of tax systems if legal tax avoidance is to be achieved.

Political risk is generally significant in both the jurisdictions in which individuals live and the places where investments are located. Family and relationship structures are often complicated. Potential claims by creditors may be less of a risk to an individual living in Europe than to a doctor practising in California, but may still be possible, particularly in view of the volatility of local markets.

It is sensible for a trust to encompass a number of different features. This not only allows the trust to meet many different planning objectives, but creates a stronger trust, better able to withstand scrutiny under the tax or other laws of a challenging jurisdiction.

When establishing any trust, there are often practical steps that should be taken to help ensure that asset protection objectives are met. In Albert's situation, given his concerns about potential claims he would be well advised to limit those to whom he provides information on the trust or trusts he creates. Although any trust Albert creates should be designed to withstand scrutiny, there is always a role for a certain amount of secrecy if asset protection is an objective. ❖

CHARITABLE GIVING

Oliver Stanley, Chairman, Rathbone Brothers plc

Oliver Stanley MA (Oxon.), Barrister, is chairman of Rathbone Brothers plc, a United Kingdom quoted private bank which has trust company subsidiaries operating in London, England; Geneva, Switzerland; Tortola, BVI, etc. His address is: University House, 11-13 Lower Grosvenor Place, London SW1W 0EX.

Phone: (+44-171) 630 5611
Fax: (+44-171) 828 2336

CHARITABLE GIVING

Oliver Stanley, Chairman, Rathbone Brothers plc

Most wealthy individuals, at some stage in their lives, make donations to charitable causes. These may vary in scale from the traditional 'penny in the blind man's hat' to the endowment of the Guggenheim Art Foundation. This article is concerned with neither of those two extremes, but is designed for donors wishing to maximise the value of their contributions, both from their own point of view and from that of the charity concerned.

SYSTEMATIC GIVING

Many donors are content to give relatively trivial amounts relatively often. But when they write out a cheque – for 100 francs, marks or dollars – to a cause they particularly support, they should remember that a large proportion of their gift will be consumed in the management, administrative and accounting expenses of dealing with it. All charitable organisations, whether for the relief of the poor or the advancement of the arts, have to maintain an infrastructure to receive, invest and distribute the funds placed in their care. It therefore follows that a systematic approach to donation will help to maximise the value of gifts made.

A system should begin with a budget. This should predict what amounts one can afford to give, in income or capital, over a stipulated period. It is helpful to divide the total into a relatively small number of units, so that each donee receives an amount substantial enough to cover administration and leave a useful balance for distribution to beneficiaries. This advice may seem obvious: it is better to give five units of £1,000 each than five thousand units of £1 each. But it is surprising how often this is overlooked!

This is only a small step towards systematising donations. The second, more important, step is to recognise that one's capacity to donate does not remain constant throughout one's life. Indeed, it can be argued that one's capacity is greatest upon one's death. This is the time when most charitable donations are made: as bequests contained in a will. The feeling is often that since one's heirs are in any case receiving their inheritance almost as a gift out the blue, if that inheritance is modestly reduced before they receive it, they will not feel too much pain! Thus it is common in English-law jurisdictions for testators to make specific legacies to the charities they wish to support: promoting the preservation of the countryside, the development of the arts or the relief of poverty, as the case may be.

However, it is misguided to confine the time of donation to death. By starting a systematic process during their lifetimes, donors can monitor how the causes they support respond to donations and to what use they put the wealth given to them.

THE CHARITABLE TRUST

To this end, there exists in English-law jurisdictions a valuable legal entity: the charitable trust. The best way to look at the charitable trust is as a staging point, half way between donors and the charities they wish to benefit. The trustees (who are legal persons) hold in trust a pool of assets which, at their discretion, they may distribute to those charities that fall within the specified class of beneficiaries. This class can be widely defined.

It is the donor who appoints the trustees to take care of distribution; who endows the trustees with the funds; and who exercises supervision over the trust (within legal limits) to ensure that his or her personal objectives are executed by the trustees. The trustees will operate accordingly during a donor's lifetime, but the trust will continue after his or her death, in perpetuity or as long as it has funds to sustain it.

There is an important supplementary advantage associated with charitable trusts: charitable trusts established under English law, with residence in English-law territories, are normally exempt from all taxation. This ensures that the funds the charitable trustees invest compound at the highest possible rate, and that when those investments receive income or realise capital gains there is no tax to pay. In the same way, distributions out of a charitable trust will not attract any form of gift or transfer taxation, so that the charitable cause receives the value gross.

There is an important set of stipulations determining whether particular purposes are charitable; this question is decided in the light of a series of precedent cases. The motives and opinions of the donor are not relevant, and the court will look for benefit to the community as a whole. The overriding test of whether an action is charitable is whether or not it is taken for 'public benefit'.

Moreover, these actions must be beneficial in a way that the law deems charitable. In the leading case of *Income Tax Commissioners vs Pemsel*, four general charitable objectives were distinguished:

(i) the relief of poverty;
(ii) the advancement of education;
(iii) the advancement of religion; and
(iv) other purposes of a charitable nature beneficial to the community, not falling under any of the three preceding heads.

It will be observed that these headings are very wide in ambit, capable of including very many worthwhile causes.

The choice of a charitable trust as a vehicle is unlikely to be made simply for tax reasons. Another reason might be that there are no suitable children or grandchildren ready to accept the responsibility of ownership and management of a family enterprise or estate, for instance. In these circumstances, the charitable trustees would serve as a substitute for family members. It is important to remember there may be no objection to the donor serving as one of a number of trustees, although a replacement would obviously need to be provided after his or her death.

THE LIECHTENSTEIN FOUNDATION

It is sometimes argued that all the functions described above are capable of being performed by a Liechtenstein foundation. This kind of foundation does, indeed, have many similarities to a charitable trust in that it exists in order to realise the aspirations and intentions of a founder, as set out in a foundation deed or statute. The objects of a foundation must be non-commercial, and the foundation capital is held, like that of a trust, in undivided form.

Scientific, artistic and cultural activities are likely to be stipulated, and a foundation may be specifically designed to provide funds for educational purposes. A foundation will have minimum capital requirements and there will be government taxes payable on this capital. Management costs will be higher, so an

English-law charitable trust is likely to prove more economical and flexible.

THE LOCATION OF THE TRUST

It is possible to set up a charitable trust in the UK itself, without any exposure to UK taxation. But there are alternatives: a trust can be set up in an overseas English-law territory (for example, the British Virgin Islands) with both family and charitable beneficiaries included. All the tax exemptions outlined above will be secured but, inevitably, the trust will not have the same purely charitable status and so may be treated differently by the tax authorities of the donor's country of residence. If that is not an issue, then this course is to be strongly recommended because it permits donors to make decisions later on about whether funds should be directed towards charitable objects or to their natural heirs.

Once the structure is in place, donation into it may be regular or occasional. Donations to selected charities may be made in accordance with a regular programme. The structure so created may prove so valuable that the estate owner can use it not only to make donations but also as an anchor, to hold resources in a state of permanent fiscal neutrality more effectively than by retaining them in his or her own name.

TRUSTEES

Finally, the economical and efficient organisation of a charitable trust depends on the capacity of the trustees. It is advisable to select an established corporate trustee in a tax-neutral, politically stable territory, regulated by English law. ❖

APPENDIX

CREDIT SUISSE PRIVATE BANKING

Philippe Bollag, Member of Senior Management, Credit Suisse Private Banking

THINK GLOBAL: ACT LOCALLY

Until a few years ago, Switzerland was the undisputed market leader in private banking services. However, competition from other financial centres has increased enormously. At the same time, globalisation and deregulation have become the watchwords for the financial markets and only organisations with a worldwide operational network can promptly and effectively meet the needs of increasingly discerning clients. Private customers are no longer satisfied with routine products and services; they expect the highest standards of professional expertise and personal commitment from their private banker.

ONSHORE AND OFFSHORE PRIVATE BANKING

According to official estimates, more than US$10,000bn in private assets is currently under management worldwide. About three-quarters of this is invested onshore in the country of residence of the owner.

This international predominance of onshore private banking has prompted Credit Suisse to extend its private banking services beyond the traditional offshore banking centres, such as Switzerland. Credit Suisse Private Banking now offers its services in Atlanta, Barcelona, Berlin, Frankfurt, Geneva, Gibraltar, Guernsey, Hong Kong, London, Los Angeles, Luxembourg, Madrid, Miami, Milan, Monte Carlo, Montreal, Munich, Nassau, New York, Nuremberg, Paris, Singapore, Tokyo, Toronto, Vancouver, Zurich and other Swiss cities as well as through a worldwide network of representative offices.

CHOOSING A PRIVATE BANKER

Finding the right partner in private banking needs special care and it is worthwhile taking the time to investigate the factors which ensure a successful relationship.

Who is the private banker?

The private banker's name and reputation are of course of prime importance. He or she should be an expert on local financial market products and have experience in international portfolio management. Comprehensive, high-quality financial advice should be offered on subjects such as tax, life insurance and international issues; and other in-house services, such as international real estate advisory and help with trust business should be available on a global basis.

How does the private banker provide the investment services?

What is the banker's performance record? He or she should be discreet, reliable and proactive, and should focus on the client's individual needs.

Where is the private banker located?

Local presence versus global presence should be considered, e.g. which private banking offshore locations will form a useful supplement to the client's onshore banking activities.

The answers to these three questions can help investors assess whether they are likely to enjoy a long-term, mutually beneficial private banking relationship. What follows is but a small selection of the private banking services offered by the Credit Suisse Group.

PRIVATE BANKING OFFSHORE

While intensifying its efforts in onshore banking, Credit Suisse is also present at all the classic offshore locations. Our private banking unit in Guernsey offers an attractive range of financial services to investors, as do our Luxembourg office and our unit in the Bahamas. These offshore services range from tailormade investment advice to trusts and funds business.

PRIVATE BANKING IN GERMANY

Private clients should be offered the opportunity to assess investment performance on an after-tax basis and compare this performance with that of other tax-efficient investment alternatives. In recent years German private investors have seen taxes on their investment income and capital rise as a result of heavy state financing requirements. In response, Credit Suisse Private Banking in Frankfurt has specialised in achieving higher after-tax returns for its private clients by offering structured financial products instead of traditional direct investments. An investment fund has been developed which is managed exclusively with tax efficiency in mind. A comprehensive individual investment service also offers private investors in Germany dividend-right securities (one of the few remaining tax-privileged products available), foreign currency hedging and a leasing fund in conjunction with project financing.

PRIVATE BANKING IN LONDON

Many private investors expect more from their private bankers than simply expertise in money and capital market investments. They expect them to be able to advise them in related areas such as real estate investment. Credit Suisse Private Banking in London can offer its clients and those of other CS Holding entities, what is called the 'International Real Estate Advisory Service in the United Kingdom and the United States property market'.

The aim of this advisory service is to look for first-class real estate investments in the United Kingdom or United States which fit easily into clients' investment portfolios and are in keeping with their individual investment goals.

Credit Suisse Private Banking in London has a coordination team of external and internal specialists covering all aspects of real estate investment, including the design and establishment of ownership structure, advice on financial and tax matters, strategic research, property management, building inspection and the establishment of trusts or holding companies.

PRIVATE BANKING IN SINGAPORE

Asia is the most dynamic economic region in the world and probably home to the next economic superpower. Asian developing nations have traded decades of political turmoil and ideological struggle for free market reforms. As a result, economies in the region have boomed. This, in turn, has opened up tremendous investment opportunities. In order to assist clients from Southeast Asia and beyond to participate in this exciting growth area, Credit Suisse Singapore Private Banking has aggressively expanded its services to cater to the needs of high net worth individuals. Discretionary portfolio management with a strong Asian flavour is available for investors of all types. In addition, Credit Suisse Singapore Private Banking provides investment advice and facilitates trading on Southeast Asia's bustling stock markets. Singapore, politically stable and with an excellent infrastructure, is the ideal gateway to Asia.

PRIVATE BANKING IN MILAN

The aim of our Italian private bankers is to offer a global service and ensure that investors always receive products that best suit their specific needs. Such products can range from traditional managed bond portfolios to non-discretionary and discretionary portfolio management for all types of investor; from foreign currency accounts (in US dollars, Deutschmarks or Swiss francs) to repos.

Strategy is tailor-made for the individual, covering business, personal and family interests and providing all the financial services a customer needs from a single source.

Our Italian customers know that their assets are in the hands of a competent, reliable, and trustworthy partner, who will take time to offer them comprehensive advice in all financial matters, who will solve their problems and who will always be at their disposal.

PRIVATE BANKING IN TOKYO

Credit Suisse Tokyo specialises in Japanese equities, bonds and commercial bills as well as in global balanced portfolios. Thanks to an outstanding investment performance, Credit Suisse has become the largest foreign trust bank in Japan. For private investors with funds of around ¥500m we can offer the same services as for our institutional clients. Japan is an ideal place for foreigners not only to have funds managed but also to have them booked. Japanese firms pay very little dividend, so the withholding-tax burden becomes insignificant. There is no capital gains tax for nonresidents.

Japan's top marginal corporate, personal and inheritance tax rates are 52%, 65% and 70%. These high rates create the need for careful structuring of corporations and estates to minimise the tax burden. Credit Suisse Tokyo specialises in global solutions that can be tailored to the needs of other nationals as well.

PRIVATE BANKING IN FRANCE

Due to its economic and cultural importance, Paris attracts French and international wealthy individuals expecting a consummate tailor-made banking service. Credit Suisse Private Banking in Paris concentrates on developing long-term relationships and is able to provide a global approach to managing assets.

Credit Suisse France offers a wide range of products, but specialises in:

- portfolio management which combines the CS group's worldwide research and local expertise of French markets;
- multi-currency accounts, including foreign exchange operations;
- estate planning for French residents supported by a newly integrated software which is updated every quarter; and
- specialist real estate and property advice.

All these services are combined with tax optimisation strategies.

Furthermore, Credit Suisse France offers a

range of specific wealth enhancement products aimed primarily at French residents, including on and offshore holdings and life insurance.

PRIVATE BANKING IN SPAIN

Gradual deregulation of investing abroad, a substantial fall in interest rates and the volatility of the peseta over the last three years have sharpened Spanish investors' appetite for investment management which can combine international expertise with knowledge of the local financial, legal, and fiscal environment.

Credit Suisse Private Banking Spain provides easy and direct access to all financial markets and products. A dedicated team of professionals designs investment policy to take advantage of both global diversification and local factors. They focus on expected return and take full account of the tax treatment of the chosen investments.

The benchmark performance for private investors in Spain, as elsewhere, is the return on domestic risk-free investment after tax. In Spain this is represented by the one-year Treasury bill, currently yielding 9.75%. Our basic task, therefore is to ensure that our customers' portfolios earn more than this.

Income tax of 56% is levied on total revenues generated, including financial income and capital gains; if taxes on net worth are included, total taxes payable can reach 70% of total revenues. Every portfolio has to be analysed from an after-tax point of view. Credit Suisse Private Banking in Spain thus advises its clients to diversify their portfolios by using tax efficient structures. Capital growth funds, for example, enjoy preferential treatment, with holders not paying income tax until divestiture actually occurs: there is also a progressive tax waiver over the life of the investment, commencing in the second year and resulting in a full tax exemption by the 15th year. For large portfolios, Credit Suisse Private Banking Spain has designed tailor-made structures that combine discretionary management and tax efficiency.

PRIVATE BANKING IN CANADA

Canadian investors are increasingly seeking access to a growing range of investment and financial services. After 40 years in the country, Credit Suisse Canada, as the premier provider of professional advice on global investing and international banking, is well positioned to respond to this need. Our philosophy has been to customise global investments to achieve consistent capital growth. We respect our clients' individual circumstances and enable them to capitalise on investment opportunities throughout the world. The solutions we provide are designed to maintain the investor's long-term purchasing power through disciplined and structured international diversification from a Canadian perspective.

To provide easy access to international markets for Canadian residents, we have created a balanced investment pooled fund. Clients seeking other sophisticated financial solutions or regular investment advice have access to our local team of seasoned investment professionals, who draw on considerable international experience and are linked to Credit Suisse's worldwide network.

PRIVATE BANKING IN HONG KONG

In 1969 Credit Suisse opened its representative office in Hong Kong, establishing a fully licensed branch in 1980. Since then, Credit Suisse Hong Kong has grown to include more than 160 employees serving both a local and an international clientele.

With over a quarter of a century of service in Hong Kong, the Hong Kong unit of Credit Suisse recently undertook significant expansion of its private banking services in the territory. The cornerstone of this expansion was the addition of highly professional, locally based portfolio management and securities administration to our existing range of services.

As Southeast Asia has grown to become one of the most demanding and sophisticated tax-favourable financial regions in the world, Credit Suisse has grown along with it. Our products, such as trusts – including asset protection, estate duty, and pre-immigration trusts – are created in conjunction with our clients in a spirit of cooperation, innovation and flexibility.

Of course, this has been the briefest of summaries of the many facets of Credit Suisse Private Banking around the world. Our private bankers, each one with specialist knowledge and a global network of research capabilities at his or her fingertips, would be more than willing to provide more detailed information on our services. ❖

Philippe Bollag is a member of senior management at Credit Suisse. He works in the Management International subdivision where he is responsible for international marketing activities. Before joining Credit Suisse Private Banking in 1994, he was head of institutional banking at Swiss Volksbank in Winterthur. Mr Bollag was involved in the creation of SOFFEX, the Swiss Options and Financial Futures Exchange. He is a graduate of the University of St Gall in business administration, economics, law and social sciences.

CREDIT SUISSE PRIVATE BANKING OFFICES

Credit Suisse
Paradeplatz 8, CH-8001 Zurich, Switzerland
Phone: (+41-1) 333 1111

Credit Suisse (Deutschland) AG
Kaiserstrasse 30, D-60020 Frankfurt am Main, Germany
Phone: (+49-69) 75380

Credit Suisse (Deutschland) AG
Haus der Schweiz, Unter den Linden 24, D-10117 Berlin, Germany
Phone: (+49-30) 201 2071

Credit Suisse (Deutschland) AG
Maximilianstrasse 23, D-80539 Munich, Germany
Phone: (+49-89) 291 770

Credit Suisse
Place Bel-Air 2, CH-1211 Geneva 70, Switzerland
Phone: (+41-22) 391 2111

Credit Suisse (Deutschland) AG
Dürrenhofstrasse 4, D-90024, Nuremberg, Germany
Phone: (+49-911) 94640

Crédit Suisse (France) SA
24 rue Cambacérès, F-75008 Paris, France
Phone: (+33-1) 4312 1000

Credit Suisse (Gibraltar) Ltd
Neptune House, Marina Bay, Gibraltar
Phone: (+350) 78399

Credit Suisse
5 Cabot Square, London EC2N 4BQ,
United Kingdom
Phone: (+44-171) 888 8000

Credit Suisse (Guernsey) Ltd
Helvetia Court, PO Box 368, St Peter Port,
Guernsey GY1 3YJ, Channel Islands
Phone: (+44-481) 719 000

Credito Svizzero
via Turati 9, 1-20121 Milan, Italy
Phone: (+39-2) 63741

Crédit Suisse (Luxembourg) SA
56 Grand-Rue, BP 40, L-2010 Luxembourg
Phone: (+352) 460 0111

Crédit Suisse (France)
Le Park Palace, 27 Avenue de la Costa,
BP 155, Monte Carlo, MC-98003 Monaco
Phone: (+33-93) 152 727

Credit Suisse
Palais Corso, Mahlerstrasse 10,
A-1010 Vienna, Austria
Phone: (+43-1) 512 2989

Credit Suisse (Moscow) Ltd
B. Palashevsky Pereulok 15A,
Moscow 103 104, Russian Federation
Phone: (+7-95) 973 2333
Phone: (+7-501) 513 0029

Credit Suisse
1 Turab Tula Street, 700003 Tashkent,
Republic of Uzbekistan
Phone: (+7-3712) 89 1166

Credit Suisse
Paseo de Recoletos 17, CP 14608,
E-28004 Madrid, Spain
Phone: (+34-1) 522 1020

Credit Suisse
Avenida Diagonal 605-7, CP 16071,
E-08080 Barcelona, Spain
Phone: (+34-3) 419 2028

Credit Suisse
Tower 49, 40th Floor,, 12 East 49th Street,
New York, NY 10017, USA
Phone: (+1-212) 238 2000

Credit Suisse
633 West Fifth Street, 64th Floor,
Los Angeles, CA 90071, USA
Phone: (+1-213) 955 8200

Credit Suisse
191 Peachtree Street NE, Suite 3500,
Atlanta, GA 30303, USA
Phone: (+1-404) 577 6100

Credit Suisse
AT&T Building, 40th Floor, 227 West Monroe
Street, Chicago, IL 60606, USA
Phone: (+1-312) 630 0086

Credit Suisse
1100 Louisiana, Suite 4750, Houston,
TX 77002, USA
Phone: (+1-713) 751 0300

Credit Suisse
50 California Street, Suite 2940,
San Francisco, CA 94111, USA
Phone: (+1-415) 391 9590

Credit Suisse
601 Brickell Key Drive, Suite 1000,
Miami, FL 33131, USA
Phone: (+1-305) 995 6500

Credit Suisse Canada
Credit Suisse Centre, Suite 1300, 525 University
Avenue, Toronto, Ontario M5G 2K6, Canada
Phone: (+1-416) 351 3500

Credit Suisse Canada
1250 René Lévesque Boulevard West, Suite 3935,
Montreal, Quebec H3B 4W8, Canada
Phone: (+1-514) 933 8774

Credit Suisse Canada
400 Burrard Street, Suite 1610, Vancouver,
British Columbia V6C 3A6, Canada
Phone: (+1-604) 684-4009

Credit Suisse (Bahamas) Ltd
Rawson Square, PO Box N-4928,
Nassau, Bahamas
Phone: (+1-809) 322 8345

Credit Suisse
Hong Kong Macau Center, 12th Floor,
1 Chaoyangmen Bei Da Jie, Beijing 100 027,
People's Republic of China
Phone: (86-10) 501 2534

Credit Suisse
International Trade Center, 18th Floor, Room 1808, 2200 Yanan Road West,
Shanghai 200335, People's Republic of China
Phone: (+86-21) 275 7050

Credit Suisse
23/F Three Exchange Square,
8 Connaught Place, Central, Hong Kong
Phone: (+852) 841 4800

Credit Suisse Trust and Banking Co. Ltd
CS Tower, 11-30 Akasaka 1-chome, Minato-ku,
CPO Box 535, Tokyo 107, Japan
Phone: (+81-3) 3589 3838

Credit Suisse
No. 2 Yodoko Building, Floor 6,
2-15 Bakuromachi 4-chome, Chuo-ku,
Osaka 541, Japan
Phone: (+81-6) 243 0121

Credit Suisse
Ankuk Fire and Marine Insurance Building,
7th Floor, 87, 1-ka Uichi-ro, Chung-ku,
CPO Box 4295, Seoul 100-191, Korea
Phone: (+82-2) 771 8231

Credit Suisse
80 Raffles Place #49-01, UOB Plaza 1,
PO Box 776, Singapore 9015
Phone: (+65) 538 6322

Credit Suisse
Hung Tai Century Tower 11F-5, Room E,
156 Min Sheng East Road, Sec. 3,
PO Box 87-770, Taipei, Taiwan
Phone: (+886-2) 717 0174

Latin America

Credit Suisse Representacion Buenos Aires
Esmeralda 130, Piso 11,
1035 Buenos Aires, Argentina
Phone: (+54-1) 334 3618

Credit Suisse Representaçâo e Assessoria Financeira S/C Ltda
Rua 7 de Setembro No. 99-8 andar,
CP 1250-ZC-00,
20050-005 Rio de Janeiro, Brazil
Phone: (+55-21) 242 6157

Credit Suisse Representaçâo e Assessoria Financeira S/C Ltda
Av. Paulista 1274, 13 andar, Conj. 34,
CP 6943, 01310-100 Sao Paulo, Brazil
Phone: (+55-11) 287 1133

Credit Suisse Consultas y Asesorias Limitada
Av. Andrés Bello 2777, Las Condes,
Santiago, Chile
Phone: (+56-2) 233 1383

Credit Suisse Representacion para Colombia
Centro Comercial 93, Calle 93, No. 14-20,
Suites 410-412, Apartado 093152,
Bogotá, Colombia
Phone: (+57-1) 218 6700

Credit Suisse Rep. Office
Campos Eliseos No. 345, Piso 9,
Col. Chapultepec Polanco,
PO Box 10581, 11560 Mexico DF
Phone: (+52-5) 202 5525

CS Consultaciones y Servicios SRL
Rincón 477 Piso 4, Esc. 402, PO Box 1330,
Montevideo, Uruguay
Phone: (+598-2) 960 159

Credit Suisse Rep. Office
Edificio Cavendes, Piso 7, Ofic. 704,
Av. Francisco de Miranda, Apartado 62522,
Caracas 1060, Venezuela
Phone: (+58-2) 283 6422

Middle East and Africa

Credit Suisse Rep. Office
32 Haroon Street, Dokki, Gizeh, Cairo, Egypt
Phone: (+20-2) 360 7865

Credit Suisse Rep. Office
Dhabi Tower, Hamdan Street, PO Box 47060,
Abu Dhabi, UAE
Phone: (+971-2) 325 048

Credit Suisse Rep. Office
Intercontinental Plaza, Suite 14H,
PO Box 8811, Dubai, UAE
Phone: (+971-4) 274 780

Credit Suisse Rep. Office
Chamber of Commerce & Industry Building,
3rd Floor, King Faisal Road, PO Box 5100,
Manama, State of Bahrain
Phone: (+973) 250 677

Credit Suisse Rep. Office
9th Floor, Sandton City Office Tower, 5th Street,
2196 Sandown, Republic of South Africa
Phone: (+27-11) 884 6741

Australia

Credit Suisse
31 Queen Street, 20th Floor, GPO Box 2844 EE,
Melbourne, Victoria 3001, Australia
Phone: (+61-3) 9614 4488